Zero Waste Management Technologies

Rouf Ahmad Bhat · Gowhar Hamid Dar ·
Younis Ahmad Hajam
Editors

Zero Waste Management Technologies

Springer

Editors
Rouf Ahmad Bhat
Department of School Education
Government of Jammu and Kashmir
Srinagar, Jammu and Kashmir, India

Gowhar Hamid Dar
Department of Environmental Science
Government Degree College Kulgam,
Higher Education Department
Srinagar, Jammu and Kashmir, India

Younis Ahmad Hajam
Sant Baba Bhag Singh University
Jalandhar, Punjab, India

ISBN 978-3-031-57274-6 ISBN 978-3-031-57275-3 (eBook)
https://doi.org/10.1007/978-3-031-57275-3

This Springer imprint is published by the registered company Springer Nature Switzerland AG
The registered company address is: Gewerbestrasse 11, 6330 Cham, Switzerland

In a world where foes and friends entwine,
Enemy and environment, a tale to define.
Humans stand as stewards of this earthly plane,
To conquer challenges, we must break the chain.
Zero waste management, our mission's decree,
Turn enemy into ally, set our world free.
For when we unite, in this common endeavour,
We protect the environment, cherish it forever.

—Ibtihaaj Bilal

Preface

In a world where the delicate balance of our planet's ecosystems is under threat and our finite resources are being depleted at an alarming rate, the concept of "zero waste" has emerged as an encouragement to protect the environment. This book delves deep into the heart of this movement, exploring the myriad strategies that empower individuals, communities, and industries to transition towards a more sustainable and harmonious way of living. The pages that follow are not just about waste reduction; they embody a philosophy that challenges the very foundations of our consumer-driven society. The journey towards zero waste is not a mere trend or a passing fad—it is a profound shift in perspective, a transformative approach that demands a re-evaluation of how we produce, consume, and dispose. In "Zero Waste Strategies", we embark on an exploration of innovative concepts, pragmatic methodologies, and inspiring success stories from around the globe. We will navigate through the complex web of challenges and opportunities, from reimagining product design to redefining supply chains, from advocating for policy changes to fostering a cultural shift in attitudes towards consumption and disposal. This book is a treasure trove of insights, a toolkit of ideas, and a source of inspiration for everyone seeking to contribute to a world where waste is minimized, resources are conserved, and the health of our planet is safeguarded for future generations. Whether you are an environmental enthusiast, a curious learner, a business leader, or a policymaker, "Zero Waste Strategies" offers something for you. It is an invitation to explore a new way of thinking, a call to action to embrace sustainability as a core value, and a reminder that our collective choices today shape the world we pass on to tomorrow. As you journey through these pages, may you be inspired to not only adopt Zero Waste Strategies in your own life but to also champion them in your communities and beyond. The time to act is now, and the power to make a difference lies within each of us. Together, let us embark on this transformative path towards a future where waste is but a distant memory, and the concept of "waste not, want not" resonates with renewed vigour. Welcome to "Zero Waste Strategies". Your adventure towards a more sustainable world starts here.

Srinagar, India — Dr. Rouf Ahmad Bhat
Srinagar, India — Dr. Gowhar Hamid Dar
Jalandhar, India — Dr. Younis Ahmad Hajam

Acknowledgments

The editors would like to acknowledge the efforts of all those who helped in making this publication possible in one way or another. The editors would like to put on record the efforts of the following for rendering tremendous support while drafting this book and making it possible to happen.

Dr. Hummara Sultan
Mrs. Rafia Rasool Gowhar
Master Mohammad Zaeem Gowhar
Master Bhat Mohammad Qais
Master Ahmad Wasif Qadri
Master Bhat Mohammad Baasim
Master Bhat Mohammad Sibtain Arshid
Mr. Naveed Qayoom
All the Authors and Supporting staff of Springer Nature

About This Book

This volume *Zero Waste Management Technologies* highlights cutting-edge research on zero waste management and the associated effects of waste on the environment. Predominantly, it focuses on the challenges of dealing with the amassed production of waste and the cumulative impact of increasing waste on the biosphere. Different sections of this book focus on the comprehensive overview of the technological advancements driving the zero waste movement. Furthermore, it explores innovations in waste reduction, recycling, and repurposing, from a global perspective, examining the diverse cultural, social, and economic factors influencing the adoption of Zero Waste Strategies worldwide. In addition, it discusses the challenges and opportunities inherent in promoting a unified global effort towards sustainable resource management. The book serves as a comprehensive roadmap for those seeking to embrace the transformative power of zero waste strategies, making it essential reading for individuals, communities, and industries passionate about fostering sustainable resource management and preserving the planet for future generations.

Key Features:

- Highlights the issues and challenges in achieving a zero waste society.
- Offering a roadmap for policymakers, business setups, and individuals to overcome the hurdles and providing insights into the potential future trajectories of the zero waste movement.
- Investigation with respect to powerful technologies regarding waste to revenue and waste to energy.
- Providing advanced biotreatment technologies for the reduction of the quantum of waste.

Contents

About the Editors

Dr. Rouf Ahmad Bhat (Ph.D.) has pursued his doctorate at Sher-e-Kashmir University of Agricultural Sciences and Technology Kashmir (Division of Environmental Sciences) and presently working in the Department of School Education, Government of Jammu and Kashmir. Dr. Bhat has been teaching graduate and postgraduate students of environmental sciences for the past three years. He is an author of more than 55 research articles (h-index 30; i-index 44; total citation 2500) and 40 book chapters and has published more than 40 books with international publishers (*Springer, Elsevier, CRC Press Taylor and Francis, Apple Academic Press, John Wiley, IGI Global*). He has his specialization in limnology, toxicology, phytochemistry, and phytoremediation. Dr. Bhat has presented and participated in numerous state, national, and international conferences, seminars, workshops, and symposiums. Besides, he has worked as an associate environmental expert in World Bank-funded Flood Recovery Project and also the environmental support staff in the Asian Development Bank (ADB)-funded development projects. He has received many awards, appreciations, and recognition for his services to the science of water testing, air, and noise analysis. He has served as an editorial board member and a reviewer of reputed international journals. Dr. Bhat is still writing and experimenting with diverse capacities of plants for use in aquatic pollution remediation.

Dr. Gowhar Hamid Dar (Ph.D.) is currently working as Assistant Professor in Environmental Science at Sri Pratap College, Cluster University Srinagar, Department of Higher Education (J&K). He has a Ph.D. in Environmental Science with specialization in Environmental Microbiology (Fish Microbiology, Fish Pathology, Industrial Microbiology, Taxonomy, and Limnology). He has been teaching postgraduate and graduate students for the past many years at Postgraduate Department of Environmental Science, Sri Pratap College, Cluster University Srinagar. He has more than 70 research articles (**h**-index **16**; **i**-index **22**; total citation **>950)** in international and national journals of repute and more than 20 books with international publishers (***Springer, Elsevier, CRC Press Taylor and Francis, Apple Academic Press, John Wiley, IGI Global***) to his credit. Moreover, he is supervising a number of students for the completion of degrees. He has been working on the isolation, identification, and characterization of microbes, their pathogenic behaviour and impact of pollution on development of diseases in fish fauna for the last several years. He has received many awards and appreciations for his services towards the science and development. Besides, he also acts as a member of various research and academic committees. Further, Dr. Dar is Principal Investigator and Co-Principal Investigator for different R&D projects sanctioned by Government of India and Government of Jammu and Kashmir.

Dr. Younis Ahmad Hajam holds a doctorate in Zoology from Guru Ghasidas Vishwavidayalaya (A Central University), Bilaspur, Chhattisgarh, India. He is currently working as Assistant Professor in the Department of Life Sciences and Allied Health Sciences, Sant Baba Bhag Singh University, Padhiana, Jalandhar, Punjab, India. Formerly, Dr. Younis Ahmad had worked as an assistant professor and head of Department of Zoology, Career Point University, Hamirpur, Himachal Pradesh, India. He is teaching undergraduate and postgraduate and doctorate students. Dr Younis Ahmad has published 42 research papers in leading national and international journals, 45 book chapters, and four (10) national proceeding. Dr. Younis Ahmad has published 8 books as an editor in Springer, Apple Academic

Press, NOVA Science Publishers, Taylor and Francis, etc. He has guided 5 M.Sc. students under thesis mode and 33 M.Sc. students under project mode. Dr. Younis Ahmad has completed one major research project (PI) and three projects under HP Chief Minister Startup Scheme. Dr. Younis Ahmad has also served as Junior Research Fellow (JRF) at Zoology, Guru Ghasidas Central University, Bilaspur, Chhattisgarh, India. He has presented his research work in more than 35 national and international conferences; Dr. Younis Ahmad has attended more than 30 national and international conferences and 5 workshops. He has organized various conferences, seminars, workshops, and webinars at the national and international levels. Dr. Younis Ahmad has also delivered invited lectures in international conference. He has received national award for his contributions in biological science. He is an editorial member of various international journals and is also a member of the Asian Council of Science editors.

Chapter 1
A Comprehensive Review on the Development of Zero Waste Management

Zulaykha Khurshid, Md. Osama Zubair, and Humaira

Abstract A revolutionary and proactive strategy to address the escalating global waste predicament and corresponding negative environmental effects is Zero Waste Management. The notion of zero waste management and its importance in facilitating sustainable development goals are thoroughly explored in this chapter. We start by defining zero waste management and its fundamental tenets, which emphasize waste reduction, reuse, and recycling within a framework of the circular economy. Analysis of the strong connection between zero waste management and sustainable development is done, demonstrating the advantages of implementing such practices on the social, economic, and environmental fronts. The hierarchy of waste management systems is shown by key ideas, using the waste hierarchy concept. We study the technological advancements that are advancing zero waste management, including waste-to-energy technology, sophisticated recycling methods, and intelligent waste management systems. The chapter also explores the crucial legal and policy foundations that support and advance zero waste programmes at the municipal, state, and federal levels. The social and behavioural components of zero waste management are also covered, with a focus on the need of community involvement and education. The chapter also discusses new trends, problems, and possibilities in the quest for a more sustainable waste management ecosystem, providing insightful information on the prospects for zero waste management in future. The thorough analysis provided in this chapter helps readers gain a greater knowledge of zero waste management and its potential to usher in a more affluent and environmentally friendly future. Societies may create the conditions for a sustainable and regenerative future by implementing this comprehensive strategy and rethinking garbage as a precious resource.

Z. Khurshid (✉)
Department of Environmental Science/Centre of Research for Development, University of Kashmir, Srinagar, J&K 190006, India
e-mail: Zulaykhazehra@gmail.com

Md. O. Zubair
Centre for Nanoscience and Nanotechnology, Jamia Millia Islamia, New Delhi 110025, India

Humaira
Centre of Research for Development, University of Kashmir, Srinagar, J&K 190006, India

R. A. Bhat et al. (eds.), *Zero Waste Management Technologies*,
https://doi.org/10.1007/978-3-031-57275-3_1

Keywords Zero waste management · Circular economy · Sustainable development · Zero waste practices · Extended producer responsibility · Waste to energy

1 Introduction

The consequences of waste generation and environmental degradation are only becoming worse (Abbas et al., 2019; Rios et al., 2019). Zero waste management has emerged as a potent and revolutionary solution. This comprehensive and sustainable approach has drawn international attention and received broad support from politicians, companies, communities, and individuals alike in the face of growing worries about resource depletion, pollution, and climate change (Awasthi, 2023; Bhat et al., 2021: Polettini, 2012). Zero waste management's primary goal is the elimination of waste via the use of a thorough, integrated system that places an emphasis on resource conservation and recovery. Zero waste promotes a circular economy where materials are continually reused, recycled, or renewed to minimize their environmental effect, in contrast to typical linear waste management, which uses a "take-make-dispose" approach (Ibrahim, 2020; Lehmann, 2011). This style of thinking is in line with the pressing need to change our disposable culture into one that prioritizes sustainability, effectiveness, and long-term resilience. According to Tang et al. (2022), Usapein and Chavalparit (2014) and Mancini et al. (2019), expanded producer responsibility, waste reduction, product redesign, and promoting environmentally friendly consumer choices are only a few of the many acts covered by the zero waste principles. The understanding that waste is not an unavoidable outcome of human activity but rather an effect of ineffective systems and unsustainable practises is at the core of this paradigm shift (Bhat et al., 2012). Zero waste management aims to disrupt the current quo and transform how society views and manages its wasted materials by reframing garbage as a valuable resource (Diliberto et al., 2020; Luidold, 2019). Numerous communities, businesses, and cities around the globe have started along the path to zero waste, setting ambitious goals and putting creative plans into practise to reduce waste and optimize resource use (Eriksen et al., 2019; Lai et al., 2022). Zero waste provides benefits for economic growth, social well-being, and a shared commitment to building a greener, more sustainable future in addition to protecting the environment (Khurshid & Khurshid, 2022). By examining zero waste management's key principles, methodologies, policy frameworks, and the different factors that influence its successful implementation, this chapter aims to provide an in-depth analysis of the practise in this context (Larrain et al., 2022; Zaman & Ahsan, 2019). By examining the environmental, financial, and social advantages of switching to zero waste systems, we hope to show how significantly this strategy can contribute to the development of a more sustainable and circular society (Diliberto et al., 2020; Mancini et al., 2019; Tang et al., 2022; Usapein & Chavalparit, 2014). Through fostering stakeholder engagement and funding state-of-the-art waste treatment technologies, this chapter will highlight the critical role that zero waste management plays in constructing a responsible and sustainable future for our planet (Awasthi, 2023; Eriksen et al., 2019; Ibrahim, 2020; Khurshid & Khurshid, 2022; Lai et al., 2022; Larrain et al., 2022; Lehmann, 2011; Luidold, 2019; Polettini, 2012; Zaman & Ahsan, 2019). People from all walks of life are united by the paradigm-changing concept of zero waste management in their shared desire to preserve the Earth's resources for future generations. Together, we set out on a journey to a world without waste—a world where garbage is no longer considered as a nuisance but rather as a chance for renewal, creativity, and a more in tune way of life with our great planet. Zero waste management is presently at the top of agendas for sustainable development due to the urgent need to find answers to the world's environmental concerns.

2 Definition of Zero Waste Management

The term "zero waste management" is an interdisciplinary and transformational strategy that aims to eliminate or drastically reduce trash creation while implementing eco-friendly procedures for handling the leftover garbage (Bertanza et al., 2018; Lange, 2021). This paradigm shift, which embraces the concepts of a circular economy, questions the linear "take-make-dispose" model of resource consumption and places a focus on waste reduction, material recovery, and sustainable production and consumption patterns (Ibrahim, 2020; Yao et al., 2022) (Fig. 1).

Zero waste management is fundamentally based on the understanding that waste is not an inevitable byproduct of human activity but rather results from inefficiencies in the existing system (Khurshid, 2019). The end goal is to completely rethink the waste management process to maximize resource recovery, reduce the environmental effect, and move us towards a more sustainable future (Abbas et al., 2019; Flowers & Linderman, 2003). Zero waste management employs a comprehensive strategy made up of several activities to achieve this lofty goal. These approaches include of reducing trash production at the source, encouraging responsible consumer behaviour, implementing state-of-the-art waste treatment technology, encouraging expanded producer responsibility, and educating the general public about good waste management practises (Lehmann, 2011). The transformational viewpoint of considering waste as a useful resource, as opposed to the conventional perception of waste as a burden, is a basic principle behind zero waste management (Mahongnao et al., 2023). Zero waste management, which embraces this paradigm change, deliberately uses cutting-edge methods like recycling, composting, and other recovery strategies to maximize the potential of waste materials (Pietzsch et al., 2017). The system becomes intrinsically circular and supports sustainable and regenerative practises by reintegrating garbage into the economic cycle as valuable resources (Bolton & Rousta, 2019; Portugaise et al., 2023).

Zero waste management is essentially a proactive and dynamic response to the problems of waste creation and environmental deterioration (Bhat et al., 2018b). Societies may envisage and work towards a world without waste by totally adopting

Fig. 1 Facets of zero waste management

this comprehensive and holistic approach. In order to lead mankind towards a more robust and peaceful cohabitation with our planet, this future will be characterized by judicious resource conservation, minimized environmental impact, and a persistent adherence to the principles of sustainability.

3 Zero Waste Management and Sustainable Development

By addressing the environmental, social, and economic components of sustainability, zero waste management plays a vital part in proceeding towards sustainable development goals (Brancoli & Bolton, 2019; Leal Filho et al., 2019). Adopting zero waste practices is indispensable to decreasing the undesirable consequences related to garbage disposal with increase in waste generation rates (Bhat et al., 2013, 2014; Masud et al., 2019). Societies and organizations may considerably cut their carbon footprints, conserve natural resources, and protect ecosystems by swapping to a zero waste management approach. Endorsement for the circular economy is one of the crucial pillars of zero waste management. In order to make the most of resource efficiency and cut waste generation, this paradigm change places a significant importance on reducing, reusing, and recycling materials in a closed-loop arrangement (Corsten et al., 2013; Rios et al., 2019). Zero waste management approaches reduce energy usage and greenhouse gas emissions during manufacturing by curtailing the extraction of raw materials (Gutberlet, 2015; Yang, 2022). Moreover, zero waste management supports sustainable production and consumption patterns by advocating societies, organizations, and industrial sectors to reassess their consumption arrays and implement eco-friendly alternates (Shukhla & Khan, 2022). This eco-friendly approach fosters the designing and manufacture of long-term, repairable, and recyclable stuffs (Haapala et al., 2013). Also, efficient waste categorization and assemblage systems allow for the recovery and recycling of valuable materials thereby minimizing their termination as useless garbage. Reduction in the waste being discarded into the landfills and incinerators can help in conservation of natural resources besides minimizing the discharge of hazardous pollutants into the environment, and curtail exploitation of raw materials (Bartl, 2011; Rios et al., 2019; Shukla & Khan, 2022). Zero waste management has other recompenses for the society as well as the economy besides environment (Pires & Martinho, 2019). Zero waste management practices lift local economies and improve standard of living particularly in those areas which are dependent on this waste by yielding employment opportunities in the collection, recycling and resource recovery dealings (Silva et al., 2020; Zaman & Ahsan, 2019). By taking considerable measures for decreasing waste generation and improving on resource recovery, many businesses and marginal communities can earn money from the sale of recycled materials. Besides this, the establishment of innovative business models and technological advancements for management of waste promotes economic growth and generates additional market prospects (Abbas et al., 2019; Khurshid & Khurshid, 2022). Zero waste management also follows the circular economy approach principles by making the economy more robust, sustainable, and lesser reliant on limited resources (Eriksen et al., 2019; Silva et al., 2020). However, working towards zero waste management objectives at the international, national as well as at municipal levels necessitates the amalgamation and advancement of zero waste activities through all-inclusive waste management policies (Anawar et al., 2019; Pietzsch et al., 2017). By filling customer demands for sustainability and environmental accountability, implementing zero waste management practices provides ventures for several sectors to simplify production processes, minimize material waste, and develop their status (Corsten et al., 2013; Usapein & Chavalparit, 2014). By tackling waste-related problems and adopting responsible resource management, zero waste management acts as a driving force for achieving sustainable development goals. Governments, businesses, communities, and individuals must work collectively with the aim of embracing and putting into practice the zero waste management principles in order to achieve this objective. With a focus on resource efficiency, waste reduction, and social cohesion with the environment, it offers a direction to a green and prosperous future.

4 Key Concepts and Principles of Zero Waste Management

The goal of zero waste management is to minimize waste generation, maximize resource recovery, and ultimately eliminate the need for landfills and incineration. It is based on a number of fundamental ideas and tenets that guide how it is used (Pires & Martinho, 2019; Yang et al., 2023). Let us examine these ideas and guidelines in further detail:

i. **Waste prevention**: The primary goal of zero waste management is to prevent waste from being generated in the first place. This involves promoting sustainable production and consumption practices, encouraging the design of products with extended lifespans and recyclability, and minimizing packaging waste. By focusing on waste prevention, the overall volume of waste can be significantly reduced (Abbas et al., 2019).
ii. **Circular economy**: Zero waste management aligns with the principles of a circular economy, where resources are kept in use for as long as possible through recycling, reusing, and repurposing. Instead of the traditional linear "take-make-dispose" model, a circular economy seeks to close the loop, creating a system where waste is seen as a valuable resource (Quicker et al., 2020; Yang, 2022).
iii. **Source separation and collection**: An essential aspect of zero waste management is the separation of waste at the source. This involves encouraging households, businesses, and industries to sort their waste into different categories such as recyclables, organic waste, and non-recyclables. Source separation facilitates efficient recycling and resource recovery processes (Zlamparet et al., 2018).
iv. **Extended producer responsibility (EPR)**: EPR is a key principle that holds producers responsible for the entire life cycle of their products, including the management of waste generated from those products. By implementing EPR programmes, manufacturers are incentivized to design products with a focus on recyclability and reusability (Franco-García et al., 2019; Tsai et al., 2013).
v. **Recycling and resource recovery**: Zero waste management prioritizes recycling and resource recovery as a means to minimize waste sent to landfills and incinerators. By recycling materials, valuable resources are conserved, reducing the need for new raw materials and lowering environmental impacts (Rios et al., 2019).
vi. **Composting**: Organic waste, such as food scraps and yard trimmings, is a significant component of municipal waste. Composting is an essential process in zero waste management, where organic waste is converted into nutrient-rich compost, which can be used to enrich soil and support sustainable agriculture (Hameed et al., 2021; Zlamparet et al., 2018).
vii. **Reuse and repair**: Encouraging the reuse and repair of products is fundamental to zero waste management. Instead of discarding items, the goal is to extend their lifespan by repairing and refurbishing them, reducing the overall demand for new products (Bolton & Rousta, 2019; Polettini, 2012).

viii. **Redesigning products and packaging**: Zero waste management advocates for product and packaging design that prioritizes sustainability. This involves using materials that are easily recyclable, reducing single-use plastics, and adopting innovative packaging solutions that minimize waste (Abbas et al., 2019; Bhat et al., 2018a; Hickel et al., 2013).

ix. **Education and awareness**: Public education and awareness play a crucial role in zero waste management. By informing and engaging communities about the importance of waste reduction and sustainable practices, individuals can make informed choices and actively participate in waste reduction efforts (Korhonen & Koivisto, 2007; Ma et al., 2017).

x. **Collaboration and partnerships**: Achieving zero waste requires collaboration and partnerships between various stakeholders, including governments, businesses, non-governmental organizations, and communities. By working together, different sectors can share knowledge, resources, and best practices to advance zero waste initiatives (Pires & Martinho, 2019).

xi. **Continuous improvement and innovation**: Zero waste management is an evolving process that requires continuous improvement and innovation. As technologies and practices advance, there are always opportunities to enhance waste management systems and achieve higher levels of waste diversion and resource recovery (Bertanza et al., 2018; Haapala et al., 2013).

By embracing these key concepts and principles, zero waste management presents a transformative vision for a more sustainable and circular future. It is a collective effort that involves the commitment of individuals, businesses, policymakers, and communities to reshape waste management practices and create a world where waste is minimized, resources are conserved, and environmental impacts are reduced.

5 5Rs of Zero Waste Management

The 5Rs of zero waste management are a group of guiding principles that emphasize waste prevention and reduction upfront. It aims to reduce the amount of waste produced and advance a more sustainable method of resource management (Tang et al., 2022). These guidelines serve as the basis for zero waste practises and offer a framework for people, organizations, and communities to embrace more ethical waste management practises (Abbas et al., 2019; Franco-García et al., 2019). The 5R's are:

i. **Refuse**: Refusing what is unnecessary is the first and most important step in zero waste management. This entails paying attention to the goods we use and the waste they produce. We may greatly minimize waste at the source by saying "no" to single-use and throwaway goods, superfluous packaging, and non-recyclable materials. Rejecting things that are not necessary or do not support our sustainability ideals reduces the amount of trash produced overall (Faibil et al., 2023; Peng et al., 2018).

ii. **Reduce**: It focuses on reducing material and product use. We may reduce the amount of trash created by being conscious of our shopping behaviour and choosing items with little packaging, robust design, and a longer lifespan. Consumption reduction reduces waste while also conserving energy, raw materials, and other production-related resources (Lange, 2021; Masud et al., 2019).
iii. **Reuse**: A crucial component of zero waste management is reuse of materials wherever possible. Instead of throwing things away after a single usage, we should look for methods to make them last longer. This may be accomplished by giving goods a new life through repair, renovation, or repurposing. Reusable items like water bottles, bags, and containers serve to lessen the demand for disposable alternatives and cut down on waste production (Eriksen et al., 2019; Meidiana et al., 2022; Silva et al., 2017).
iv. **Recycle**: Recycling is the act of turning garbage into new goods or materials in order to prevent it from going to a landfill or being burned. Even while recycling is a crucial part of waste management, it should only be used as a last resort if the first three Rs have been followed. Recycling materials must be properly sorted and separated in order to be processed and reused in the production of new goods (Larrain et al., 2022; Xue et al., 2013).
v. **Repurpose**: Repurposing or upcycling items that may no longer serve their original purpose is the fifth R in zero waste management. We can come up with inventive methods to give things or materials new uses rather than throwing them away. Repurposing objects encourages resourcefulness and creativity in addition to reducing waste (Agarwal & Agarwal, 2021; Flowers & Linderman, 2003) (Fig. 2).

6 Concept of Waste Hierarchy

Waste hierarchy is a key and frequently used concept that offers crucial direction for environmentally friendly waste management techniques. It describes a methodical process to rank waste treatment choices according to their sustainability and adverse effects on the environment. The waste hierarchy is arranged with the most desirable and eco-friendly practices at the top and the least appealing possibilities at the bottom (Fig. 3).

The overarching concept of the waste hierarchy is to "avoid and reduce waste". This level emphasizes how crucial it is to stop waste production at the source and reduce the quantity of waste created. Adopting proactive methods and sustainable practises that lower total demand for raw materials, energy, and resources is the objective here. Designing products with minimal packaging, encouraging sensible consumption patterns, and pressuring firms to use eco-friendly manufacturing techniques are all examples of waste avoidance measures. Societies may greatly reduce the demand for waste management and the accompanying environmental responsibilities by concentrating on trash avoidance and reduction (Gutberlet, 2015; Lesar et al.,

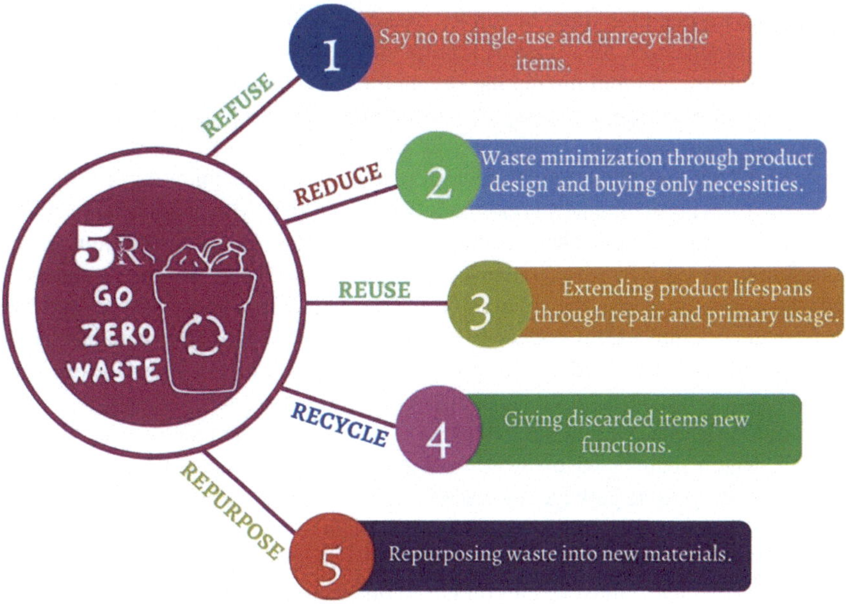

Fig. 2 5Rs of zero waste management

Fig. 3 Concept of waste hierarchy in zero waste management

2018). As we ascend the waste hierarchy, we come across the idea of "reuse waste". This level emphasizes the value of repurposing goods and resources, prolonging their useful lives, and minimizing the demand for new product development. Reuse projects entail restoring, repairing, or reusing objects to keep them from ending up in the trash. Reusing more minimizes not just the amount of trash that has to be disposed off but also the usage of precious resources and greenhouse gas emissions that come from producing new goods (Anawar et al., 2019; Mahongnao et al., 2023). The waste hierarchy's third level is "recycle waste". Recycling is essential to waste management since it involves gathering, processing, and reprocessing resources to make new goods. Recycling relieves strain on natural ecosystems by converting waste materials into useful resources. It also conserves energy and raw materials. Recycling also prevents garbage from going to landfills, which reduces the environmental impact of waste disposal and helps to promote a more sustainable and circular economy (Park, 2021; Tsai et al., 2013). The next step in the hierarchy is "recover energy", which focuses on using some waste materials' potential for energy. By turning non-recyclable garbage into useable energy sources through energy recovery techniques like waste-to-energy incineration, we can lessen our dependency on fossil fuels and reduce greenhouse gas emissions. Energy recovery supports waste management initiatives, but it must be careful not to undermine attempts to reduce trash or recycle (Mancini et al., 2019; Quicker et al., 2020). The waste hierarchy's fifth level is "treat waste". The recovery of valuable materials from organic waste is emphasized at this stage. Anaerobic digestion and composting are two examples of treatment processes that turn organic waste into valuable resources like compost that is rich in nutrients and biogas. In addition to lowering greenhouse gas emissions from landfills, these procedures offer a sustainable option for handling organic waste (Shukla & Khan, 2022). The final step in the waste hierarchy is "dispose of waste". Landfilling is a disposal strategy that is least desired and should only be used in cases where other waste management techniques are impractical. Landfills may have a big influence on the ecosystem, contaminating the soil and water, emitting greenhouse gases, and causing aesthetic blight. According to Abbas et al. (2019) and Goriaeva et al. (2000), the waste hierarchy seeks to reduce dependency on disposal techniques and place an emphasis on sustainable waste management solutions. Individuals, businesses, and governments may manage waste more sustainably by adhering to the waste hierarchy. Prioritizing waste avoidance, reuse, recycling, and resource recovery helps create a circular economy where resources are saved and waste production is reduced, which in turn minimizes environmental consequences. The trash hierarchy supports the objective of building a greener and more resilient future by directing us towards a more sustainable and responsible approach to waste management.

7 Innovative Technologies Driving Zero Waste Management

In order to achieve zero waste goals, technological advancements are especially important in influencing the landscape of waste management practises. In order to maximize resource recovery, lessen environmental effect, and promote sustainable waste management practises, the integration of cutting-edge technology is more important as the world's garbage output rises. We examine a variety of cutting-edge technologies in this part that have been developed to aid in the creation of zero waste management systems. These developments include waste-to-energy options, sophisticated recycling methods, intelligent waste management systems, and the incorporation of blockchain and artificial intelligence. The path to a world with zero waste is made more attainable and revolutionary by embracing these technical advancements.

i. **Waste-to-energy (WtE) technologies**: The goal of waste-to-energy technology is to transform non-recyclable garbage into usable energy sources, such as fuel, heat, or electricity (Mancini et al., 2019). There are various ways to turn garbage into energy, including pyrolysis, gasification, and incineration:

   - **Incineration**: This process involves burning waste at high temperatures in controlled environments. The heat generated during incineration can be used to produce steam, which drives turbines to generate electricity. Incineration significantly reduces the volume of waste and can be an effective solution for handling non-recyclable materials (Quicker et al., 2020).
   - **Gasification and pyrolysis**: Thermal processes called gasification and pyrolysis turn organic waste into synthetic gas (syngas) and charcoal, respectively. While biochar may be used as a soil improvement in agriculture, syngas can be utilized as a fuel for power generation or other industrial purposes (Mancini et al., 2019).

     WtE technologies pose questions about emissions and air pollution even though they can assist to lessen the environmental effects of waste disposal and offer a source of renewable energy. The environmental sustainability of WtE technologies depends on well-planned and run facilities with strong emission control systems.

ii. **Advanced recycling methods**: The goal of innovations in material recycling is to recover valuable resources from difficult-to-recycle complex items and materials. For instance, valuable metals like gold and palladium are recovered from discarded devices through electronic waste (e-waste) recycling technology. Furthermore, improvements in textile recycling allow for the extraction of fibres from used textiles in order to produce new textiles. Modern recycling methods put a strong emphasis on enhancing the efficacy and efficiency of recycling procedures, which results in better rates of material recovery and less trash production. Here are a few of them:

- **Mechanical-biological treatment (MBT)**: To separate recyclables from mixed garbage, MBT combines mechanical sorting and biological processing. The procedure entails shredding the trash to mechanically separate recyclables, followed by biological treatment, such as composting or anaerobic digestion, for the residual organic fraction (Ameli et al., 2019; Larrain et al., 2022).
- **Chemical recycling**: Recycling of feedstock, commonly referred to as chemical recycling, employs chemical procedures to disassemble plastic waste into its monomers, or fundamental components. Closing the loop and lowering the need for virgin plastics is possible by using these monomers to create new plastics (Lai et al., 2022; Park, 2021; Tsai et al., 2013).
- **Plasma gasification**: Plasma gasification is an advanced thermal conversion technology that uses high-temperature plasma to convert waste into a syngas. The syngas can be used for energy production, and the process can handle a wide range of waste materials, including hazardous waste.

  Advanced recycling techniques play a crucial role in diverting valuable resources from landfills and promoting a more circular economy where waste is viewed as a resource rather than a burden.

iii. **Smart waste management systems**: More effective garbage collection and processing are made possible by the use of smart waste management systems, which also promote more environmentally friendly waste management techniques. Technology and data analytics are used by smart waste management systems to improve garbage collection, transportation, and processing:

- **Sensor-based waste sorting**: Different forms of waste may be automatically identified and sorted using sensors that are mounted in trash cans or at waste sorting facilities. Due to this technology, recycling procedures are more effective and recyclable materials are less likely to be contaminated (Park, 2021; Tsai et al., 2013).
- **Route optimization**: The routes garbage trucks use to gather waste are optimized using sophisticated algorithms that analyse data on waste generating trends. As a consequence, waste management services use less fuel, emit fewer pollutants, and incur less expense (Lahane & Kant, 2022; Quicker et al., 2020).
- **Internet of things (IoT) integration**: Internet of things (IoT) integration: IoT gadgets may be used to track the amount of waste in bins, allowing prompt collections and avoiding overflow. Additionally, real-time monitoring of waste management procedures is made possible by IoT integration, which enables the detection and correction of inefficiencies (Ameli et al., 2019; Lange, 2021).

iv. **Biological nutrient recovery (BNR)**: BNR focuses on removing useful nutrients from organic waste and wastewater, such as phosphate and nitrogen. These nutrients can be recovered and reused, which eliminates the need for synthetic fertilizers and closes the nutrient loop. They are crucial for agricultural fertilizers. In addition to recovering nutrients, BNR technologies including struvite precipitation, ammonia recovery, and anaerobic digestion also generate biogas, a sustainable energy source (Lahane & Kant, 2022; Quicker et al., 2020).

v. **Zero waste manufacturing and design**: Optimizing product design to reduce waste generation during production and throughout the product's life cycle is the main goal of zero waste manufacturing and design practices. Utilizing eco-friendly materials, minimizing packaging, creating goods that can be disassembled, and applying closed-loop manufacturing techniques are all part of this. Industries may transition to zero waste production and sustainable consumption habits by using circular design concepts (Larrain et al., 2022; Lesar et al., 2018).

vi. **Distributed waste management solutions**: Smaller-scale waste treatment systems are placed closer to the point of waste creation as part of decentralized waste management solutions. Examples include home composting systems, neighbourhood recycling facilities, and anaerobic digesters for the disposal of organic waste. According to Agarwal and Agarwal (2021) and Mahongnao et al. (2023), decentralization can lower transportation costs, promote community engagement, and open up prospects for the recovery of local resources.

vii. **Artificial intelligence (AI) and machine learning**: To enhance workflows and decision-making, waste management operations are increasingly utilizing AI and machine learning technologies. Large volumes of data can be analysed by these technologies to forecast trash creation trends, improve sorting procedures, and spot possibilities for waste reduction. In addition to encouraging correct garbage separation, AI-driven smart bins can also boost recycling practises amongst the general population (Ameli et al., 2019; Lange, 2021).

viii. **Blockchain technology**: It is being investigated as a way to improve traceability and transparency in garbage management. It may be used to monitor waste streams, confirm recycling and disposal procedures, and make sure waste rules are being followed. Blockchain can promote a more effective and responsible waste management ecosystem and assist stakeholders develop confidence in one another (Bolton & Rousta, 2019; Yang et al., 2023).

ix. **Efficient product packaging**: Innovative packaging techniques seek to minimize waste production, especially in the consumer products industry. This includes supporting reusable packaging models, using compostable or biodegradable packaging materials, and designing packaging that uses as little material as possible (Meidiana et al., 2022; Park, 2021) (Fig. 4).

The path to zero waste is made more realistic and efficient by combining these technological advancements with current waste management practises and legislation. However, it is critical to understand that technology cannot reach zero waste targets on its own; it must be supported by robust regulatory frameworks, widespread public involvement, and a change in perspective towards the circular economy. For the subject of zero waste management to advance, ongoing research, cooperation amongst many stakeholders and constant technology advancement is essential.

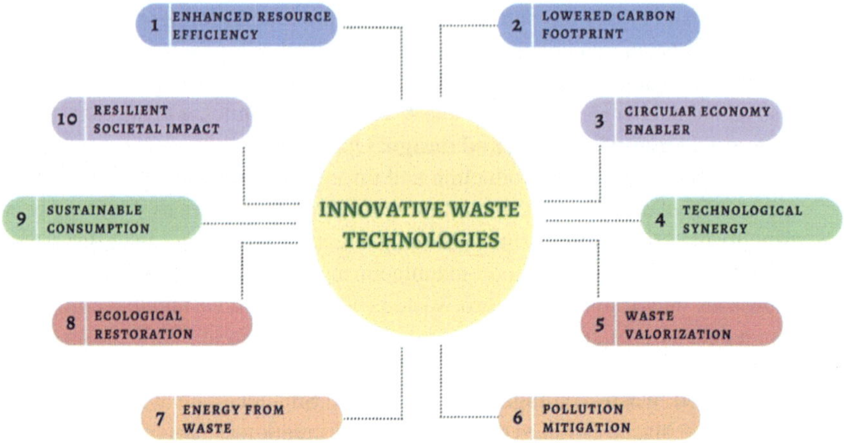

Fig. 4 Advantages of implementing innovative technologies for zero waste management

8 Policy and Legislative Framework for Zero Waste Management

A robust policy and legal framework that offers a clear roadmap, defines goals, and develops laws to direct waste management practises is necessary for the effective implementation of zero waste management. This framework is essential for encouraging sustainable practises, changing behaviour, and fostering an environment that supports resource recovery and waste reduction (Quicker et al., 2020; Rios et al., 2019). The main components of the policy and legal framework for zero waste management are explained in detail below:

i. **Waste management guidelines**: National waste management plans have been formed by several nations, outlining their vision and objectives for resource recovery and waste management. Specific goals for trash diversion, recycling rates, and landfill reduction are frequently included in these programmes. National waste management regulations set the foundation for regional and local implementation of zero waste projects and offer a strategic direction (Haapala et al., 2013; Zhang et al., 2022).
ii. **Waste hierarchy integration**: The inclusion of the waste hierarchy is a crucial component of the policy framework. The waste hierarchy places waste avoidance and reduction at the top of the list, followed by reuse, recycling, energy recovery, and disposal in order of environmental sustainability. Governments emphasize the value of resource recovery and waste reduction by incorporating the waste hierarchy into waste management policy (Faibil et al., 2023; Quicker et al., 2020).
iii. **Extended producer responsibility (EPR)**: Producers are held responsible for the environmental effect of their goods throughout their lifecycles, including the management of post-consumer waste, under the crucial policy approach

known as environmental product responsibility (EPR). Manufacturers are encouraged to build items that are simpler to recycle, repair, or reuse thanks to EPR rules. Additionally, it encourages the development of systems for the collecting and recycling of certain goods and materials, such as packaging, electronics, and batteries (Peng et al., 2018; Portugaise et al., 2023; Tsai et al., 2013).

iv. **Waste collection and separation arrangements**: The policy and regulatory framework must include efficient garbage collection and separation systems. To guarantee that distinct waste streams, such as recyclables, organic garbage, and non-recyclables, are collected separately, governments must enact regulations that encourage source separation at homes, companies, and enterprises. As a result, recycling and resource recovery may be done effectively (Park, 2021; Shukhla & Khan, 2022).

v. **Fixed recycling programmes**: Numerous jurisdictions have put in place obligatory recycling programmes that mandate that companies and households separate recyclables from the waste stream. These initiatives set recycling goals and may apply sanctions for failure to meet them. Governments might promote more involvement and raise recycling rates by making recycling mandatory (Tandon, 2020; Zaman & Ahsan, 2019).

vi. **Monetary enticements and provision**: Financial incentives can be included into the policy framework to promote recycling and waste minimization. For organizations and communities that make investments in waste reduction infrastructure and recycling facilities, these incentives may take the form of grants, subsidies, tax credits, or rebates. The market for recycled materials can also be developed with financial assistance.

vii. **Guidelines on landfilling and incineration**: Policies might establish stringent rules and charge higher costs for waste disposal in an effort to deter the dumping of garbage in landfills and incinerator plants. By creating financial disincentives for disposing of garbage in landfills or incinerators, these policies promote resource recovery and waste reduction (Zlamparet et al., 2018).

viii. **Civic responsiveness and training activities**: There may be provisions in the policy framework for public awareness and education efforts. These campaigns seek to educate the public on the value of recycling, waste reduction, and good waste management techniques. Public education can encourage behavioural changes and boost involvement in garbage reduction initiatives (Portugaise et al., 2023).

ix. **Scrutinizing and reporting means**: Governments can set up reporting and monitoring systems to keep tabs on the development of zero waste objectives. In order to evaluate the success of policies and pinpoint areas for improvement, regular data collection and reporting on trash generation, recycling rates, and landfill diversion are necessary (Abbas et al., 2019; Bertanza et al., 2018).

x. **Collaboration and stakeholder participation**: Collaboration between governmental bodies, corporations, non-governmental organizations, communities, and other stakeholders is essential for the achievement of zero waste management. To progress zero waste projects, the policy framework should promote stakeholder involvement, public-private collaborations, and knowledge exchange (Pires & Martinho, 2019).

xi. **Novelty and exploration provision**: Policies can encourage the development of new waste management techniques and technologies. Governments may encourage technical breakthroughs that improve trash diversion, recycling effectiveness, and resource recovery by investing in research and development (Tandon, 2020).

The basis of effective zero waste management is a strong policy and legal framework. It offers the required instructions, rewards, and rules to encourage resource recovery, waste reduction, and environmentally friendly waste management techniques. Governments may strive towards attaining their zero waste objectives and promoting a more sustainable future for the world when these efforts are backed by comprehensive legislation.

9 Social and Behavioural Aspects of Zero Waste Management

Initiatives for zero waste management are successfully implemented when social and behavioural factors are taken into account. These factors include people's attitudes, beliefs, values, and behaviours towards waste management practises in communities and organizations. Promoting waste reduction, recycling, and sustainable consumption habits requires an understanding of and attention to social and behavioural variables (Silva et al., 2017; Yang, 2022). As it relates to the social and behavioural components of zero waste management, the following major considerations should be noted:

i. **Public education and awareness**: Promoting public support and engagement in zero waste management requires increasing public understanding of its significance. People can be made aware of the negative effects that trash has on the environment, the advantages of recycling and composting, and the necessity of reducing waste at the source through educational programmes. People are more likely to adopt sustainable waste management practises when they are given knowledge and encouraged to feel responsible (Tang et al., 2022; Tsai et al., 2013).

ii. **Community involvement**: Successfully implementing zero waste projects depends on community involvement. Participation in the community promotes communication of thoughts, issues, and suggestions, fostering a sense of ownership and group accountability. Communities can collaborate to identify problems with waste management, provide solutions, and put into action programmes that are suited to their unique requirements and preferences (Faibil et al., 2023; Polettini, 2012).

iii. **Strategies for changing behaviour**: Effective behaviour modification techniques may motivate people and companies to embrace zero waste procedures. These tactics can include offering recycling incentives, promoting recyclable goods, putting trash sorting programmes in place, and making recycling facilities more accessible. To promote sustainable waste management practises, behaviour modification campaigns might make use of social norms, peer pressure, and rewarding behaviour (Peng et al., 2018).
iv. **Cultural and societal norms**: Individual behaviours can be influenced by cultural norms and societal attitudes towards trash. Implementing zero waste practises may be made simpler in some cultures where resource conservation and waste reduction may already be firmly engrained. In contrast, efforts may be required to change societal norms and values away from disposability in societies where this is the norm (Mancini et al., 2019).
v. **Intergroup cooperation**: Collaboration amongst many stakeholders, including governments, corporations, non-governmental organizations, and neighbourhood associations, is necessary for zero waste management. Designing and implementing comprehensive waste management policies and programmes requires effective communication and coordination between these groups (Portugaise et al., 2023).
vi. **Policy and regulations**: The legal and policy frameworks can have an impact on social and behavioural characteristics. By introducing supporting policies like extended producer responsibility, landfill bans, and recycling objectives, governments may play a significant role in advancing zero waste management. Regulations have the power to foster the adoption of sustainable waste management practises by both people and companies (Larrain et al., 2022; Yang, 2022).
vii. **Infrastructure and accessibility**: People's decisions on how to dispose of their garbage can be influenced by the accessibility and availability of waste management infrastructure, such as composting and recycling facilities. Recycling rates may be hampered by a lack of easy recycling choices. For trash diversion to be effective, it is essential to make sure that appropriate garbage collection and processing facilities are available to all areas (Ameli et al., 2019; Luidold, 2019).
viii. **Social equality**: When managing zero waste, social equality issues need to be taken into account. The availability and affordability of waste management programmes and services for all socioeconomic categories must be guaranteed. To address possible discrepancies in waste management access and benefits, targeted measures may be required (Zlamparet et al., 2018).

For the effective adoption of zero waste practises, it is essential that the social and behavioural components of waste management are addressed right now. Societies may transition to a more sustainable and circular waste management system by raising public awareness, involving communities, encouraging behaviour change, and enacting supporting regulations. A more environmentally conscious society will be developed via the use of socially inclusive and behaviourally informed tactics.

10 Future Outlook of Zero Waste Management

As the world shifts in favour of sustainable practises, the prospects for zero waste management are bright. According to Xue et al. (2013) and Ma et al. (2017), governments, businesses, and communities are becoming more aware of the pressing need to solve waste problems and transition to more circular and sustainable waste management systems. Future prospects for zero waste management are defined by a few important factors:

i. **Technological developments**: Future waste management practises will be heavily influenced by technology. Technologies for sorting, processing, and recycling garbage will advance resource recovery, minimize waste transported to landfills, and lessen the impact of waste disposal on the environment. Robotics, machine learning, and artificial intelligence combined will enhance sorting precision, streamline garbage collection routes, and create more effective waste-to-energy systems (Ameli et al., 2019; Faibil et al., 2023).
ii. **Circular economy shift**: Zero waste management will be fueled by the move towards a circular economy. According to Zlamparet et al. (2018), businesses will emphasize product durability, repairability, and recyclability by adopting circular design concepts. This change will result in more closed-loop systems, which reduce waste output and save resources by reusing, refurbishing, or recycling items and materials (Luidold, 2019).
iii. **Policy and regulation**: To promote the adoption of zero waste practises, strong policy backing and legal frameworks are necessary. Governments will implement waste management plans that put an emphasis on recycling, reduce trash, and create extended producer responsibility (EPR) rules. EPR will encourage sustainable design and waste reduction by holding producers responsible for managing the end-of-life of their goods (Portugaise et al., 2023; Silva et al., 2020).
iv. **Community engagement and public awareness**: Promoting a zero waste culture will need significant community involvement and public awareness. People will be empowered by educational programmes and awareness campaigns to make ecologically responsible decisions, such as limiting the use of single-use plastics, properly sorting garbage, and supporting sustainable goods and companies (Zlamparet et al., 2018).
v. **Waste-to-energy and resource retrieval**: Waste-to-energy technology and resource recovery techniques will keep developing, providing sustainable energy options while keeping garbage out of landfills. The generation of energy from organic waste will be further improved by advances in gasification, pyrolysis, and anaerobic digestion, which will also open up new potential for biogas and biofuel applications (Yao et al., 2022; Zaman & Ahsan, 2019).

vi. **Collaboration and information sharing**: Sharing information, best practises, and experiences in zero waste management will need close cooperation between governments, businesses, academic institutions, and NGOs. International collaboration helps governments learn from successful waste management projects throughout the world and accelerates development (Bartl, 2011; Faibil et al., 2023; Polettini, 2012).
vii. **Sustainable consumption arrays**: The zero waste movement will be significantly fuelled by consumer behaviour. Growing consumer demand for eco-friendly goods and packaging will encourage companies to adopt green practises and stop waste before it starts (Silva et al., 2020).
viii. **Innovation and start-ups**: By creating new technologies and business models that support waste reduction, recycling, and upcycling, innovative start-ups and entrepreneurs will contribute to zero waste management. These endeavours will encourage waste management companies to adopt sustainable practises (Ameli et al., 2019).
ix. **Zero waste capitals and societies**: To eliminate waste, cities and communities will embrace ambitious zero waste targets. To execute comprehensive trash reduction plans, boost recycling rates, and develop circular waste management systems, local governments will work with citizens, companies, and organizations (Luidold, 2019; Ma et al., 2017).
x. **Combating climate change**: Zero waste management is consistent with more general measures to combat climate change. Zero waste practises help the world's efforts to fight climate change and lessen environmental degradation by lowering waste creation and greenhouse gas emissions from landfills and incinerators (Faibil et al., 2023; Polettini, 2012).

A cleaner, more sustainable future with less waste and more resource efficiency is presented by zero waste management. Societies may get closer to reaching zero waste targets and building a more sustainable future for future generations by embracing technology advancements, adopting circular economy ideas, enacting helpful legislation, and encouraging community involvement.

11 Conclusion

This chapter emphasized the critical importance of a comprehensive and sustainable approach to waste management. It explores the key concepts and principles of zero waste management and provides a roadmap for minimizing waste generation, maximizing resource recovery and moving towards a circular economy. Focusing on prevention, reuse, recycling and responsible disposal of waste, the waste hierarchy serves as a framework for making informed waste management decisions. By prioritizing waste prevention and adopting innovative recycling and resource recovery strategies, society can significantly reduce its environmental footprint and promote a more sustainable future. The concept of zero waste management fits seamlessly into the circular economy principle, where waste is seen as a valuable resource rather than a burden. This paradigm shift challenges traditional linear consumption patterns and promotes a systematic approach that emphasizes product longevity and recyclability. The integration of innovative technologies such as waste-to-energy conversion technology and advanced recycling technology promises to optimize resource recovery and reduce waste sent to landfills. Furthermore, zero waste management goes beyond technical and infrastructure solutions. The importance of social and behavioural aspects, public awareness, education and engagement are essential elements in promoting waste reduction behaviour and fostering a culture of sustainability has been highlighted. Furthermore, the principle of extended producer responsibility emphasizes the joint responsibility of producers and consumers in waste management. Zero waste management encourages the development of environmentally friendly and recyclable products by holding manufacturers accountable for the entire product lifecycle. Political and legal frameworks play an important role in the implementation of zero waste practices. Supportive policies, regulations and incentives can create an enabling environment for reducing waste and conserving resources. Harnessing collective efforts and resources towards the common goal of zero waste requires cooperation amongst various sectors such as governments, businesses, NGOs and communities. To pave the way for a more sustainable and waste conscious society, it is imperative that policymakers, industry, communities and individuals jointly adopt the principles of zero waste agriculture. In this way, we can minimize waste, conserve resources and create a future together where environmental protection is the cornerstone of the global progress towards sustainability. After all, zero waste management is not just an aspiration. This is a viable and transformative approach to achieving sustainable development. Towards a more resilient and sustainable future where waste is minimized, resources are valued and environmental impact is reduced by adopting principles of waste prevention, recycling and circular economy you can work on it. The road to zero waste requires commitment, innovation and collaboration at all levels and is worth working for the well-being of our planet and future generations.

References

Abbas, J., Mahmood, S., Ali, H., Ali Raza, M., Ali, G., Aman, J., Bano, S., & Nurunnabi, M. (2019). The effects of corporate social responsibility practices and environmental factors through a moderating role of social media marketing on sustainable performance of business firms. *Sustainability, 11*(12), 3434. https://doi.org/10.3390/su11123434

Agarwal, V., & Agarwal, S. (2021). Performance profiling of solid lubricant for eco-friendly sustainable manufacturing. *Journal of Manufacturing Processes, 64*, 294–305. https://doi.org/10.1016/j.jmapro.2020.12.070

Ameli, M., Mansour, S., & Ahmadi-Javid, A. (2019). A simulation-optimization model for sustainable product design and efficient end-of-life management based on individual producer responsibility. *Resources, Conservation and Recycling, 140*, 246–258. https://doi.org/10.1016/j.resconrec.2018.02.031

Anawar, H. M., Strezov, V., & Adyel, T. M. (2019). The reuse and recycling of coal mining waste with zero-waste approach by technological development and integrated management for sustainable growth and benefits. *Sustainable and Economic Waste Management,* 31–46. https://doi.org/10.1201/9780429279072

Awasthi, A. K. (2023). Zero waste: A potential strategy for sustainable waste management. *Waste Management & Research: The Journal of the International Solid Wastes and Public Cleansing Association, ISWA, 41*(6), 1061–1062. https://doi.org/10.1177/0734242X231170264

Bartl, A. (2011). Barriers towards achieving a zero waste society. *Waste Management, 31*(12), 2369–2370. https://doi.org/10.1016/j.wasman.2011.09.013

Bertanza, G., Ziliani, E., & Menoni, L. (2018). Techno-economic performance indicators of municipal solid waste collection strategies. *Waste Management, 74,* 86–97. https://doi.org/10.1016/j.wasman.2018.01.009

Bhat, R. A., Dar, G. A., Jehangir, A., Bhat, B. A., & Yousuf, A. R. (2012). Municipal solid waste generation and present scenario of waste management during Yatra season in Pahalgam: A tourist health resort of Kashmir valley. *International Journal of Current Research, 4,* 004–009.

Bhat, R. A., Dar, S. A., Dar, D. A., & Dar, G. H. (2018a). Municipal solid waste generation and current scenario of its management in India. *International Journal of Advance Research in Science and Engineering, 7*(2), 419–431.

Bhat, R. A., Dervash, M. A., Mehmood, M. A., & Hakeem, K. R. (2018b). Municipal solid waste generation and its management, a growing threat to fragile ecosystem in Kashmir Himalaya. *American Journal of Environmental Sciences.* https://doi.org/10.3844/ajessp.2018

Bhat, R. A., Kamili, A. N., & Bandh, S. A. (2013). Characterisation and composition of municipal solid waste (MSW) generated in Yusmarg: A health resort of Kashmir valley. A glance at the world. *Waste Management, 33,* 774–777.

Bhat, R. A., Nazir, R., Ashraf, S., Ali, M., Bandh, S. A., & Kamili, A. N. (2014). Municipal solid waste generation rates and its management at Yusmarg forest ecosystem, a tourist resort in Kashmir. *Waste Management and Research, 32,* 165–169.

Bhat, R. A., Singh, D. V., Qadri, H., Dar, G. H., Dervash, M. A., Bhat, S. A., Unal, B. T., Ozturk, M., Hakeem, K. R., & Yousaf, B. (2021). Vulnerability of municipal solid waste: An emerging threat to aquatic ecosystems. *Chemosphere, 287,* 132223.https://doi.org/10.1016/j.chemosphere.2021.132223

Bolton, K., & Rousta, K. (2019). Solid waste management toward zero landfill: A Swedish model. In *Sustainable resource recovery and zero waste approaches* (pp. 53–63). Elsevier. https://doi.org/10.1016/B978-0-444-64200-4.00004-9

Brancoli, P., & Bolton, K. (2019). Life cycle assessment of waste management systems. In *Sustainable resource recovery and zero waste approaches* (pp. 23–33). Elsevier. https://doi.org/10.1016/B978-0-444-64200-4.00002-5

Corsten, M., Worrell, E., Rouw, M., & Van Duin, A. (2013). The potential contribution of sustainable waste management to energy use and greenhouse gas emission reduction in the Netherlands. *Resources, Conservation and Recycling, 77,* 13–21. https://doi.org/10.1016/j.resconrec.2013.04.002

Diliberto, C., Meux, E., Diliberto, S., Garoux, L., Marcadier, E., Rizet, L., & Lecomte, A. (2020). A zero-waste process for the management of MSWI fly ashes: Production of ordinary Portland cement. *Environmental Technology, 41*(9), 1199–1208. https://doi.org/10.1080/09593330.2018.1525434

Eriksen, M. K., Christiansen, J. D., Daugaard, A. E., & Astrup, T. F. (2019). Closing the loop for PET, PE and PP waste from households: Influence of material properties and product design for plastic recycling. *Waste Management, 96,* 75–85. https://doi.org/10.1016/j.wasman.2019.07.005

Faibil, D., Asante, R., Agyemang, M., Addaney, M., & Baah, C. (2023). Extended producer responsibility in developing economies: Assessment of promoting factors through retail electronic firms for sustainable e-waste management. *Waste Management & Research: The Journal of the*

International Solid Wastes and Public Cleansing Association, ISWA, 41(1), 117–142. https://doi.org/10.1177/0734242X221105433

Flowers, A. D., & Linderman, K. (2003). Hazardous waste disposal: A waste-fuel blending approach. *Production and Operations Management, 12*(3), 307–319. https://doi.org/10.1111/j.1937-5956.2003.tb00205.x

Franco-García, M. L., Carpio-Aguilar, J. C., & Bressers, H. (2019). *Towards zero waste, circular economy boost: waste to resources* (pp. 1–8). Springer International Publishing.

Goriaeva, O. Y., Skorohodov, V. I., Naboychenko, S. S., & Verhaege, M. A. (2000). A new technology for the processing of precious metal containing secondary raw materials. In *Recycling of metals and engineered materials* (pp. 731–737). https://doi.org/10.1002/9781118788073.ch65

Gutberlet, J. (2015). Cooperative urban mining in Brazil: Collective practices in selective household waste collection and recycling. *Waste Management, 45*, 22–31. https://doi.org/10.1016/j.wasman.2015.06.023

Haapala, K. R., Zhao, F., Camelio, J., Sutherland, J. W., Skerlos, S. J., Dornfeld, D. A., Jawahir, I. S., Clarens, A. F., & Rickli, J. L. (2013). A review of engineering research in sustainable manufacturing. *Journal of Manufacturing Science and Engineering, 135*(4), 041013. https://doi.org/10.1115/1.4024040

Hameed, M., Bhat, R. A., Pandit, B. A., Ramzan, S., Dijoo, Z. K., & Wani, M. A. (2021). Qualitative assessment of compost engendered from municipal solid waste and green waste by indexing method. *Journal of the Air & Waste Management Association*. https://doi.org/10.1080/10962247.2021.1959466

Hickel, R., Brüshaver, K., & Ilie, N. (2013). Repair of restorations—Criteria for decision making and clinical recommendations. *Dental Materials, 29*(1), 28–50. https://doi.org/10.1016/j.dental.2012.07.006

Ibrahim, S. (2020). Waste management in Balikpapan city based on circular economy. In *Waste management as economic industry towards circular economy* (pp. 11–16). https://doi.org/10.1007/978-981-15-1620-7_2

Khurshid, Z. (2019). Management of urban solid waste in India. *International Journal of Research and Analytical Reviews (IJRAR), 6*(1), (E-ISSN 2348-1269, P- ISSN 2349-5138).

Khurshid, Z., & Khurshid, R. (2022). Environmental degradation as a multifaceted consequence of human development. In *Environmental biotechnology: Sustainable remediation of contamination in different environs.* New York, NY, USA: CRC Press.

Korhonen, H., & Koivisto, E. M. (2007). Playability heuristics for mobile multi-player games. In *Proceedings of the 2nd International Conference on Digital Interactive Media in Entertainment and Arts* (pp. 28–35). https://doi.org/10.1145/1306813.1306828

Lahane, S., & Kant, R. (2022). Investigating the sustainable development goals derived due to adoption of circular economy practices. *Waste Management, 143*, 1–14. https://doi.org/10.1016/j.wasman.2022.02.016

Lai, W. L., Sharma, S., Roy, S., Maji, P. K., Sharma, B., Ramakrishna, S., & Goh, K. L. (2022). Roadmap to sustainable plastic waste management: A focused study on recycling PET for triboelectric nanogenerator production in Singapore and India. *Environmental Science and Pollution Research International, 29*(34), 51234–51268. https://doi.org/10.1007/s11356-022-20854-2

Lange, J. P. (2021). Managing plastic waste—Sorting, recycling, disposal, and product redesign. *ACS Sustainable Chemistry & Engineering, 9*(47), 15722–15738. https://doi.org/10.1021/acssuschemeng.1c05013

Larrain, M., Billen, P., & Van Passel, S. (2022). The effect of plastic packaging recycling policy interventions as a complement to extended producer responsibility schemes: A partial equilibrium model. *Waste Management, 153*, 355–366. https://doi.org/10.1016/j.wasman.2022.09.012

Leal Filho, W., Saari, U., Fedoruk, M., Iital, A., Moora, H., Klöga, M., & Voronova, V. (2019). An overview of the problems posed by plastic products and the role of extended producer responsibility in Europe. *Journal of Cleaner Production, 214*, 550–558. https://doi.org/10.1016/j.jclepro.2018.12.256

Lehmann, S. (2011). Optimizing urban material flows and waste streams in urban development through principles of zero waste and sustainable consumption. *Sustainability, 3*(1), 155–183. https://doi.org/10.3390/su3010155

Lesar, B., Humar, M., & Hora, G. (2018). Quality assessment of recycled wood with and without non-wooden materials from selected recycling companies in Europe. *Waste Management, 79*, 362–373. https://doi.org/10.1016/j.wasman.2018.08.002

Luidold, S. (2019). Recycling of technologic metals: Fundamentals and technologies. In *Sustainable resource recovery and zero waste approaches* (pp. 223–238). Elsevier. https://doi.org/10.1016/B978-0-444-64200-4.00016-5

Ma, Y., Yin, Y., & Liu, Y. (2017). A holistic approach for food waste management towards zero-solid disposal and energy/resource recovery. *Bioresource Technology, 228*, 56–61. https://doi.org/10.1016/j.biortech.2016.12.090

Mahongnao, S., Sharma, P., & Nanda, S. (2023). Conversion of waste materials into different by-products of economic value. In *Waste management and resource recycling in the developing world* (pp. 665–699). Elsevier. https://doi.org/10.1016/B978-0-323-90463-6.00030-0

Mancini, E., Arzoumanidis, I., & Raggi, A. (2019). Composting and anaerobic digestion of organic urban waste: A systematic literature review of life cycle assessment case studies. In *Advances in waste-to-energy technologies* (pp. 249–264). CRC Press. https://doi.org/10.1201/9780429423376

Masud, M. H., Akram, W., Ahmed, A., Ananno, A. A., Mourshed, M., Hasan, M., & Joardder, M. U. H. (2019). Towards the effective E-waste management in Bangladesh: A review. *Environmental Science and Pollution Research International, 26*(2), 1250–1276. https://doi.org/10.1007/s11356-018-3626-2

Meidiana, C., Kurniawan, T. A., Yudono, A., & Surjono, S. (2022). Community-based waste management model in optimizing waste reduction: Waste bank practices in Indonesia. In *Modern challenges and approaches to humanitarian engineering* (pp. 98–113). IGI Global. https://doi.org/10.4018/978-1-7998-9190-1.ch006

Park, S. (2021). Assessing the recycling performance of extended producer responsibility (EPR) programs: A case study of the steel can packaging recycling system in South Korea. *Journal of the Air & Waste Management Association, 71*(5), 586–596. https://doi.org/10.1080/10962247.2020.1866120

Peng, B., Tu, Y., Elahi, E., & Wei, G. (2018). Extended producer responsibility and corporate performance: Effects of environmental regulation and environmental strategy. *Journal of Environmental Management, 218*, 181–189. https://doi.org/10.1016/j.jenvman.2018.04.068

Pietzsch, N., Ribeiro, J. L. D., & de Medeiros, J. F. (2017). Benefits, challenges and critical factors of success for zero waste: A systematic literature review. *Waste Management (New York, N.Y.), 67*, 324–353. https://doi.org/10.1016/j.wasman.2017.05.004

Pires, A., & Martinho, G. (2019). Waste hierarchy index for circular economy in waste management. *Waste Management, 95*, 298–305. https://doi.org/10.1016/j.wasman.2019.06.014

Polettini, A. (2012). Waste and climate change: Can appropriate management strategies contribute to mitigation? *Waste Management, 32*(8), 1501–1502. https://doi.org/10.1016/j.wasman.2012.04.017

Portugaise, M. K., Jóhannsdóttir, L., & Murakami, S. (2023). Extended producer responsibility's effect on producers' electronic waste management practices in Japan and Canada: Drivers, barriers, and potential of the urban mine. *Discover Sustainability, 4*(1), 8. https://doi.org/10.1007/s43621-023-00124-y

Quicker, P., Consonni, S., & Grosso, M. (2020). The zero waste utopia and the role of waste-to-energy. *Waste Management & Research: The Journal of the International Solid Wastes and Public Cleansing Association, ISWA, 38*(5), 481–484. https://doi.org/10.1177/0734242X20918453

Rios, F. C., Grau, D., & Chong, W. K. (2019). Reusing exterior wall framing systems: A cradle-to-cradle comparative life cycle assessment. *Waste Management, 94*, 120–135. https://doi.org/10.1016/j.wasman.2019.05.040

Shukla, S., & Khan, R. (2022). Sustainable waste management approach: A paradigm shift towards zero waste into landfills. In *Advanced organic waste management* (pp. 381–395). Elsevier. https://doi.org/10.1016/B978-0-323-85792-5.00006-X

Silva, A., Rosano, M., Stocker, L., & Gorissen, L. (2017). From waste to sustainable materials management: Three case studies of the transition journey. *Waste Management (New York, N.Y.), 61*, 547–557. https://doi.org/10.1016/j.wasman.2016.11.038

Silva, A. L. P., Prata, J. C., Walker, T. R., Campos, D., Duarte, A. C., Soares, A. M., Barcelò, D., & Rocha-Santos, T. (2020). Rethinking and optimising plastic waste management under COVID-19 pandemic: Policy solutions based on redesign and reduction of single-use plastics and personal protective equipment. *Science of the Total Environment, 742*, 140565. https://doi.org/10.1016/j.scitotenv.2020.140565

Tandon, N. (2020). Creating "zero-waste campuses". In *Sustainable waste management: Policies and case studies: 7th IconSWM—ISWMAW 2017, volume 1* (pp. 183–186). Springer Singapore. https://doi.org/10.1007/978-981-13-7071-7_16

Tang, C., Xue, Y., Wu, H., Irfan, M., & Hao, Y. (2022). How does telecommunications infrastructure affect eco-efficiency? Evidence from a quasi-natural experiment in China. *Technology in Society, 69*, 101963. https://doi.org/10.1016/j.techsoc.2022.101963

Tsai, T. H., Wu, S. J., & Hwang, H. (2013). Waste recycling policies under extended producer responsibility: Take-back mandate versus deposit-refund. *Global Journal of Economics, 2*(02), 1350005. https://doi.org/10.1142/S2251361213500055

Usapein, P., & Chavalparit, O. (2014). Development of sustainable waste management toward zero landfill waste for the petrochemical industry in Thailand using a comprehensive 3R methodology: A case study. *Waste Management & Research: The Journal of the International Solid Wastes and Public Cleansing Association, ISWA, 32*(6), 509–518. https://doi.org/10.1177/0734242X14533604

Xue, M., Li, J., & Xu, Z. (2013). Management strategies on the industrialization road of state-of-the-art technologies for e-waste recycling: The case study of electrostatic separation—A review. *Waste Management & Research, 31*(2), 130–140. https://doi.org/10.1177/0734242X12465464

Yang, G., Zhang, Q., Zhao, Z., & Zhou, C. (2023). How does the "zero-waste city" strategy contribute to carbon footprint reduction in China? *Waste Management (New York, N.Y.), 156*, 227–235. https://doi.org/10.1016/j.wasman.2022.11.032

Yang, J. (2022). Zero waste theory and practice. In *From zero waste to material closed loop: The way towards circular economy* (pp. 103–157). Springer Nature Singapore. https://doi.org/10.1007/978-981-16-7683-3_8

Yao, Q., Kuai, L., & Wang, C. L. (2022). How frontline employees' communication styles affect consumers' willingness to interact: The boundary condition of emotional ability similarity. *Journal of Retailing and Consumer Services, 68*, 103082. https://doi.org/10.1016/j.jretconser.2022.103082

Zaman, A., & Ahsan, T. (2019). *Zero-waste: Reconsidering waste management for the future.* Routledge.

Zhang, S., Tan, Q., Xu, G., & Li, J. (2022). Unveiling characteristics and trend of zero waste research: A scientometric perspective. *Environmental Science and Pollution Research International, 29*(29), 44391–44403. https://doi.org/10.1007/s11356-021-18048-3

Zlamparet, G. I., Tan, Q., Stevels, A. B., & Li, J. (2018). Resource conservation approached with an appropriate collection and upgrade-remanufacturing for used electronic products. *Waste Management, 73*, 78–86. https://doi.org/10.1016/j.wasman.2017.11.053

Chapter 2
Advanced Waste Management Strategies for Sustainable Development

Syed Maqbool Geelani and Aminoddin Haji

Abstract As waste generation rises globally, implementing zero waste strategies through advanced technologies and systems is becoming an urgent priority. This article provides a comprehensive overview of key components of modern zero waste management. It examines waste reduction approaches, collection methods, separation techniques, biological treatment processes, thermal and chemical conversion technologies, recycling systems, composting, policy tools, and public engagement tactics. For each aspect, current methods are explained, implementation challenges are discussed, and future directions are considered. Employing this range of solutions can help communities maximize landfill diversion, resource recovery, and environmental sustainability. The article aims to provide a technical yet accessible portrait of the path to zero waste.

Keywords Recycling · Composting · Waste to energy · Circular economy · Source separation

1 Introduction

Solid waste management is a major challenge facing urban areas worldwide. As populations grow and consumption rises, the amount of waste generated continues to increase. Managing this waste responsibly and sustainably has become an urgent priority (Bhat et al., 2012). Zero waste management represents a holistic approach aimed at reducing waste generation and recovering as many materials from the waste stream as possible (Zaman & Lehmann, 2011). Implementing zero waste strategies requires utilizing various technologies and systems that together enable communities to maximize diversion from landfills (Czekała et al., 2023a). This article provides an

S. M. Geelani (✉)
Department of Environmental Science, Sri Pratap College, Cluster University, Srinagar 190001, India
e-mail: geelani111@gmail.com

A. Haji
Department of Textile Engineering, Yazd University, Yazd, Iran

R. A. Bhat et al. (eds.), *Zero Waste Management Technologies*,
https://doi.org/10.1007/978-3-031-57275-3_2

Fig. 1 Comparison of two ecosystems: **a** environment with waste, **b** environment without waste

in-depth look at key technologies and infrastructure that make up modern zero waste management. It covers waste reduction strategies, collection methods, separation and sorting systems, biological and chemical conversion technologies, waste-to-energy systems, recycling methods, composting, and innovations in policy, regulation, and public engagement (Fatimah et al., 2020). For each aspect, current technologies are explained, implementation challenges discussed, and future directions explored. Taken together, advancing these systems can help cities around the world make progress toward zero waste goals and circular economies (Vu et al., 2020). A glimpse of waste impact on aesthetic quality of different ecosystems is depicted in Fig. 1.

1.1 Waste Minimization and Reduction

The first priority of zero waste management is to reduce the amount of waste generated in the first place (Noori et al., 2022). Less material entering the waste stream means lower management costs and environmental impacts (Ngoc & Schnitzer, 2009). Waste reduction starts with smarter product and packaging design (Blumbergs et al., 2022).

1.1.1 Strategies for Waste Minimization Include

Design for durability: Making products that are long-lasting, repairable, and easy to maintain. This reduces breakdowns and the need for replacement.

Design for reuse: Creating products and packaging meant to be used multiple times and enabling take-back systems. Reusable decreases single-use disposables.

Use of recycled materials: Incorporating post-consumer recycled materials into manufacturing reduces virgin resource demands and makes use of available recycled stocks (Saeedi et al., 2023).

Minimal and recyclable packaging: Eliminating excess packaging, using recyclable and recycled packaging materials, and enabling easy separation boosts recovery (Saeedi et al., 2023).

Eliminating toxic materials: Substituting or phasing out materials containing heavy metals, chemicals, and other toxics prevents hazardous waste.

Dematerialization: Relying more on virtual services, digital products, and technologies like 3D printing to cut down on physical production.

Product stewardship: When manufacturers are responsible for end-of-life management, it incentivizes design improvements.

Consumer education: Teaching consumers about waste impacts and sustainable purchasing helps drive demand for minimization.

Government policies like extended producer responsibility laws, disposal bans, procurement standards, and eco-design regulations all encourage waste reduction by shifting responsibility and setting top-down targets (Khetriwal et al., 2009). But design innovation by manufacturers and brand owners aided by consumer pressure is essential for driving change. The various reliable waste management strategies that promote sustainable development are depicted in Fig. 2.

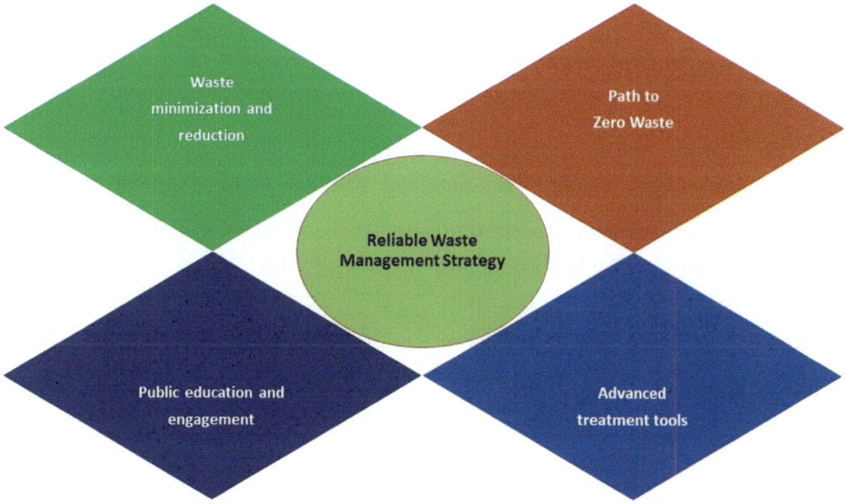

Fig. 2 Advanced waste management sustainable strategies

2 Waste Collection

An effective solid waste collection system is the necessary first step in recovering materials after waste generation (Al-Salem et al., 2009). Collection methods must be convenient, efficient, and suited to the types and volumes of waste produced (Bhat et al., 2014; Rawat et al., 2013). Common collection streams include:

Municipal solid waste (MSW): Household waste collected curbside, at drop-off sites, or at transfer stations. MSW is a mix of organics, recyclables, and other materials (Al Sa'di, 2009; Bhat et al., 2018a).

Commercial and industrial waste: Businesses and industrial facilities generate significant waste requiring collection tailored to their needs (Bhat et al., 2013). On-site separation of recyclables is encouraged.

Construction and demolition debris (C&D): Building projects generate large quantities of materials including wood, drywall, metals, masonry, glass, and plastics. This often allows source separation.

Hazardous and special wastes: Toxic, radioactive, electronic, medical, and other regulated wastes require specialized collection and management under strict guidelines (Ostad-Ali-Askari, 2022).

Organics: Food and yard waste collection enables composting. Many cities now provide separate green bins for these materials.

2.1 Key Collection Methods Include

Curbside pickup: MSW collection trucks make rounds through neighborhoods picking up bins left at the curb. Recycling and organics can have separate bins.

Drop-off centers: Residents transport materials to central sites with bins for various waste and recycling streams. This requires consumer effort.

Deposit return schemes: A deposit is paid when purchasing bottles, batteries, electronics, etc. which is refunded upon return to designated collection points.

Buy-back centers: Informal networks of scrap collectors buy recyclable materials from consumers and businesses for resale to processors. These help supplement formal recycling.

On-demand pickup: It is requested as-needed collection for bulky items, yard waste, etc. via phone or online scheduling. It is useful for extra service beyond routine collection.

Self-haul: Residents and businesses transport waste and recyclables to transfer stations and processing sites themselves. This is common for C&D debris.

Routing optimization, automated trucks, RFID tracking, and smart bin sensors are improving collection efficiency. But designing accessible systems and building public participation remains critical.

3 Sorting and Separation

Collected solid waste contains a heterogeneous mix of materials. To enable recycling and recovery, the waste stream must be separated into constituent parts (Dalrymple et al., 2007). This occurs both at the source and through centralized processing infrastructure.

3.1 Key Sorting and Separation Techniques Include

Source separation: Households, schools, businesses, etc. segregate recyclables and waste into different bins for collection. Less contamination improves recovery quality.

Dual-stream collection: Recyclables are separated into fiber (paper, cardboard, etc.) and non-fiber materials (glass, plastic, metal) streams during curbside pickup.

Single-stream collection: All recyclables are collected together in one bin for sorting at a recovery facility. It is more convenient for users but with higher contamination.

Drop-off separation: Recycling drop-off sites have dedicated containers for common materials like glass, plastic, paper, and metals for users to self-sort (Matuszewska et al., 2022).

Deconstruction: Selective demolition to dismantle buildings and infrastructure keeps valuable metals, wood, concrete, glass, and other materials separated.

Material recovery facilities (MRFs): They utilize conveyors, screens, magnets, optical sorters, and manual sorting to categorize commingled recyclables after collection for processing.

Dirty MRFs: It can separate limited recyclables from mixed solid waste, but with high contamination and high residue disposal.

Construction and demolition (C&D) processing facilities: Sites with crushers, screens, magnets, and picking stations to recover ingredients from C&D debris for recycling.

Waste-to-energy (WTE) processing: Front-end separation at WTE plants removes bulky, hazardous, and recyclable materials from MSW combustion feedstock (Torres-Lozada et al., 2023).

Automated sorting technology is rapidly improving recovery rates and economics from single-stream collections. But low-tech source separation remains vital where capabilities are limited.

4 Biological Treatment Methods

Organic waste which includes food scraps, yard debris, paper, and other biodegradable materials makes up a substantial portion of the MSW stream (Davis & Song, 2006). Many of these materials can be biologically processed to recover value and avoid landfill disposal (Bhat et al., 2018b; Narayana, 2009).

Composting: Aerobic decomposition of organics by microorganisms produces a stabilized, nutrient-rich soil amendment (Hajam et al., 2023). It can be conducted in home bins, community sites, and large industrial facilities (Hameed et al., 2021).

Anaerobic digestion: It breaks down organics without oxygen via microbial action to produce biogas for energy generation and a digestate that can be composted or used as fertilizer (Wierzbowska et al., 2021).

Feeding programs: Unserved or unsold food can be recovered from businesses and consumers to provide meals for those in need through partnerships with food banks and shelters.

Animal feed: Some food scraps can be processed into animal feed instead of being wasted. This is now being expanded beyond traditional farm uses.

Waste-to-biofuels: Emerging technologies use chemical, microbial, and enzymatic processes to convert organics into various transportation biofuels.

Land application: Applying sludge biosolids, manures, food waste, and other organics to land can improve soil health. It must be done responsibly to avoid contamination.

Black soldier fly farming: It allows organic waste to be converted into an insect-based animal feed while producing valuable frass fertilizer as a byproduct.

Proper source separation of organics coupled with local processing infrastructure minimizes waste while creating environmental and social benefits through resource recovery.

5 Advanced Thermal and Chemical Conversion

Innovative technologies that use thermal, chemical, or biochemical processes enable conversion of residual waste and plastics that remain after recycling into fuels, chemicals, and energy (Ellis et al., 2021).

Waste-to-energy (WTE): Waste incineration generates electricity and steam heat. It reduces landfill volumes but has air pollution concerns.

Gasification: Heating waste with low oxygen produces syngas that can be used to create ethanol, diesel, and other fuels after conditioning (Maluf Filho et al., 2022).

Pyrolysis: Heating plastic waste in the absence of oxygen produces pyrolysis oil and gas for fuels, waxes, and chemical feedstocks (Matuszewska et al., 2022).

Hydrothermal processing: Using hot compressed water breaks down organics into biofuel precursors. It is effective for wet waste.

Plastic chemical recycling: It breaks down polymers into monomers, oils, gases or feedstocks for producing new plastics through depolymerization, solvolysis, etc. (Matuszewska et al., 2022).

Fuel synthesis/refining: Combining waste-derived syngas or bio-oils with catalytic refining and other chemical processes creates usable transportation fuels.

Microwave decomposition: Using microwave energy and heat, carbon-based waste materials can be broken down into a syngas suitable for making liquid fuels.

These technologies show promise for getting more value out of non-recyclable waste. However, high capital costs remain a barrier to widespread adoption, along with variability in feedstock composition.

6 Recycling Systems and Markets

Development of closed-loop recycling systems requires advanced sorting and reprocessing technologies along with businesses willing to incorporate recycled materials into new products (Hagelüken & Goldmann, 2022).

Material recovery facilities (MRFs): Quickly sort, decontaminate, and densify recyclable materials from collection programs to meet specifications for remanufacturing (Boom, 2022).

Plastics recycling: Cleaning, melting, and pelletizing enables reintroduction of PET, HDPE, LDPE, PP, PVC, PS, and other resins into new plastic products.

Paper and cardboard recycling: Hydropulping, screening, cleaning, and deinking reprocesses fibers into recycled paper stock. New boxes contain high recycled content.

Glass beneficiation: Processing to remove contaminants enables crushed recycled glass to displace mined minerals like feldspar in glassmaking furnaces.

Metal separation: Shredding, magnetic, and eddy-current separation allow high-quality recovery of ferrous and non-ferrous metals for smelting into new metals.

Textile recycling: Processing recovers fibers for incorporation into insulation, stuffing, rags, or lower-grade fabrics. It can also be chemically depolymerized for virgin-like fibers.

Organics recycling: Turning food, yard, and paper waste into compost, biogas, and other products completes the recovery cycle of biological materials (Czekała et al., 2023b).

Strong end markets for recycled materials locally drive quality and make the economics work (Gregson et al., 2015). Better product design for recyclability and use of recycled content also help push progress toward circularity (Tecchio et al., 2017).

7 Composting Systems

Diverting organic waste from landfills through controlled aerobic decomposition into compost enables both waste reduction and useful product generation (Czekała et al., 2022; Levis et al., 2010). Compost has growing applications:

Soil amendment: Compost is added to soil mixes and potting blends to provide nutrients, improve texture and drainage, and enhance water retention.

Landscaping: It is used as mulch, mixed into turf, and added to topsoil for gardens, urban plantings, golf courses, roadsides, etc.

Erosion control: It can be applied to stabilize embankments and prevent water runoff.

Landfill cover: It provides an alternative to soil for daily landfill cover and closure cap material.

Storm water filtration: It acts as a sorbent for filtration and bio-retention in rain gardens, swales, green roofs, etc. to remove pollutants.

Agriculture: It supplies nutrients and organic matter when applied to benefit crops and pasturelands.

Silviculture: It is used for conditioning soils around trees and in nurseries to grow seedlings.

Waste-to-energy (WTE): It can serve as a feedstock for generating biogas through anaerobic digestion (Czekała et al., 2023b).

Composting requires proper feedstock preparation, optimized microbial decomposition, continuous aeration and mixing, temperature control, moisture management, sanitation, site maintenance, and quality testing of finished products (Vigneswaran et al., 2016). Both small-scale localized and large-scale industrial composting facilities and processes make these diverse applications possible (Lin et al., 2011).

8 Policy and Regulation

Achieving zero waste management goals requires effective policies, regulations, and economic incentives enacted by local and national governments (Stavi & Lal, 2015).

Recycling mandates: Requirements for households and businesses to recycle certain materials, with fines for non-compliance. Helps establish new social norms.

Landfill bans: Preventing readily recyclable materials and organics from disposal accelerates adoption of alternatives.

Incineration taxes: Makes WTE disposal more costly to incentivize higher recycling rates. Revenues support waste programs.

Virgin material taxes: Increases costs of new resource extraction, which evens the playing field for recycled material use.

Pay-as-you-throw: User fees for waste collection based on amount generated create financial incentive to reduce and recycle.

Extended producer responsibility: Makes manufacturers responsible for products and packaging at end-of-life, encouraging better design.

Recycled content mandates: Require minimum levels of recycled materials in manufactured goods, which grows demand.

Compost procurement: Policies directing governments to purchase and use locally produced compost for projects.

Green public procurement: Allows factors like recyclability, recycled content, and waste reduction to drive purchasing decisions by agencies. Effective policies require monitoring and enforcement. Life-cycle assessments and cost-benefit analyses help identify optimal targets and phased timelines (Ahamed et al., 2016). Policy portfolios should utilize multiple reinforcing levers tailored to local context.

9 Public Education and Engagement

Gaining participation from residents, schools, businesses, institutions, and other stakeholders is essential for zero waste programs to work at a community scale (Phillips et al., 2011). This requires multifaceted education and engagement.

Community campaigns: Events, advertising, social media, and door-to-door visits promote programs and proper separation behaviors through fun and accessible messaging.

School curricula: Teaching the next generation about recycling, waste impacts, and sustainability starts habits early. Programs can include facility audits.

Business outreach: Tools like waste audits and technical assistance help businesses assess current practices and transition to meeting new requirements.

Institutional partnerships: Local reuse organizations, environmental groups, industry associations, and social services collaborate to maximize reach and synchronize efforts.

Feedback mechanisms: Apps, cart hang-tags, and other feedback on set-out errors help identify where more instruction is needed in a transparent non-punitive way.

Incentive programs: Contests, tax breaks, challenges, and rewards provide motivations to participate above just obligations.

Community design: It ensures convenient settings like parks, streets, offices, schools, markets, and events.

10 Recognition Programs Access to Recycling and Organics Collection in Public

Certifications, awards, and public acknowledgment for individuals, schools, and businesses achieving zero waste goals help motivate leadership. Ongoing public education paired with program evaluation and adaptation is key for both launching new initiatives successfully as well as maintaining engagement long-term.

11 The Path to Zero Waste

While once treated as simply a disposal problem, municipal solid waste management has evolved into a system aimed at maximizing sustainability through waste reduction, materials recovery, and technological innovation (Anomanyo, 2004). Achieving zero waste will take continued progress across all components of the system. Stronger product and packaging design, advanced collection systems, recycling and composting infrastructure, waste conversion technologies, coordinated policies, and public participation together can enable communities to minimize their environmental footprint while benefiting from the productive use of recovered resources (Rogoff, 2013). Collaborative efforts by multiple stakeholders from residents to private industry to government are needed to bring zero waste strategies from niche applications into the mainstream, ultimately transitioning societies from the take-make-waste model to a circular economy. Ongoing research, transparency, education, and adaptable policies will help guide the systems and culture change this transition requires (Bhat et al., 2021).

12 Conclusion

Transitioning from current waste practices to zero waste systems represents a complex challenge but also an enormous opportunity. Achieving this shift will require continued innovation and improvement across all aspects of solid waste management. This includes creative product and process redesign to minimize waste generation, advanced collection and sorting technologies for high recovery, biological and chemical solutions to process residuals, robust local markets for recycled materials, and effective policies and public engagement to enable implementation. No single solution will be sufficient. Rather, integrating a portfolio of complementary technological and social strategies tailored to local context is needed to bring zero waste goals into reality. With persistent effort and collaboration among committed stakeholders, cities and nations can gradually replace the prevailing take-make-waste economic model with circular systems that maximize sustainability. The environmental and economic benefits of doing so will only continue to grow over time as resource constraints increase. The technologies and knowledge exist—we must now take advantage of them to make zero waste the new standard.

References

Ahamed, A., Yin, K., Ng, B. J. H., Ren, F., Chang, V. C., & Wang, J. Y. (2016). Life cycle assessment of the present and proposed food waste management technologies from environmental and economic impact perspectives. *Journal of Cleaner Production, 131*, 607–614.

Al Sa'di, M. G. M. (2009). *Reuse-recycling options for municipal solid waste in Zahret Al-Finjan landfill* [Doctoral dissertation].

Al-Salem, S. M., Lettieri, P., & Baeyens, J. (2009). Recycling and recovery routes of plastic solid waste (PSW): A review. *Waste Management, 29*(10), 2625–2643.

Anomanyo, E. D. (2004). *Integration of municipal solid waste management in Accra (Ghana): Bioreactor treatment technology as an integral part of the management process*. Lund University.

Bhat, R. A., Dar, G. A., Jehangir, A., Bhat, B. A., & Yousuf, A. R. (2012). Municipal solid waste generation and present scenario of waste management during Yatra season in Pahalgam: A tourist health resort of Kashmir valley. *International Journal of Current Research, 4*, 004–009.

Bhat, R. A., Dar, S. A., Dar, D. A., & Dar, G. H. (2018a). Municipal solid waste generation and current scenario of its management in India. *International Journal of Advance Research in Science and Engineering, 7*(2), 419–431.

Bhat, R. A., Dervash, M. A., Mehmood, M. A., & Hakeem, K. R. (2018b). Municipal solid waste generation and its management, a growing threat to fragile ecosystem in Kashmir Himalaya. *American Journal of Environmental Sciences*. https://doi.org/10.3844/ajessp.2018

Bhat, R. A., Kamili, A. N., & Bandh, S. A. (2013). Characterisation and composition of municipal solid waste (MSW) generated in Yusmarg: A health resort of Kashmir valley. A glance at the world. *Waste Management, 33*, 774–777.

Bhat, R. A., Nazir, R., Ashraf, S., Ali, M., Bandh, S. A., & Kamili, A. N. (2014). Municipal solid waste generation rates and its management at Yusmarg forest ecosystem, a tourist resort in Kashmir. *Waste Management and Research, 32*, 165–169.

Bhat, R. A., Singh, D. V., Qadri, H., Dar, G. H., Dervash, M. A., Bhat, S. A., Unal, B. T., Ozturk, M., Hakeem, K. R., & Yousaf, B. (2021). Vulnerability of municipal solid waste: An emerging threat to aquatic ecosystems. *Chemosphere, 287*, 132223. https://doi.org/10.1016/j.chemosphere.2021.132223

Blumbergs, E., Serga, V., Shishkin, A., Goljandin, D., Shishko, A., Zemcenkovs, V., Markus, K., Baronins, J., & Pankratov, V. (2022). Selective disintegration-milling to obtain metal-rich particle fractions from E-waste. *Metals, 12*, 1468.

Boom, Y. J. (2022). *Environmental challenges in incorporating recycled plastics in asphalt pavements* [Doctoral dissertation]. RMIT University.

Czekała, W., Drozdowski, J., & Łabiak, P. (2023a). Modern technologies for waste management: A review. *Applied Sciences, 13*(15), 8847. https://doi.org/10.3390/app13158847

Czekała, W., Janczak, D., Pochwatka, P., Nowak, M., & Dach, J. (2022). Gases emissions during composting process of agri-food industry waste. *Applied Sciences, 12*, 9245.

Czekała, W., Pulka, J., Jasiński, T., Szewczyk, P., Bojarski, W., & Jasiński, J. (2023b). Waste as substrates for agricultural biogas plants: A case study from Poland. *Journal of Water and Land Development, 56*, 1–6.

Dalrymple, I., Wright, N., Kellner, R., Bains, N., Geraghty, K., Goosey, M., & Lightfoot, L. (2007). An integrated approach to electronic waste (WEEE) recycling. *Circuit World, 33*(2), 52–58.

Davis, G., & Song, J. H. (2006). Biodegradable packaging based on raw materials from crops and their impact on waste management. *Industrial Crops and Products, 23*(2), 147–161.

Ellis, L. D., Rorrer, N. A., Sullivan, K. P., Otto, M., McGeehan, J. E., Román-Leshkov, Y., Wierckx, N., & Beckham, G. T. (2021). Chemical and biological catalysis for plastics recycling and upcycling. *Nature Catalysis, 4*(7), 539–556.

Fatimah, Y. A., Govindan, K., Murniningsih, R., & Setiawan, A. (2020). Industry 4.0 based sustainable circular economy approach for smart waste management system to achieve sustainable development goals: A case study of Indonesia. *Journal of Cleaner Production, 269*, 122263.

Gregson, N., Crang, M., Fuller, S., & Holmes, H. (2015). Interrogating the circular economy: The moral economy of resource recovery in the EU. *Economy and Society, 44*(2), 218–243.

Hagelüken, C., & Goldmann, D. (2022). Recycling and circular economy—Towards a closed loop for metals in emerging clean technologies. *Mineral Economics, 35*(3–4), 539–562.

Hajam, Y. A., Kumar, R., & Kumar, A. (2023). Environmental waste management strategies and vermi transformation for sustainable development. *Environmental Challenges*, 100747.

Hameed, M., Bhat, R. A., Pandit, B. A., Ramzan, S., Dijoo, Z. K., & Wani, M. A. (2021). Qualitative assessment of compost engendered from municipal solid waste and green waste by indexing method. *Journal of the Air & Waste Management Association*. https://doi.org/10.1080/10962247.2021.1959466

Khetriwal, D. S., Kraeuchi, P., & Widmer, R. (2009). Producer responsibility for e-waste management: Key issues for consideration–learning from the Swiss experience. *Journal of Environmental Management, 90*(1), 153–165.

Levis, J. W., Barlaz, M. A., Themelis, N. J., & Ulloa, P. (2010). Assessment of the state of food waste treatment in the United States and Canada. *Waste Management, 30*(8–9), 1486–1494.

Lin, B. B., Chappell, M. J., Vandermeer, J., Smith, G., Quintero, E., Bezner-Kerr, R., Griffith, D. M., Ketcham, S., Latta, S. C., McMichael, P., & McGuire, K. L. (2011). Effects of industrial agriculture on climate change and the mitigation potential of small-scale agro-ecological farms. *CABI Reviews, 2011*, 1–18.

Maluf Filho, D., Coelho, S. T., & Perecin, D. (2022). Opportunities and challenges of gasification of municipal solid waste (MSW) in Brazil. *Energies, 15*, 2735.

Matuszewska, A., Owczuk, M., & Biernat, K. (2022). Current trends in waste plastics' liquefaction into fuel fraction: A review. *Energies, 2022*(15), 2719.

Narayana, T. (2009). Municipal solid waste management in India: From waste disposal to recovery of resources? *Waste Management, 29*(3), 1163–1166.

Ngoc, U. N., & Schnitzer, H. (2009). Sustainable solutions for solid waste management in Southeast Asian countries. *Waste Management, 29*(6), 1982–1995.

Noori, M. A., Al-Janabi, T. A., & Hussien, S. A. S. (2022). Solid waste recycling and management cost optimization algorithm. *Bulletin of Electrical Engineering and Informatics, 11*, 2225–2232.

Ostad-Ali-Askari, K. (2022). Management of risks substances and sustainable development. *Applied Water Science, 12*(4), 65.

Phillips, P. S., Tudor, T., Bird, H., & Bates, M. (2011). A critical review of a key waste strategy initiative in England: Zero waste places projects 2008–2009. *Resources, Conservation and Recycling, 55*(3), 335–343.

Rawat, I., Kumar, R. R., Mutanda, T., & Bux, F. (2013). Biodiesel from microalgae: A critical evaluation from laboratory to large scale production. *Applied Energy, 103*, 444–467.

Rogoff, M. J. (2013). *Solid waste recycling and processing: Planning of solid waste recycling facilities and programs*. Elsevier.

Saeedi, K., Visvizi, A., Alahmadi, D., & Babour, A. (2023). Smart cities and households' recyclable waste management: The case of Jeddah. *Sustainability, 15*, 6776.

Stavi, I., & Lal, R. (2015). Achieving zero net land degradation: Challenges and opportunities. *Journal of Arid Environments, 112*, 44–51.

Tecchio, P., McAlister, C., Mathieux, F., & Ardente, F. (2017). In search of standards to support circularity in product policies: A systematic approach. *Journal of Cleaner Production, 168*, 1533–1546.

Torres-Lozada, P., Manyoma-Velásquez, P., & Gaviria-Cuevas, J. F. (2023). Prioritization of waste-to-energy technologies associated with the utilization of food waste. *Sustainability, 15*(7), 5857. https://doi.org/10.3390/su15075857

Vigneswaran, S., Kandasamy, J., & Johir, M. A. H. (2016). Sustainable operation of composting in solid waste management. *Procedia Environmental Sciences, 35*, 408–415.

Vu, H. L., Ng, K. T. W., Fallah, B., Richter, A., & Kabir, G. (2020). Interactions of residential waste composition and collection truck compartment design on GIS route optimization. *Waste Management, 102*, 613–623.

Wierzbowska, J., Sienkiewicz, S., & Załuski, D. (2021). Nitrogen fractions in soil fertilized with waste organic materials. *Agronomy, 11*, 1474.

Zaman, A. U., & Lehmann, S. (2011). Urban growth and waste management optimization towards 'zero waste city.' *City, Culture and Society, 2*(4), 177–187.

Chapter 3
Zero Waste Management System

Palaq, Raman Jasrotia, Seema Langer, and Kenjabek Rozumbetov

Abstract Zero waste management is the holistic approach which utilizes waste as a resource produced in the intermediate steps in the process of consumption of resource. It reduces the cost of product formation, protects the environment and creates additional job opportunities. This idea is being utilized in several sectors including management of waste and its treatment, manufacturing and urban development. Zero waste index is known to substitute energy, greenhouse gases, water by the resources recovered from the waste. In this chapter, challenges of industrial solid waste, e-waste, food waste, waste packaging and zero construction waste have been discussed to transform the traditional waste toward zero waste vision. Zero waste management so far has been used to form biodegradable plastic, clothes in sustainable fashion, production of biogas from organic waste. This can be achieved by using 7R strategy—reduce the amount of waste, reuse the resources, recycle resources, refuse to receive unnecessary objects, repair things for their prolonged use, rethink and refuse to think judiciously about the resources to be used and refuse to waste them. Such strategy is very useful as it does not utilize the processes like incineration that emit more toxins and pollutants thus degrading the quality of air. Zero waste management at home includes use of disposables, reusing the bottles, handmade cleaning tools, using ceramic dishes, cloth napkins, downloading music and videos instead of purchasing CDs and DVDs, using refillable pens, refillable whiteboard markers and many more. It is important to increase the awareness among the stakeholders by organizing campaigns. Moreover, sorting of the waste properly using temporary disposing sites, proper time management and improved waste logistics will improve the technical operations used in zero waste management. This strategy needs to be developed sustainably to minimize the environmental damage and protect natural resources.

Keywords Waste management · Electronic waste · Biodegradable plastic · Biogas

---

Palaq · R. Jasrotia (✉) · S. Langer
Department of Zoology, University of Jammu, Jammu, Jammu & Kashmir, India
e-mail: drraman.jasrotia@jammuuniversity.ac.in

K. Rozumbetov
Department of General Biology and Physiology, Karakalpak State University Named After Berdakh, Nukus, Uzbekistan

R. A. Bhat et al. (eds.), *Zero Waste Management Technologies*,
https://doi.org/10.1007/978-3-031-57275-3_3

1 Introduction

Rapid modernization in the community lifestyle, increasing population have greatly increased the release of solid waste throughout the world, especially in the developing nations (Bhat et al., 2014; Guerrero et al., 2013; Hajam et al., 2023; Minghua et al., 2009) thereby emerging as a global environment issue (Seng et al., 2010; Sharholy et al., 2007). Release of waste by any civilized society is an indication of its inefficiency and indicates the misallocation of the available resources. Release of waste material causes the depletion of natural resources thereby generating 70% of the waste from the global resources (Ramsar, 2012; UN-MEA, 2006). The waste siege problem is becoming more and more severe (Fan et al., 2020). Such wastes cause depletion of the natural resources by using the resources like water and energy, create pressure on the land where it is being released, cause environmental pollution and also create an extra cost for management of such waste. Billion of tons of the resources are being utilized everyday across the globe that releasing about 3.4–4 billion tons of municipal solid waste (Bhat et al., 2012; Chalmin & Gaillochet, 2009; Giljum et al., 2008). The huge quantity of solid waste has even caused a massive burden for the environmental authority to handle solid waste in a very much sustainable way (Awasthi et al., 2021). In this context, raw material consumption at this huge with rapid-mounting speed causes to theatrical geological natural resource depletion, along with causing a negative impact on environment threat through emissions from different activities (mining, transport, refining and disposal practices) (Ke et al., 2022). Management of solid waste becomes important when the society is growing from low-density population in agriculture to urban population which is of high density. Moreover, rapid growth of industries has produced numerous products which are not easily decomposed. Certain products released from various industries contain substances which due to their toxicity build up to certain high levels in nature that may pose a threat to the existence of humans by degrading natural resources like freshwater, agricultural soil, etc. (Desmond, 2006; Menikpura et al., 2012; Troschinetz & Mihelcic, 2009).

2 Need for Zero Waste Management

The concept of proper management of waste was taken into consideration long before the modern civilization came into existence. Innovative techniques like composting, recycling, landfilling and many advanced methods like incineration have been implemented for managing the solid waste (Bhat et al., 2013; Hameed et al., 2021). However, zero waste is the most holistic approach of this century that truly achieves the goal of waste management in a sustainable manner globally (Zaman & Lehmann, 2011a). The term zero waste was first used by Palmer (2004) who recovered resources from waste chemicals. This concept, however, had been gaining the public attention since the late 1990s when various organizations across the world adopted zero waste concept by setting a goal of zero waste release to landfills. No waste bill was proposed by various municipal organizations, and Canberra was the first city to officially achieve the target of zero waste (Connett, 2013; Snow & Dickinson, 2003). Zero waste trust was established in New Zealand for the first time in 1997 that supported the minimization of the waste and implemented zero waste management. Another trust also aimed of forming "closed loop material economy" wherein effort was made to minimize the waste initially later reducing it to zero level by reusing, repairing and recycling the products (Tennant-Wood, 2003). First comprehensive zero waste management plan was implemented in California by Del Norte (in 2000), and in 2001, many plans were adopted for zero waste management (Connett, 2013). In New Zealand, zero waste trust defined zero waste as a strategy that helps to redesign the ways by which the flow of resources occurs in the society. This has also been considered as an "end of pipe" that aims to minimize the release of waste and designs methods to maximize the recycling of resources thus ensuring the reusing, repairing and recycling of materials. This concept has redesigned the industries in a way that we do not visualize the environment as a supplier of material (Tennant-Wood, 2003). Designing, managing and processing of the products are done in such a way so that the volume and toxicity of the waste are minimized, thus aiming at the conservation of the resources (ZWIA, 2009). The Department of Environment in San Francisco defines zero waste as nothing to be sent for landfilling or incineration (SF, Environment). In England, it has been defined as a simplistic way to reduce the effect of waste on the environment. This has been visualized as a holistic approach that prevents the release and formation of waste thus conserving the resources and also recovering the resources from waste product (Phillips et al., 2011).Thus, zero waste management can be summed up as conserving resources by responsible production of commodities, consuming, reusing and recovering the products, packing the materials without discharging them to soil, water or air thus threatening the environment or human health.

3 Steps in Zero Waste Management

Though the idea of zero waste has been introduced in last few decades, the concept is still growing and is now being practiced in the extracting the resources, producing and managing the waste. The guidelines for zero waste management have been implemented which are categorized under four phases. During the beginning phase, integration and application of several techniques and methods for the extraction of resources, industrial ecology, cradle to cradle and the principles of green engineering have been devised for the promotion of industrial symbiosis and various up-cycling processes. Such strategies of extraction should ensure that the existing resources are utilized instead of extracting or using the virgin and non-renewable resources. All the principles of this phase should ensure minimal degradation of environment and maximum utilization of the resources by eco-effectiveness. The products which are not easy to disintegrate recycle or recover from the end-of-life product should not be manufactured (Ziout et al., 2014). In this phase, preventing and reducing the waste are possible in these segments by forming sustainable design, production and consumption (Tseng, 2013; Vallero, 2011). In the second step, it is ensured that citizens become more responsible in consumption and utilization of resources. Such sustainable practices are vital for both local and global communities. The overconsumption of the resources creates waste which can be minimized by implementing sustainable consumption practices. Earlier studies reveal that sustainable consumption and behavioral patterns leading to reduction of waste are socially oriented and exposed to the peer pressure and do not depend upon individual attitudes (Cecere et al., 2014). Third phase of zero waste management is based upon down-cycling wherein sorting of waste, accumulating and gathering, recycling, recovery of the resource and treating the waste are done for maximum resource recovery with minimum environmental degradation (Fig. 1). ZW should focus upon the economic activities by creating more job opportunities (Lee et al., 2014). Extended responsibility of the producers is vital to ensure the stability and protection of the environment.

During the final phase, regulatory policies are used in development of zero waste management by promoting zero waste programs, plans and activities and banning the burning of waste or releasing them into landfills. ZW strategies and policies should promote the recycling activities and contributing to economic growth by recycling jobs.

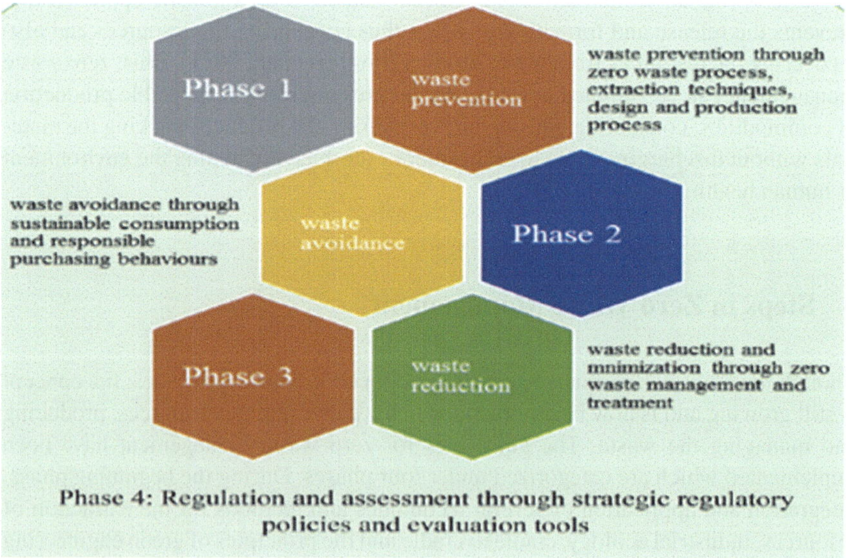

Fig. 1 Showing steps in zero waste management

4 7R's of Zero Waste Management

1. **Rethink**: Every single time, one goes to purchase something, one should give it a second thought whether it is really a necessity to buy that commodity. This in the long long will help to conserve the resources and prevent from building a pile of trash. One can also go for purchasing local products rather than using a brand; this will decrease the transportation energy and at the same time will support small entrepreneurs who are engaged in the manufacturing units.
2. **Refuse**: This is the second step of 7R strategy (Fig. 2) that helps to reduce the waste; this step aims at refusing to buy the non-recyclable products.
3. **Reduce**: Efforts should be made to minimize the harmful, toxic, wasteful and non-recyclable commodities. Less dependency on such products will lead to the least release of waste material that may find its place in a landfill. Reducing the amount of consumable will reduce the amount of packaging material thus reducing the waste.
4. **Repurpose**: Finding the use of an item in some way might prevent the disposal of an item into the landfill. However, some products cannot be reduced, reused or recycled. Old newspaper can be used in arts and crafts, certain jars can be used for holding plants, and mugs can be used as pen holders (Fig. 3). Waste yarns can be used to make the carpet rugs, cardboards and paper can be used to form greeting cards and other attractive showcases, old clothing can be used to manufacture sling bags, and lipsticks after use can be filled with solidified Vaseline mixed with flower essence that can be used as natural lip color.
5. **Reuse**: Single-use culture of the things should be avoided, instead of purchasing the new things; we should start reusing the things which we already have. This can be initiated by discarding the use of single-use eating utensils, water bottles and paper plates and replacing them with reusable articles.
6. **Recycle**: Releasing of the waste into the environment may cause the toxic effects on the environment in one way or the other (Bhat et al., 2021). Formation of new product will cause extraction of new resources that can be exhausted if overused.
7. **Rot**: All the organic waste should be separated (food, tree limbs, scraps, etc.), from rest of the waste and be allowed to undergo decomposition in a controlled environment that will ultimately improve the fertility of soil (Hameed et al.,

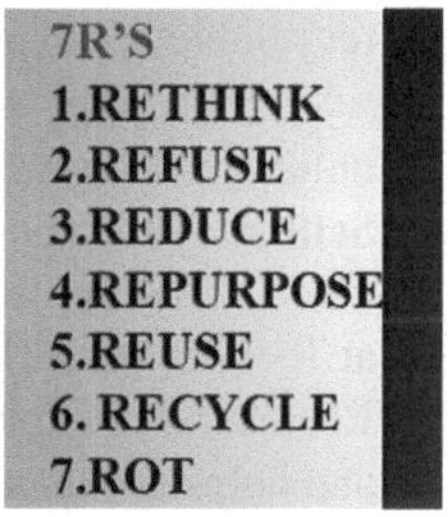

Fig. 2 7R's of zero waste management

Fig. 3 Showing the recycled products made by applying zero waste management strategies

2021). Composting is called the "black gold" as it ensures the maximum addition of nutrients to the soil.

5 Mobile/PC Applications Related to Zero Waste Management

The following four applications have been used in waste management:

1. **Pompom trash to cash**: Any waste material can be sold using this remotely recyclable application. Price of a thing is set in kg, and a request is sent for the sale on the app, and the team from pompom will collect it. Waste is sent to the industries for recycling and reusing.
2. **Sellixo**: This application helps to sell dry waste including paper, plastic or metal. This application was developed by 14-year-old girl and won 6.4 lakh at Technovation challenge in the manufacturing year.
3. **Ecolekt**: This application was developed by Bangalore-based company Jogiv and helps to manage the problem of e-waste.

4. **Encashea**: It is a Bangalore-based application in which the vendors tie up with the team of Encashea that buys scrap and lists the cost on the app. One just needs to book a slot by choosing the time.

5.1 Zero Waste Index

Zero waste index is a formula that measures the potential of unused resources to be offset by zero waste management systems that majorly aim at minimizing the depletion of natural resources. Thus, performance of zero waste city can be done by measuring the resources which are collected from environment, the consumed resources, resources which are wasted, the recycled resources, recovered resources and resources which are substituted and offset extraction of resources by waste management system. The formula for zero waste index is given in Eq. 1:

$$\text{Zero waste Index} = \frac{\sum \text{Amount of waste managed by the city} \times \text{Substitute for the system}}{\text{Cumulative waste released in the city}} \quad (1)$$

However, diversion rate of the waste does not reveal the efficiency of replacing virgin material in waste management system that is very important in conserving natural resources throughout the world. Thus, zero waste indexes are a tool that measures the virgin substitute material by waste management system. This tool at a global level measures the potential of virgin material and the decreasing rate of natural resources. It also helps to compare various management systems thus giving a broader vision for the need of unutilized materials including carbon, energy and water. ZW index acts as an indicator that helps in the measurement of the performance zero waste management systems.

5.2 Challenges of Solid Waste

1. **Industrial waste**: Such waste is released through the end product of mines, mills and factories. About 9.2 billion tons of global waste including 1.74 tons/year industrial waste was generated in 2011. In developing countries, industrial waste release is increasing at a fast rate but will decrease slowly in upcoming future (Frost & Sullivan, 2012; NBSC, 2013). The rate of recycling and collection of waste in developing countries is lower (China 67% in 2010; India has below 50% in 2010) and is therefore facing many challenges due to industrial waste generation (Bhat et al., 2018b).
2. **Environmental concerns**: The release of hazardous waste by illegal dumping seeks more attention than their adverse effects on the environment. Such dumping practices still exist as a persistent problem that poses a significant threat to human

health (Lega et al., 2012; Penelope et al., 2010). Due to enormous generation and inability to treat such wastes, illegal dumping of waste still persists especially in the developing nations of the world (Bhat et al., 2018a; Chartsbin, 2013). The waste released in the environment should always be treated, and different disposal methods should be adopted so as to cause the least adverse effects on the environment. Majority of the waste generated by developed nations is transported to developing countries by transboundary movement (Huang et al., 2012; Kojima et al., 2013) that causes environmental pollution. This is due to the fact that majority of developing nations lack the waste recycling, reusing and waste treatment facilities thus causing dumping of waste in the environment (Salihoglu, 2010; Thomas & Fannin, 2011).

3. **E-waste**: E-waste commonly termed as electronic waste includes all the electric equipments that is being dumped into the environment in the form of waste. E-waste has emerged as one of the fastest-growing waste materials during the last decade (Huisman, 2010; Zeng et al., 2013). Various electronic devices have been formed of plastics, metal, along with certain additives majority of which have very toxic effects on health of living organisms. Toxic substances like polychlorinated dibenzo-p-dioxins, flame retarders and metals like arsenic, selenium, lead, cadmium if not properly recycled, can find their way into the environment and resulting in serious environmental issues (Duan et al., 2011; Lundgren, 2012). Ingestion of dust can cause serious health hazards including silicosis (Grant et al., 2013; Qu et al., 2007). Moreover, they may also cause breathing difficulties, coughing, choking, convulsions, coma or even death (Samarasekera, 2005; Wang et al., 2006).
4. **Food waste**: All discarded food comprises the food waste and is categorized under five headings: Agricultural production, handling and post-harvest storage, processing of the food, distribution and consumption. Majority of the food waste is from developed countries compared to the developing ones which roughly are counted about 1.3 billion tons per year (Gustavsson et al., 2011; Hall et al., 2009). In industrialized nations, food waste at consumer level is about 222 million tons which is equivalent to net food production in Africa (230 million tons) (Sub-Sahara).
5. **Packaging waste**: Any abandoned product which is used for handling, protection and delivery of the goods constitutes the packaging waste (Marsh & Bugusu, 2007). According to an estimate, 420 million tons of the packaging waste have been produced (Li et al., 2013; Xie et al., 2013). Majority of the packaging waste includes paper and cardboard (40%), plastic waste (19%), glass (20%), wood (15%) and metal (Huang et al., 2007). Extravagant packaging is so prevalent in developed nations that wrapping a small piece of chocolate in three layers of plastic is an everyday practice (Ji, 2013; Zheng, 2013). Extravagant packaging utilizes more resources thus costing more money, and on being discarded, they find their way to the environment where majority of the products are not easily degraded thus causing accumulation of waste in the environment. Moreover, such packaging makes a simply commodity so heavy that it causes financial losses too.

5.3 Types of the Bins Used in Zero Waste Management

1. **Red dustbin**: Such bins are used to collect the non-biodegradable waste, commonly known as rejected dustbin, because the waste collected over here cannot be recycled. Since such wastes are non-biodegradable, they are usually treated by treatment methods like incineration. These bins are used to collect biomedical waste and are commonly observed in hospitals comprising wastes like surgical needles, knives, blades, sanitary napkins, etc. Such discarded needles, if not properly disposed, can act as a source of Hepatitis B and C.
2. **Blue dustbin**: Blue dustbins are used to carry the dry and non-biodegradable products like plastic cans, aerosol cans, broken bulbs, cardboards etc., since such material cannot be broken down by the microorganisms; hence, it is non-biodegradable. It is advisable to recycle such products as it will reduce the waste from the landfills thereby causing less pollution of the groundwater.
3. **Green dustbins**: It is commonly called an organic bin as it contains wastes of organic origin like rotten eggs, fruit and vegetable peels, garden waste, etc. Such waste can also be used as a manure as it gets converted to zero waste by composting. Such waste bins become a problem when the waste is in excess thus causing foul smell leading to human diseases on inhalation.

5.4 Zero Waste Biorefinery Using Sustainable Material Lignocellulose

There have been numerous efforts for the use of technology sustainably for the processing of lignocelluloses into value-added products including animal feeding stuffs, fine chemical, pulp and paper, biofuels, enzymes and so on (Fig. 4). There have been significant advancements in this industry when innovative studies on lignocellulose, various lignolytic enzymes, including lignin peroxidases, manganese peroxidases and laccase their formation and purification have been done using wide array of applications of biotechnology (Asgher et al., 2017; Iqbal et al., 2013). This industry has the potential to utilize all kinds of biomass like agricultural products, algae and municipal resources. Such integrated transition from petrochemical approach to sustainable biorefinery has been widely explored (Langeveld et al., 2012; McCormick & Kautto, 2013) and has the following advantages:

- It reduces dependency on petrochemical products.
- It reduces the consumption of petrol, diesel, coal and other non-renewable resources.
- It enhances the diversific use of bio-renewable resources.
- It reduces the emission of greenhouse gases.
- It safeguards and boosts the growth of greener ecosystem.

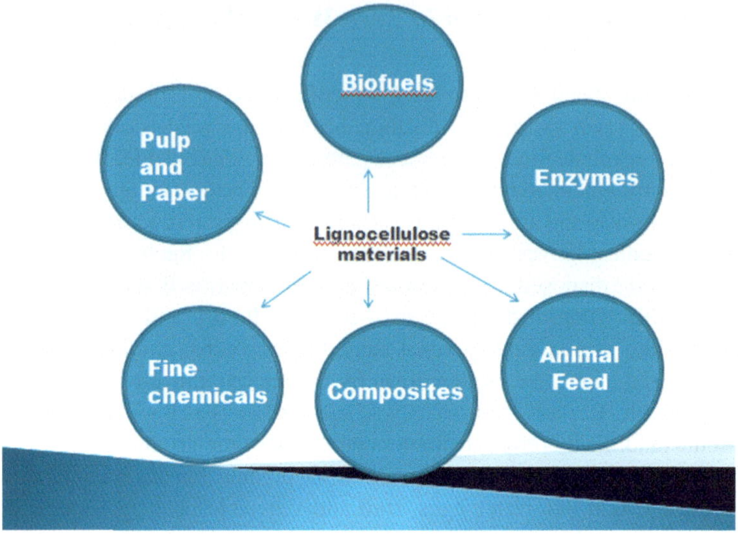

Fig. 4 Showing the use of sustainable technology for the processing of lignocelluloses into value-added products

5.5 Ten Major Goals for Achieving Zero Waste in 2031

Following are the ten goals that have to be achieved by 2031 for zero waste management:

I. To ensure the protection of environment.
II. To ensure the health and safety standards.
III. To ensure dedication in institutes and governance.
IV. To bring awareness in various communities.
V. To segregate various types of the waste released in the environment.
VI. To promote partnerships and collaborations.
VII. To make maximum use of innovative technologies and infrastructure.
VIII. Promoting educational awareness at all levels.
IX. Investing in 3R infrastructure (ecotowns, science parks, eco-industrial zones).
X. Implementation and review process.

5.6 Zero Waste in Sustainable Fashion

Designing the things using zero waste has become one of the most sustainable manufacturing methods. Some of the most innovative techniques of sewing, seam closings, seam construction enables us to have the most fashionable clothes in a sustainable way. This technique has a very bright future as it eliminates any wasteful clothing from production using whole of the bits of the cloth so that nothing is rendered waste out of it. Contrary to this, the traditional cloth makers release an average of 15% cloth as a waste (Rissanen, 2011; Townsend & Mills, 2013). The fashion industry however ensures that all the variously designed parts of a clothing fit together like a well-solved puzzle. Many old denim pants can be utilized to make sling bags when recycled properly. Many international designers like YeohleeTeng and St. James have given solutions to zero waste, and also, one of the emerging designers, Titaniainglis received "Sustainable Design Award of the Eco-Domani Fashion Foundation" in 2012. Many fabrics have been recycled by designers using organic dyes that not only eliminate zero waste completely but also help in sustainable presentation of fabric which would have increased the pile of waste clothes.

5.7 Project ZeroWIN

ZeroWIN was a project that was funded by the EC running between 2009 and 14 and was established toward zero waste in industrial network. It had 30 partners across the Europe that used their skills to allow the strategies that should be chosen for case studies in real life. This project demonstrated how technology and innovative methods can be utilized to achieve zero waste. Its partners elaborated the proposal discussing its necessity, stressing upon the fact that it can be achieved by necessary expertise in the respective field and practical skill to achieve the goals of the project. Focus was laid upon the new innovative methods and technology before aiming on the maximum use of the resources and preventing the release of waste. Report of the success through evaluation and the policies was worked out, and extensive dissemination of the results was revealed. Project ZeroWIN stressed upon wastes from car industry, photovoltaic sector and electric and electronic facilities. This project has aimed to use the best innovative technologies in addition to the already existing approaches to:

Reduce greenhouse gas emission by 30%.
Recycling of the waste upto 70%.
Reduction in freshwater utilization by 75%.

5.8 Zero Waste Management at Home

See Table 1.

Table 1 Showing the list of things at home where zero waste management strategies can be applied

Shopping bags	Instead of buying the commodities in plastic bags, carry cloth-made bags
Small quantities of packaged items	Buy the things in bulk in order to reduce the packaging material
Online shopping	Avoid shopping from online stores which utilize large amounts of plastics as packaging material
Fruits and vegetables	Buy fresh fruits and vegetables from local street hawkers that will provide livelihood to them. Purchase them in small quantity as rottening of the fruits and vegetables will increase the waste
Milk in plastic packaging	Use recyclable milk pouches, if possible, use glass bottles that will help in recycling
Sanitary napkins	Try using menstrual cups or eco-pads made from biodegradable material, mainstream sanitary napkins if used will increase the bulk of plastic waste in the environment
Lotions, creams and showering gels	Use the bigger bottles instead of purchasing numerous small ones
Leftover soap	The soap was cut to small pieces and was immersed in water which can be freezed later for a few hours and grated it to use with water as liquid soap
Wet wipes	Instead of using wipes, try using a cotton cloth
Diapers	Try using cloth diapers
Trash batteries	Try to use the rechargeable batteries
Trashing old medicines	Donate the unused medicines to a medicine store that will reduce the amount of waste released in the environment
Plastic/nylon scrubber for cleaning utensils	Use coconut husk
Plastic liners for dustbins	Use paper bags as lining material and wash the dustbins frequently
Plastic tiffin boxes	Try using steel
Footwear	Give away before it is unwearable, or get it repaired by a cobbler
Wrapping paper	Give a cloth bag instead of using a gift paper, or use a nice jute basket

6 Zero Waste Countries: Case Studies

1. **Adelaide, Australia**: South Australia with Adelaide as its capital city with a total of about 1,089,728 inhabitants occupies an urban area of 841.5 km^2 (UN-HABITAT, 2010) and accounts for 85% of the Australian population. Zero waste SA is a state government organization established under zero waste SA Act

(2004) that has improved waste management and fostered zero waste in South Australia (ZWSA, 2011). Plastic shopping bags banning was one of the main initiatives to minimize the release of waste in Adelaide. The composition of the municipal solid waste with the countries will be variable in different seasons of the year (UN-HABITAT, 2010) majority of which constitutes the construction and demolition waste. An average person generated about 681 kg of municipal solid waste in Adelaide in 2008–2009 of which 46% was recycled, 8% was composted and 46% found its way to landfill.

2. **Kamikatsu, Japan**: Kamikatsu, Tokushima Prefecture, a town located along the upper reaches of Katsuura River, focused on zero waste and has emerged as the first municipal city in Japan to adapt Zero waste policy. The residents of the town sorted the waste into 45 types under 13 categories and managed to recycle 81% of the refuse in 2016. 1580 residents of this city had some 800 households that gathered into 55 communities of variable sizes that dot the landscape at an altitude of 100–800 m. They took the concept of recycling seriously and categorized their whooping waste into various categories: aluminum cans, steel cans, spray cans, paper cartons, cardboard, newspapers, paper flyers, magazines and so on. In 2003, this city set a goal of eliminating waste by 2020 without putting any of it to landfills or incinerators and thus became the first municipal city in Japan to adopt zero waste policy in which 80% of the trash was either recycled or reused. This was difficult initially due to the opposition to this idea, but later benefits were realized, and now looking at its recycling facility, there are monitors that look at town's recycling facility by using monitors that ensure whether the trash has been properly cleaned or not. Such trash is then put into 34 bins before they are recycled, and specific stickers were put on such bins depicting the type of the waste that has to be added to such bins before getting the final recycled product. Also, a specific feature of this city was "Kuru kuru shop" where the residents can bring in and take recycled products free of cost. Also, there are factories where local women use items out of discarded products like turning old clothes, purses, blankets or pillow covers to make stuffed dolls or designer bags. Reports claim that Japan saves about two-thirds of the money which was earlier used to burn the trash.

International Day of Zero Waste 2023: The United Nations General Assembly on 14th December 2022 formally recognized the importance of zero waste initiatives and proclaimed 30th March as International Day of Zero Waste, to be observed annually beginning in 2023. In line with the theme for International Day of Zero Waste 2023—"**Achieving sustainable and environmentally sound practices of minimizing and managing waste**," the Ministry of Housing and Urban Affairs organized "**Swachhotsav**"—International Day of Zero Waste: Rally for Garbage Free Cities in New Delhi.

7 Conclusions

Zero waste is a noble approach of managing the evil of waste in the twenty-first century, wherein many workers have proposals, plans, policies, strategies and have implemented them to attain the zero waste goals. Presently, ZW has targeted to zero landfills by diverting it to some recycled product; however, 100% diversion is practically not possible in waste management systems. Hence, studies to convert existing systems to zero waste systems are vital for maximizing the attainment of ZW goals. The study has recognized the scope, relevance and implementation of prior studies and goals in Zero waste management systems. This has also revealed that professionals in this area are applying ZW in various areas though the scale and scope of zero waste are very vast. However, few of the studies have addressed ZW design, production, sustainable consumption, ZW evaluation and assessment. According to this review, zero waste concepts have been extensively utilized in various phases of production and waste management systems. Countries will be able to attain ZW goals by forming zero waste strategy and integration and promotion of zero waste initiatives. Hence, priority has to be laid upon finding areas of zero waste and developing guidelines on national zero waste; thus, the article will be a boon for policymakers to form zero waste guidelines based on evidences.

References

Asgher, M., Ahmad, Z., & Iqbal, H. M. N. (2017). Bacterial cellulose-assisted de-lignified wheat straw-PVA based bio-composites with novel characteristics. *Carbohydrate Polymers, 161*, 244–252. https://doi.org/10.1016/j.carbpol.2017.01.032. PMID:28189235.

Awasthi, A. K., Cheela, V. R. S., D'Adamo, I., et al. (2021). Zero waste approach towards a sustainable waste management. *Resources. Environment and Sustainability, 3*, 100014.

Bhat, R. A., Dar, G. A., Jehangir, A., Bhat, B. A., & Yousuf, A. R. (2012). Municipal solid waste generation and present scenario of waste management during Yatra season in Pahalgam: A tourist health resort of Kashmir valley. *International Journal of Current Research, 4*, 004–009.

Bhat, R. A., Dar, S. A., Dar, D. A., & Dar, G. H. (2018a). Municipal solid waste generation and current scenario of its management in India. *International Journal of Advance Research in Science and Engineering, 7*(2), 419–431.

Bhat, R. A., Dervash, M. A., Mehmood, M. A., & Hakeem, K. R. (2018b). Municipal solid waste generation and its management, a growing threat to fragile ecosystem in Kashmir Himalaya. *American Journal of Environmental Sciences*. https://doi.org/10.3844/ajessp.2018

Bhat, R. A., Kamili, A. N., & Bandh, S. A. (2013). Characterisation and composition of municipal solid waste (MSW) generated in Yusmarg: A health resort of Kashmir valley. A glance at the world. *Waste Management, 33*, 774–777.

Bhat, R. A., Nazir, R., Ashraf, S., Ali, M., Bandh, S. A., & Kamili, A. N. (2014). Municipal solid waste generation rates and its management at Yusmarg forest ecosystem, a tourist resort in Kashmir. *Waste Management and Research, 32*, 165–169.

Bhat, R. A., Singh, D. V., Qadri, H., Dar, G. H., Dervash, M. A., Bhat, S. A., Unal, B. T., Ozturk, M., Hakeem, K. R., & Yousaf, B. (2021). Vulnerability of municipal solid waste: An emerging threat to aquatic ecosystems. *Chemosphere, 287*, 132223. https://doi.org/10.1016/j.chemosphere.2021.132223

Cecere, G., Mancinelli, S., & Mazzanti, M. (2014). Waste prevention and social preferences: The role of intrinsic and extrinsic motivations. *Ecological Economics, 107*, 163–176. https://doi.org/10.1016/j.ecolecon.2014.07.007

Chalmin, P., & Gaillochet, C. (2009). From waste to resource: An abstract of world waste survey 2009. http://www.veolia-environmentalservices.com/veolia/ressources/files/1/927,753,Abstract_2009_GB-1.pdf

Chartsbin. (2013). Global illegal waste dumping by country. http://chartsbin.com/view/576

Connett, P. (2013). *The zero waste solution*. Chelsea Green Publishing.

Desmond, M. (2006). Municipal solid waste management in Ireland: Assessing for sustainability. *Irish Geography, 39*(1), 22–33.

Duan, H., Li, J., Liu, Y., Yamazaki, N., & Jiang, W. (2011). Characterization and inventory of PCDD/Fs and PBDD/Fs emissions from the incineration of waste printed circuit board. *Environmental Science & Technology, 45*, 6322–6328.

Fan, X., Yu, B., Chu, Z., Chu, X., Huang, W. C., & Zhang, L. (2020). A stochastic frontier analysis of the efficiency of municipal solid waste collection services in China. *Science of the Total Environment, 743*, 140707.

Frost, & Sullivan. (2012). The global industrial waste recycling & services markets. In *TEKES Growth Workshop in Helsinki on 2nd October 2012*. https://tapahtumat.tekes.fi/uploads/c8ffe124/TekesGG_Workshop_021012_global_industrial_waste_presentation-9175.pdf

Giljum, S., Lutz, C., Jungnitz, A., Bruckner, M., & Hinterberger, F. (2008). *Global dimensions of European natural resource use; First results from the global resource accounting model (GRAM)*. Sustainable Europe Research Institute (SERI) and Institute for Economic Structures Research (GWS). http://www.petre.org.uk/pdf/Giljum%20et%20al_GRAMresults_petrE.pdf

Grant, K., Goldizen, F. C., Sly, P. D., Brune, M. N., Neira, M., Van den Berg, M., & Norman, R. E. (2013). Health consequences of exposure to E-waste: A systematic review. *The Lancet Global Health*. https://doi.org/10.1016/S2214-109X(13)70101-3

Guerrero, L. A., Maas, G., & Hogland, W. (2013). Solid waste management challenges for cities in developing countries. *Waste Management, 33*, 220–232.

Gustavsson, J., Cederberg, C., Sonesson, U., Van Otterdijk, R., & Meybeck, A. (2011). *Global food losses and food waste*. Food and Agriculture Organization of the United Nations.

Hajam, Y. A., Kumar, R., & Kumar, A. (2023). Environmental waste management strategies and vermi transformation for sustainable development. *Environmental Challenges*, 100747.

Hall, K. D., Guo, J., Dore, M., & Chow, C. C. (2009). The progressive increase of food waste in America and its environmental impact. *PLoS ONE, 4*(11), e7940.

Hameed, M., Bhat, R. A., Pandit, B. A., Ramzan, S., Dijoo, Z. K., & Wani, M. A. (2021). Qualitative assessment of compost engendered from municipal solid waste and green waste by indexing method. *Journal of the Air & Waste Management Association*. https://doi.org/10.1080/10962247.2021.1959466

Huang, Y., Bird, R. N., & Heidrich, O. (2007). A review of the use of recycled solid waste materials in asphalt pavements. *Resources, Conservation and Recycling, 52*(1), 58–73.

Huisman, J. (2010). *WEEE recast: From 4 kg to 65%: The compliance consequences*. United Nations University.

Iqbal, H. M. N., Kyazze, G., & Keshavarz, T. (2013). Advances in the valorization of lignocellulosic materials by biotechnology: An overview. *BioResources, 8*(2), 3157–3176. https://doi.org/10.15376/biores.8.2.3157-3176

Ji, H. Q. (2013). Application of low carbon materials in packaging. *Advanced Materials Research, 651*, 462–465.

Ke, J., Cai, K., Yuan, W., et al. (2022). Promoting solid waste management and disposal through contingent valuation method: A review. *Journal of Cleaner Production, 379*, 134696.

Kojima, M., Yoshida, A., Sasaki, S., & Chung, S. (2013). Lessons learned from illegal transboundary movement of hazardous waste in Asia. In *International trade in recyclable and hazardous waste in Asia* (p. 149).

Langeveld, H., Sanders, J., & Meeusen, M. (2012). *The biobased economy: Biofuels. Materials, and chemicals in the post-oil era*. Earthscan.

Lee, J., Pedersen, A. B., & Thomsen, M. (2014). The influence of resource strategies on childhood phthalate exposure—The role of REACH in a zero waste society. *Environment International, 73*, 312–322. https://doi.org/10.1016/j.envint.2014.08.003

Lega, M., Ceglie, D., Persechino, G., Ferrara, C., & Napoli, R. M. A. (2012). Illegal dumping investigation: A new challenge for forensic environmental engineering. *WIT Transactions on Ecology and the Environment, 163*, 3.

Li, L., Diederick, R., Flora, J. R., & Berge, N. D. (2013). Hydrothermal carbonization of food waste and associated packaging materials for energy source generation. *Waste Management, 33*(11), 2478–2492.

Lundgren, K. (2012). *The global impact of E-waste: Addressing the challenge*. International Labour Organization.

Marsh, K., & Bugusu, B. (2007). Food packaging—Roles, materials, and environmental issues. *Journal of Food Science, 72*(3), R39–R55.

McCormick, K., & Kautto, N. (2013). The bioeconomy in Europe: An overview. *Sustainability, 5*(6), 2589–2608. https://doi.org/10.3390/su5062589

Menikpura, S. N. M., Gheewala, S. H., & Bonnet, S. (2012). Framework for life cycle sustainability.

Minghua, Z., Xiumin, F., Rovetta, A., Qichang, H., Vicentini, F., Bingkai, L., Giusti, A., & Yi, L. (2009). Municipal solid waste management in Pudong new area, China. *Waste Management, 29*, 1227–1233.

NBSC (National Bureau of Statistic of China). (2013). The China statistical yearbook (2003–2011). http://data.stats.gov.cn/workspace/index?m¼hgnd

Penelope, C., Tara, M., Kim, W., & Alicia, P. (2010). *Illegal dumping and crime prevention: A case study of ash road*. Liverpool Council.

Phillips, P. S., Tudor, T., Bird, H., & Bates, M. (2011). A critical review of a key waste strategy initiative in England: Zero waste places projects 2008–2009. *Resources, Conservation and Recycling, 55*(3), 335–343.

Qu, W., Bi, X., Sheng, G., Lu, S., Fu, J., Yuan, J., & Li, L. (2007). Exposure to polybrominated diphenyl ethers among workers at an electronic waste dismantling region in Guangdong, China. *Environment International, 33*, 1029–1034.

Ramsar. (2012). The Ramsar Convention on wetlands. Background and context to the development of principles and guidance for the planning and management of urban and peri-urban wetlands (COP11 DR11) [Online]. Retrieved May 06, 2012, from http://www.ramsar.org/pdf/cop11/doc/cop11-doc23-e-urban.pdf

Rissanen, T. (2011). *Zero waste fashion repatterned exhibition catalogue*. Columbia College Chicago. Retrieved July 5, 2013, from http://www.colum.edu/ADGallery/Exhibits/2010_zero waste/zerowastecatalog_final.pdf

Salihoglu, G. (2010). Industrial hazardous waste management in Turkey: Current state of the field and primary challenges. *Journal of Hazardous Materials, 177*(1), 42–56.

Samarasekera, U. (2005). Electrical recycling exposes workers to toxic chemicals. *Lancet, 366*, 1913–1914.

Seng, B., Kaneko, H., Hirayama, K., & Katayama-Hirayama, K. (2010). Municipal solid waste management in Phnom Penh, capital city of Cambodia. *Waste Management & Research, 29*, 491–500.

Sharholy, M., Ahmad, K., Vaishya, R. C., & Gupta, R. D. (2007). Municipal solid waste characteristics and management in Allahabad, India. *Journal of Waste Management, 27*, 490–496.

Snow, W., & Dickinson, J. (2003). *The road to zero waste: Strategies for sustainable communities*. Auckland. http://www.zerowaste.co.nz/assets/Reports/roadtozerowaste150dpi.pdf

Tennant-Wood, R. (2003). Going for zero: A comparative critical analysis of zero waste events in southern New South Wales. *Australasian Journal of Environmental Management, 10*(1), 46–55.

Thomas, J. K., & Fannin, D. (2011). The transboundary trade of hazardous wastes, 2000–2006. *Environmental Justice, 4*(1), 55–62.

Townsend, K., & Mills, F. (2013). Mastering zero: How the pursuit of less waste leads to more creative pattern cutting. *International Journal of Fashion Design, Technology, and Education, 6*(2), 104–111. https://doi.org/10.1080/17543266.2013.793746

Troschinetz, A. M., & Mihelcic, J. R. (2009). Sustainable recycling of municipal solid waste in developing countries. *Waste Management, 29*(2), 915–923.

Tseng, M.-L. (2013). Modeling sustainable production indicators with linguistic preferences. *Journal of Cleaner Production, 40*, 46–56. https://doi.org/10.1016/j.jclepro.2010.11.019

UN-HABITAT. (2010). *Solid waste management in the world's cities. Water and sanitation in the world's cities report 2010*. Earthscan.

UN-MEA. (2006). The UN millennium ecosystem assessment report [Online]. Retrieved May 06, 2012, from http://www.publications.parliament.uk/pa/cm200607/cmselect/cmenvaud/77/77.pdf

Vallero, D. A. (2011). Green engineering and sustainable design aspects of waste management (chap. 2). In T. M. Letcher & D. A. Vallero (Eds.), *Waste* (pp. 11–21). Academic Press.

Wang, L. K., Hung, Y. T., Lo, H. H., & Yapijakis, C. (2006). *Hazardous industrial waste treatment*. CRC Press.

Xie, M., Qiao, Q., Sun, Q., & Zhang, L. (2013). Life cycle assessment of composite packaging waste management—A Chinese case study on aseptic packaging. *International Journal of Life Cycle Assessment, 18*(3), 626–635.

Zeng, X., Li, J., Stevels, A. L. N., & Liu, L. (2013). Perspective of electronic waste management in China based on a legislation comparison between China and the EU. *Journal of Cleaner Production, 51*, 80–87.

Zheng, B. J. (2013). Green packaging materials and modern packaging design. *Applied Mechanics and Materials, 271*, 77–80.

Ziout, A., Azab, A., & Atwan, M. (2014). A holistic approach for decision on selection of end-of-life products recovery options. *Journal of Cleaner Production, 65*, 497–516. https://doi.org/10.1016/j.jclepro.2013.10.001

ZWIA. (2009). *Zero waste definition*. Zero Waste International Alliance. Retrieved November 12, 2013, from http://zwia.org/standards/zw-definition/

ZWSA. (2011). *ZeroWaste SA: About us* [Online]. Retrieved May 31, 2011, from http://www.zerowaste.sa.gov.au/about-us

Chapter 4
Zero Waste in the Mining: Reflections and Proposal for an Emerald Mine in Brazil

Nilza Carolina Ramos Silva Carvalho, José Ângelo Sebastião Araujo dos Anjos, and Fábio Carvalho Nunes

Abstract In recent decades, researchers have proposed a new development model, the circular economy (CE). The new model seeks to develop more efficient technologies with zero waste, including in mining. The discussion about zero waste in mining is just starting, but there are already proposals based on the principles of reduction, reuse, recycling, restoration, and re-establishment of uses (5Rs). In this chapter we present some examples, emphasizing a case study in an emerald mine in Brazil. In addition, we discussed the challenges of the mineral sector, such as the need to develop new technologies and the need for the emergence of new value chains and better regulation of the sector.

Keywords Circular economy · Zero waste · Mining · Anthropogenic circularity

1 Introduction

The linear economy model causes numerous environmental problems, due to the disorderly consumption of natural resources, pollution, and the increase in waste production. To mitigate or avoid many of these problems, it is necessary to adopt an economy model that effectively recovers degraded areas, reduces or avoids waste, increases the useful life, and reuses products whenever possible (Murray et al., 2017). Researchers call this new model circular economy (CE).

Mining has significant impacts on the environment, as its current exploration, production, and waste management systems are based on extract-produce-discard.

N. C. R. S. Carvalho · J. Â. S. A. dos Anjos
Departamento de Geologia, Instituto de Geociências, Universidade Federal da Bahia, Rua Barão de Jeremoabo, S/N, Ondina CEP: 40.170-290, Brazil

F. C. Nunes (✉)
Academic Department, Federal Institute Baiano (IF BAIANO), Campus Alagoinhas, R. Manoel Romão - Alagoinhas Velha, Alagoinhas, BA 48030530, Brazil
e-mail: fabio.nunes@ifbaiano.edu.br

R. A. Bhat et al. (eds.), *Zero Waste Management Technologies*,
https://doi.org/10.1007/978-3-031-57275-3_4

However, there is potential for the application of the circular economy model in the sector, which could reconcile economic development with environmental protection (Lieder & Rashid, 2016).

The waste generated in mining represents a global concern, due to the amount and variability of waste, which tend to grow due to the increase in population. The global population has reached more than 7.6 billion inhabitants and could reach 8.5 billion in 2030, 9.7 billion in 2050, and 10.9 billion in 2100 (UN, 2019). Demand for mineral resources increases with population growth and the desire for a better life; however, resources are finite and need better management (Nunes et al., 2023).

In recent decades, researchers have advocated the implementation of an economy based on circularity and zero waste, including in mining. Using the keywords "circular" and "economy" and "mining" in the Scopus database, it is noted that between 2005 and 2022, 819 articles were published in journals, with research intensifying from 2017 (Fig. 1). There is, therefore, an increase in the interest of researchers on the subject. When the keywords "zero waste" and "mining" are used, the number of publications decreases to 280, and when "circular" and "economy" and "mining" and "zero waste" are used, 41 publications appear, but it follows the trend of increase (Figs. 2 and 3).

Discussions on circular economy and zero waste in mining are still starting, so there is a clear potential to be explored (Nunes et al., 2023). Therefore, the main objective of this chapter is to reflect on the paradigm shift in the sector, challenges

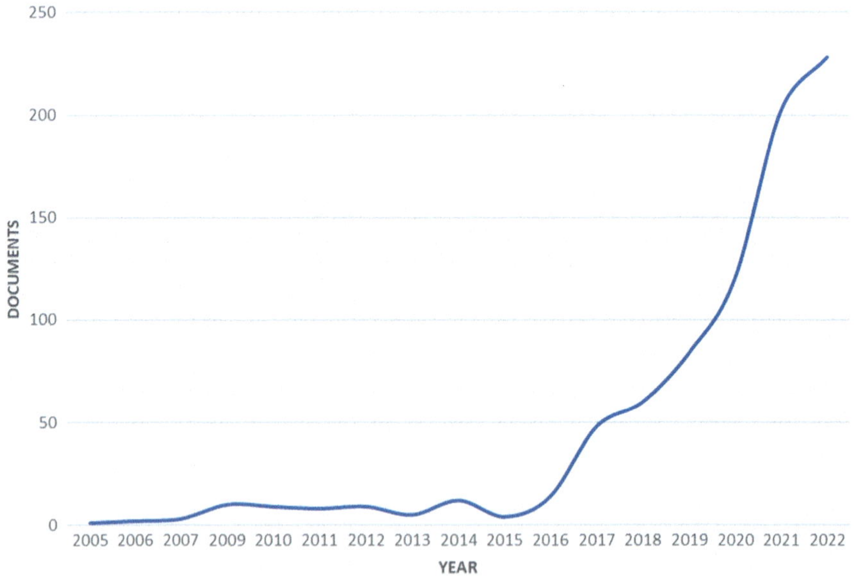

Fig. 1 Number of documents (819) on "circular economy" and "mining" in the Scopus database until 2022

Fig. 2 Number of documents (309) on "zero waste" and "mining" in the Scopus database until 2022

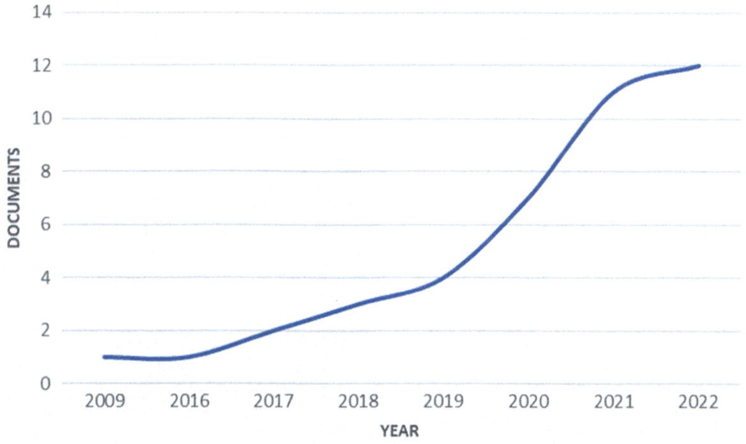

Fig. 3 Number of documents (41) on "circular economy" and "mining" and "zero waste" in the Scopus database until 2022

and possibilities, presenting some examples present in the literature and emphasizing a case study carried out in an emerald mine in Brazil.

2 Mining Waste: A Global Problem

In recent decades, there occurred a growth in the use of mineral resources; however, there has been a decrease in the quality and concentration of ores (Esposito et al., 2015). Most of these resources are finite, and the entire production chain of mining, from extraction to disposal, has created several negative environmental impacts (Edraki et al., 2014; Kinnunen & Kaksonen, 2019).

Mining waste consists of the remaining fine-grained rock (1 and 600 μm), dissolved metals, and processing reagents after valuable minerals have been extracted (Edraki et al., 2014). About 5–7 billion tons of mining tailings are generated every year in the world (Xu et al., 2019) in large, medium, and small enterprises, which are deposited in dams, tailings piles (Fig. 4), or disposed of in an inadequate in the environment (Fig. 7).

It is believed that billions of tons of mining tailings have been deposited worldwide, and this number is growing (Schoenberger, 2016; Wang et al., 2014). The global demand for raw materials has increased in recent decades and is expected to continue increasing, due to: (a) the increasing demand due to the deterioration or depletion of high-grade ore deposits (Wang et al., 2014); (b) the reheating of post-COVID-19 production chains; and (c) new post-Russia-Ukraine post-conflict geopolitical demands, with emphasis on the energy and war sectors. The new global demands will certainly produce more and more mineral residues.

It is important to mention that mining of low-grade ore is becoming the norm (Jones & Boger, 2012), and in this case, sub-micrometer grinding and classification

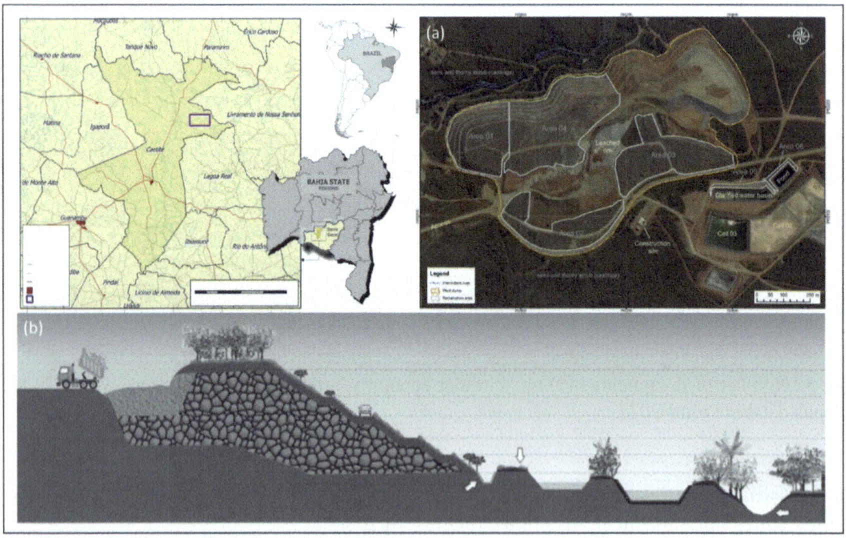

Fig. 4 Example of uranium mine tailings piles, Caetité, Brazil. *Source* Adapted from Alves et al. (2018)

become unavoidable, which makes it difficult to separate tailings in wastewater and favors the alarming accumulation of mineral fluids (Wang et al., 2014).

According to Mohanty et al. (2010), for each ton of metal extracted from the ore, between 2 and 12 tons of overburden material is removed as waste. Such an estimate is frightening and signals the need for better use and reuse in mining. After the selective extraction and processing of the desired material, all waste is released into tailing ponds, natural or constructed; however, they are often inappropriately released into the environment (Samal et al., 2013).

Wetland waste discard remains a proven technology widely adopted by the mining industry. However, the approach did not bring a definitive solution (Watson et al., 2010), which signals the need to rethink the mining waste management process. Furthermore, over the years there have been many dam failures and disastrous spills (Aires et al., 2018; Escobar, 2015; Fernandes et al., 2016; Ferreira et al., 2021; Garcia et al., 2017; Jones & Boger, 2012; Prasad et al., 2021; Nunes et al., 2022).

Mining waste are usually very acidic or alkaline, saline, and poor in nutrients and have a high content of heavy metals (Álvarez-Rogel et al., 2021). The leachate also has high levels of heavy metals, which can contaminate continental and marine ecosystems and affect human health (Alves et al., 2023; Anjos & Sánchez, 2021; Kan et al., 2021).

The circular economy has been seen as an alternative to enhance the use of mineral waste, aiming at high efficiency, mitigating and/or avoiding environmental problems. Changing the way we think, act, and behave requires us to overcome typical industrial efficiency solutions, which are still based on linear models (Luthra, 2022), that is, extraction-production-use-disposal.

The vision of the mineral sector is changing, seeking to reduce, reuse, and recycle mineral waste, but also to recover and restore uses in degraded areas (Nunes et al., 2023), favoring anthropogenic circularity (Zeng, 2023). Anthropogenic circularity is an emerging area of interdisciplinary development of earth sciences, technology science, and management science (Zeng & Li, 2021) that is gaining strength in the face of the challenges of the humanity to continue developing, but without compromising the environment. The challenges are many, but the change in mentality is taking place in step with technological innovations.

3 Is Zero Waste in Mining Possible?

Research on circular economy and zero waste in mining is still in its infancy, and the vast majority of articles are theoretical. However, the practices suggested or applied point to the positive possibility of adopting a new model in mining, based on circularity and zero waste production. The main researches are in the coal sector and based in China (Nunes et al., 2023), but they are also observed in other sectors, such as copper, tungsten, zinc, phosphorus, and rare earth mining (Table 1).

Table 1 Articles on circular economy and zero waste in mining

Sector	References
Coal	Mo et al. (2005), Xu and Zhang (2006), Si (2006), Xie et al. (2009), Zeng et al. (2009), Sun (2009), Ru-yin and Xiao-ting (2009), Wang and Wu (2010), Zhao et al. (2012), Yu (2017), Liu et al. (2019), Pactwa et al. (2020), Le and Dong (2021)
Copper	Kotarska et al. (2018), Figueiredo et al. (2018)
Potassium	Yakovleva et al. (2021)
Tungsten	Figueiredo et al. (2018)
Zinc	Figueiredo et al. (2018)
Phosphor	Geissler et al. (2018), Barquet et al. (2020)
Rare earth	Figueiredo et al. (2018), Favot and Massarutto (2019)
Red mud	Paramguru et al. (2004)

Articles published on circular economy and zero waste in mining point to a series of challenges, which are interconnected and can be organized into the following categories:

(a) energy efficiency;
(b) efficiency of material resources;
(c) circular economy models;
(d) decision support;
(e) political support.

The elaborated knowledge, reflections, and strategies pointed out by the authors mentioned in Table 1 were the basis for the proposal of circularity and zero waste in Emerald mining, located in Bahia, Brazil, which will be presented in the following item.

3.1 Zero Waste Model in Emerald Mining

Brazil's mineral resources comprise a production of 72 mineral substances, 23 of which are metallic, with iron being the most important, 45 non-metallic, and 4 energetic, which generates annual revenues of more than 30 billion dollars, more than 35% of the balance of the Brazilian trade balance (IBRAM, 2021) (Fig. 5). Mining in Brazil is mainly formed by small and microcompanies (87%) (IBRAM, 2019; Nunes et al., 2022), among them cooperatives, which increases the importance of alternatives that aim to avoid waste, increase profits, recover degraded areas, and reuse waste more cheaply and effectively.

According to Brazilian legislation, the mining activity is considered a legal form of wealth extraction, provided it meets certain rules. Law No. 11,685 of 2008 conceptualizes the prospector as anyone of Brazilian nationality that acts directly in the process of mineral extraction, either individually or in an associative way.

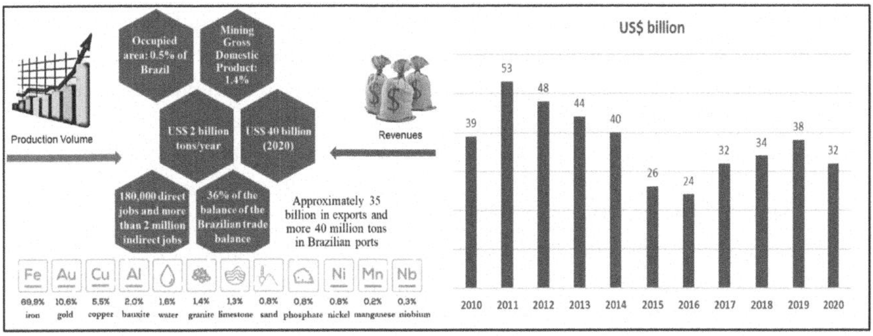

Fig. 5 Brazilian mineral sector, main substances and impacts on the economy. Adapted Nunes et al. (2022)

The mining is the place where mineral resources are extracted, with immediate use, as it does not depend on previous research work due to its nature, location, dimension, and economic use. The mining stands out in the Brazilian economy since the early eighteenth century and states of Minas Gerais, Goiás, Mato Grosso, and Bahia constitute traditional spaces of mining activities.

Emerald is the third most valuable gem on the planet, behind only the diamond and the ruby. The first occurrences of emerald in Brazil were discovered in the Bahia State of Bahia, in 1912. However, the country was only one of the largest emerald producers in the world from the 1960s, after the discovery of several deposits of high gemological quality in Minas Gerais, Bahia, and Goiás (Cornejo & Bartorelli, 2010).

Brazil, from the Federal Constitution of 1988, prioritized the granting of Lavras to prospectors linked to cooperative systems. The object was to favor the formalization of the activity and, especially, to block illegal activities.

The emerald mine studied is located in the district of Carnaíba de Cima, Pindobaçu municipality, Bahia State, Brazil. It is explored by the Mineral Cooperative of Bahia (MCB) by the underground mining method and with access to the mineralized layers from pits and galleries.

The rudimentary techniques used are causing many negative environmental impacts. Waste generated by mining fronts is discarded on slopes, forming extensive dams in valleys and urban areas.

Inadequate disposition of emerald mining residues may cause, among other impacts, the degradation of rivers and groundwater. In mining, there are two types of solid waste: sterile and tailings. The National Mining Agency (NMA) from Brazil, on December 2, 2021, published Resolution 85 that provides for the procedures for the use of tailings and steriles. The agency understands that the steriles are the fresh materials discarded before ore processing, and the tailings are all materials discarded during or after processing of ore.

The initial step of our research was to recognize the emerald exploration, proposing as a starting point a volumetric and spatial analysis of the waste piles (tailings and sterile). This analysis can leverage or decrease investments in a particular technology, as well as recognition in terms of volume and spatiality.

In the emerald mining two economically viable products were identified: (a) emerald itself and (b) the concentrate of molybdenum, extracted by flotation of the molybdenite, found associated with quartz.

The removal of high-value emeralds generates a large volume of rock, which are transported to surface, where they will be manually drawn up by the Quijilas (Fig. 6), name of people who manually select the waste and then they discard the steriles.

Molybdenite's selection is sent to a warehouse external to the mining, where it is subjected to the flotation process to remove the molybdenum concentrate (40%). After flotation there is a thin tailing, which is even discarded on site (Fig. 6).

After analyzing all stages of emerald mining, a circular economy proposal and zero residue for the mine were elaborated, which can be observed in Fig. 7. In the proposed model the cycle begins with the mining of two products: emerald and molybdenite. High-added emeralds are selected in underground mines, and rock residues are arranged on surface to manual mining. After manual mining, waste should be physically and chemically analyzed for possible agricultural uses. If the wastes are inadequate it should be sent for the recovery of the areas degraded, where bio-georemediation techniques will be used.

Molybdenite extraction begins with subsurface selection and subsequent destination for the flotation shed. Flotation generates several wastes that can be used in

Fig. 6 Craft production of molybdenite in the district of Carnaíba de Cima, Bahia, Brazil

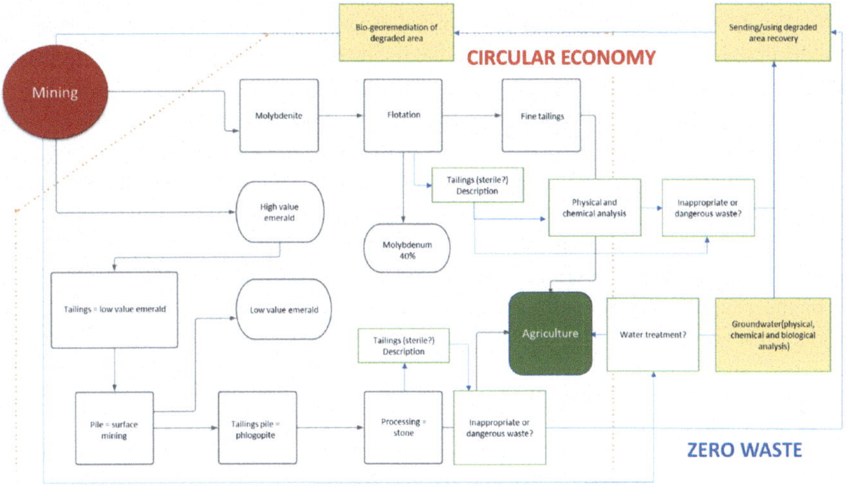

Fig. 7 Proposal of circular economy and zero waste in emerald mining

agriculture after physical and chemical analysis, because molybdenite's tailings is rich in K, Mg, Fe, S, and Mo. However, if the waste is inadequate, it should be sent for the recovery of degraded areas using bio-georemediation techniques.

Finally, it is important to signal that groundwater and wastewater should also undergo quality analysis (physical, chemical, and biological) for agricultural use or environmental recovery.

4 Final Considerations: Possibilities and Challenges

Most geological materials extracted so far have been transformed into products and waste, a linear economy (Zeng & Li, 2021). As a result, according to Elhacham et al. (2020), the global mass produced by man exceeded all living biomass. Technological support for implementing circular economy in mining should be environmentally friendly (or green technology), whose characteristics are less pollutant emission, reasonable use of resources and energy, and recycle more waste (Guo et al., 2007).

Circular economy is essentially an ecological economy, as it aims to develop proposals and techniques for efficient use of resources, low waste, and problems, as it is based on the principles of 3Rs (reduce, reuse, and recycle) (Pactwa et al., 2020; Zhao et al., 2012), but also in recovering and restoring degraded areas (5Rs) (Nunes et al., 2023). The circular economy employs the law of industrial ecology to achieve the efficiency of resources and with minimal or zero waste.

The circular and zero waste economy put a lot of concern in chemistry to make innovative products, using ecological renewable raw material. In addition, the substances with which these products are made will be increasingly treated as a secondary resource and not simply discarded as garbage (Clark et al., 2016). Several authors dedicated to the subject suggest that the new economy with zero waste should be based on the principles and practices of anthropogenic circularity.

Anthropogenic circularity seeks to create a flow of closed-circuit material, usually linking green chemistry, reverse supply chain, and industrial ecology at different scales (Cooper et al., 2020; Heisel & Rau-Oberhuber, 2020). Green chemistry is considered as the design of results and processes that eliminate the use and generation of hostile substances, aiming to reduce or prevent the use of dangerous substances and minimize the waste generated by chemical reactions (Anastas & Lankey, 2000).

In the reverse supply chain, some old, high-quality scrap can be reused as a product, for example:

(a) The ordinary old scrap can be dismantled for remanufacturing (Zeng & Li, 2021).
(b) Remanufactured products can be updated to the same quality standards (Savaskan et al., 2004).
(c) Some discarded materials may be disassembled for recycling and inserted in the original production processes (Zeng & Li, 2021).
(d) Waste can be recovered as raw materials using green chemistry (Li et al., 2015).

In industrial ecology processes seek to imitate natural ecological systems, where connections between organisms allow them to live and consume each other and residue from each other (ALLENBY, 2006). Industrial ecology should think of the production system as a large web, where the waste of an industry, for example, can serve as a raw material for another industry (Lifset, 2014).

Anthropogenic circularity has plenty of room to achieve a win-win goal, which is to reduce the criticality of mineral resources and at the same time contribute to sustainable development. The barriers and the great challenges for an economy with zero waste are due to the inadequacy of the life cycle, the insight governance, and the emergence of the discipline of anthropogenic circularity. Material flow analysis and life cycle assessment are the central methodologies to identify hidden problems. Mineral processing and foundry, as well as end of life management, are indicated as critical priority areas to improve anthropogenic circularity (Zeng, 2023).

In this chapter, we present a proposal for economic circularity for an emerald mine managed by small entrepreneurs, which aims to maximize profits and generate zero waste. We hope that our contribution will be applied to the context to which it was thought and/or adapted to other places. Certainly the incorporation of anthropogenic circular principles and practices will help improve it and strengthen ideas and practices around the world for a healthier world.

References

Aires, U. R. V., Santos, B. S. M., Coelho, C. D., da Silva, D., & Calijuri, M. L. (2018). Changes in land use and land cover as a result of the failure of a mining tailings dam in Mariana, MG, Brazil. *Land Use Policy, 70*, 63–70. https://doi.org/10.1016/j.landusepol.2017.10.026

Allenby, B. (2006). The ontologies of industrial ecology. *Progress in Industrial Ecology, 3*, 28–40. https://cspo.org/legacy/library/110215F0XT_lib_AllenbyPIEontolo.pdf

Álvarez-Rogel, J., Peñalver-Alcalá, A., Jiménez-Cárceles, F. J., Tercero, M. C., & González-Alcaraz, M. N. (2021). Evidence supporting the value of spontaneous vegetation for phytomanagement of soil ecosystem functions in abandoned metal(loid) mine tailings. *CATENA, 201*, 105191. https://doi.org/10.1016/j.catena.2021.105191

Alves, L. J., et al. (2018). Uranium mine waste phytostabilization with native plants—A case study from Brazil. In M. N. V. Prasad, P. J. de Campos Favas, & S. K. Maiti (Org.), *Bio-geotechnologies for mine site rehabilitation* (Vol. 1, 1st ed., pp. 299–322). Elsevier.

Alves, et al. (2023). Reflections on the uraniferous province of lagoa real: Routes, dispersion and impacts on health. In Vithanage, M. & Prasad, M. N. V. (Org.), *Medical geology: En route to one health* (Vol. 1, 1st ed., pp. 273–290). Wiley.

Anastas, P. T., & Lankey, R. L. (2000). Life cycle assessment and green chemistry: The yin and yang of industrial ecology. *Green Chemistry, 2*, 289–295. https://pubs.rsc.org/en/content/articlelanding/2000/gc/b005650m#!divCitation

Anjos, J. A. S. A., & Sánchez, L. E. (2021). *Diagnóstico ambiental do meio físico da Plumbum Mineração e Metalurgia* (1st ed., 416 pp.). Geologia Ambiental e Médica do Estado da Bahia.

Barquet, K., Järnberg, L., Rosemarin, A., & Macura, B. (2020). Identifying barriers and opportunities for a circular phosphorus economy in the Baltic Sea region. *Water Research, 171*(15), 115433. https://doi.org/10.1016/j.watres.2019.115433

Clark, J. H., Farmer, T. J., Herrero-Davila, L., & Sherwood, J. (2016). Circular economy design considerations for research and process development in the chemical sciences. *Green Chemistry, 18*, 3914–3934. https://pubs.rsc.org/en/content/articlelanding/2016/gc/c6gc00501b

Cooper, D. R., Ryan, N. A., Syndergaard, K., & Zhu, Y. (2020). The potential for material circularity and independence in the U.S. steel sector. *Journal of Industrial Ecology, 24*, 748–762. https://doi.org/10.1111/jiec.12971

Cornejo, C., & Bartorelli, A. (2010). *Minerals & precious stones of Brazil*. Solaris Cultural Productions.

Edraki, M., Baumgartl, T., Manlapig, E., Bradshaw, D., Franks, D. M., & Moran, C. J. (2014). Designing mine tailings for better environmental, social and economic outcomes: A review of alternative approaches. *Journal of Cleaner Production, 84*, 411–420. https://doi.org/10.1016/j.jclepro.2014.04.079

Elhacham, E., Ben-Uri, L., Grozovski, J., Bar-On, Y., & Milo, M. R. (2020). Global human-made mass exceeds all living biomass. *Nature, 588*, 442–444. https://doi.org/10.1038/s41586-020-3010-5

Escobar, H. (2015). Mud tsunami wreaks ecological havoc in Brazil. *Science, 350*, 1138–1139. https://doi.org/10.1126/science.350.6265.1138

Esposito, M., Tse, T., & Soufani, K. (2015). Is the circular economy a new fast-expanding market? *Thunderbird International Business Review, 59*(1), 9–14. https://doi.org/10.1002/tie.21764

Favot, M., & Massarutto, A. (2019). Rare-earth elements in the circular economy: The case of yttrium. *Journal of Environmental Management, 240*, 504–510. https://doi.org/10.1016/j.jenvman.2019.04.002

Fernandes, G. W., Goulart, F. F., Ranieri, B. D., Coelho, M. S., Dales, K., Boesche, N., Bustamante, M., Carvalho, F. A., Carvalho, D. C., Dirzo, R., Fernandes, S., Galetti, P. M., Millan, V. E. G., Mielke, C. J., Ramirez, L., Neves, A., Rogass, C., Ribeiro, S. P., Scariot, A., & Soares-Filho, B. (2016). Deep into the mud: Ecological and socio-economic impacts of the dam breach in Mariana, Brazil. *Natureza & Conservação, 14*(2), 35–45. https://doi.org/10.1016/j.ncon.2016.10.003

Ferreira, G. W. D., Ribeiro, B. T., Weindorf, D. C., Teixeira, B. I., Chakraborty, S., Li, B., Guilherme, L. R. G., & Scolforo, J. R. S. (2021). Assessment of iron-rich tailings via portable X-ray fluorescence spectrometry: The Mariana dam disaster, southeast Brazil. *Environmental Monitoring and Assessment, 193*, 203. https://doi.org/10.1007/s10661-021-08982-7

Figueiredo, J. A., Vila, M. C., Martins, D., Futuro, A., & Fiúza, A. (2018). Design optimization of a tailings reprocessing: Tungsten and zinc recovery. In *International Multidisciplinary Scientific Geo Conference Surveying Geology and Mining Ecology Management, SGEM* (Vol. 18(2.1), pp. 293–300). https://doi.org/10.5593/sgem2018/2.1/S07.037

Garcia, L. C., Ribeiro, D. B., Roque, F. O., Ochoa-Quintero, J. M., & Laurance, W. F. (2017). Brazil's worst mining disaster: Corporations must be compelled to pay the actual environmental costs. *Ecological Applications, 27*, 5–9. https://doi.org/10.1002/eap.1461

Geissler, B., Hermann, L., Mew, M. C., & Steiner, G. (2018). Striving toward a circular economy for phosphorus: The role of phosphate rock mining. *Minerals, 8*(9), 395. https://doi.org/10.3390/min8090395

Guo, M., Hu, S. C., & Liu, X. H. (2007). Improving multipurpose utilization of mineral resources by developing mining recycling economy. In *Conservation and utilization of mineral resources* (pp. 1–5). https://doi.org/10.3969/j.issn.1001-0076.2007.03.001

Heisel, F., & Rau-Oberhuber, S. (2020). Calculation and evaluation of circularity indicators for the built environment using the case studies of UMAR and Madaster. *Journal of Cleaner Production, 243*, 118482. https://doi.org/10.1016/j.jclepro.2019.118482

IBRAM Brazilian Mining Institute. (2019). *Economia mineral*. https://ibram.org.br/publicacoes/page/2/#publication

IBRAM—Brazilian Mining Institute. (2021). https://ibram.org.br/publicacoes/page/2/#publication

Jones, H., & Boger, D. V. (2012). Sustainability and waste management in the resource industries. *Industrial and Engineering Chemistry Research, 51*, 10057–10065. https://doi.org/10.1021/ie202963z

Kan, X. Q., Dong, Y. Q., Feng, L., Zhou, M., & Hou, H. B. (2021). Contamination and health risk assessment of heavy metals in China's lead-zinc mine tailings: A meta-analysis. *Chemosphere, 267*, 128909. https://doi.org/10.1016/j.chemosphere.2020.128909

Kinnunen, P. H. M., & Kaksonen, A. H. (2019). Towards circular economy in mining: Opportunities and bottlenecks for tailings valorization. *Journal of Cleaner Production, 228*, 153–160. https://doi.org/10.1016/j.jclepro.2019.04.171

Kotarska, I., Mizera, B., & Stefanek, P. (2018). Mining waste in the circular economy—Idea versus reality. *E3S Web of Conferences, 41*, 02013. https://doi.org/10.1051/e3sconf/20184102013

Le, D. C., & Dong, T. B. (2021). Circular economy model and the implementation in Vietnamese coal mining industry. *Inżynieria Mineralna, 1*(2). https://doi.org/10.29227/IM-2021-02-40

Li, J., Zeng, X., & Stevels, A. (2015). Ecodesign in consumer electronics: Past, present and future. *Critical Reviews in Environmental Science and Technology, 45*, 840–860. https://doi.org/10.1080/10643389.2014.900245

Lieder, M., & Rashid, A. (2016). Towards circular economy implementation: A comprehensive review in context of manufacturing industry. *Journal of Cleaner Production, 115*, 36–51. https://doi.org/10.1016/j.jclepro.2015.12.042

Lifset, R. (2014). Speaking industrial ecology. *Journal of Industrial Ecology, 18*, 785–786. https://doi.org/10.1111/jiec.12218

Liu, X., Guo, P., & Guo, S. (2019). Assessing the eco-efficiency of a circular economy system in China's coal mining areas: Emergy and data envelopment analysis. *Journal of Cleaner Production, 206*, 1101–1109. https://doi.org/10.1016/j.jclepro.2018.09.218

Luthra, S. (2022). Resources melioration and the circular economy: Sustainability potentials for mineral, mining and extraction sector in emerging economies. *Resources Policy, 77*, 102652. https://doi.org/10.1016/j.resourpol.2022.102652

Mo, X.-P., Cheng, D.-J., Hu, S.-Y., & Zhou, Q.-H. (2005). Study on development strategy of circular economy for China's coal industry. *Modern Chemical Industry, 25*, 12–16.

Mohanty, M., Dhal, N. K., Patra, P., Das, B., & Reddy, P. S. R. (2010). Phytoremediation: A novel approach for utilization of iron-ore wastes. In D. Whitacre (Eds.), *Reviews of Environmental Contamination and Toxicology Volume 206* (pp. 29–47). Springer. https://doi.org/10.1007/978-1-4419-6260-7_2

Murray, A., Skene, K., & Haynes, K. (2017). The circular economy: An interdisciplinary exploration of the concept and application in a global context. *Journal of Business Ethics, 140*(3), 369–380. https://doi.org/10.1007/s10551-015-2693-2

Nunes, F. C., Alves, L. D. J., Prasad, M. N. V., de Carvalho, C. C. N., & Nolasco, E. C. (2022). Phytoremediation strategies for rehabilitation of soils affected by red mud: The Mariana tailing dam collapse (Minas Gerais, Brazil). *Eurasian Soil Science, 1*, 1–13. https://doi.org/10.1134/S1064229322050052

Nunes, F. C., Oliveira, L. C., Alves, L. J., Prasad, M. N. V., & Anjos, J. A. S. A. (2023). Mine waste: Contributions to the circular economy. In M. N. V. Prasad & M. Smol (Org.), *Sustainable and circular management of resources and waste towards a green deal* (Vol. 1, 1st ed., pp. 1–19). Elsevier.

Pactwa, K., Woźniak, J., & Dudek, M. (2020). Coal mining waste in Poland in reference to circular economy principles. *Fuel, 270*, 117493. https://doi.org/10.1016/j.fuel.2020.117493

Paramguru, R. K., Rath, P. C., & Misra, V. N. (2004). Trends in red mud utilization—A review. *Mineral Processing and Extractive Metallurgy Review, 26*(1), 1–29. https://doi.org/10.1080/08827500490477603

Prasad, M. N. V., Lander, J. A., & Nunes, F. C. (2021). Global remediation industry and trends. In M. N. V. Prasad (Org.), *Handbook of assisted and amendment: Enhanced sustainable remediation technology* (Vol. 1, 1st ed., pp. 3–28). Wiley.

Ru-yin, L., & Xiao-ting, Z. (2009). Negative entropy mechanism of the circular economy development countermeasures in mining area. *Procedia Earth and Planetary Science, 1*(1), 1678–1685. https://doi.org/10.1016/j.proeps.2009.09.258

Samal, S., Ray, A. K., & Bandopadhyay, A. (2013). Proposal for resources, utilization and processes of red mud in India—A review. *International Journal of Mineral Processing, 118*, 43–55. https://doi.org/10.1016/j.minpro.2012.11.001

Savaskan, R. C., Bhattacharya, S., & van Wassenhove, L. N. (2004). Closed-loop supply chain models with product remanufacturing. *Management Science, 50*(2), 239–252. http://www.jstor.org/stable/30046061

Schoenberger, E. (2016). Environmentally sustainable mining: The case of tailings storage facilities. *Resources Policy, 49*, 119–128. https://doi.org/10.1016/j.resourpol.2016.04.009

Si, J. (2006). Development pattern of circular eco-industrial park of coal enterprise. *Meitan Xuebao/Journal of the China Coal Society, 31*(4), 549–552. https://caod.oriprobe.com/articles/10779661/Development_pattern_of_circular_eco_industrial_par.htm

Sun, H. (2009). Fuzzy evaluation on circular economy development of coal mining based on improved algorithm. In *2009 International Conference on E-Learning, E-Business, Enterprise Information Systems, and E-Government*. https://doi.org/10.1109/eeee.2009.52

United Nations (UN), Department of Economic and Social Affairs, Population Division. (2019). *World population prospects 2019: Highlights* (ST/ESA/SER.A/423). https://population.un.org/wpp/Publications/Files/WPP2019_Highlights.pdf

Wang, C., Harbottle, D., Liu, Q., & Xu, Z. (2014). Current state of fine mineral tailings treatment: A critical review on theory and practice. *Minerals Engineering, 58*, 113–131. https://doi.org/10.1016/j.mineng.2014.01.018

Wang, L., & Wu, C. (2010). Construction of coal eco-industrial park based on the theory of circular economy. In *International Conference on E-Product E-Service and E-Entertainment* (pp. 1–3). https://doi.org/10.1109/ICEEE.2010.5661466

Watson, A. H., Corser, P. G., Pardo, E. E. G., Christian, T. E. L., & Vandekeybus, J. (2010). A comparison of alternative tailings disposal methods—The promises and realities. In R. Jewell & A. B. Fourie (Eds.), *Mine Waste 2010: Proceedings of the First International Seminar on the*

Reduction of Risk in the Management of Tailings and Mine Waste (pp. 499–514). Australian Centre for Geomechanics. https://doi.org/10.36487/ACG_rep/1008_41_Watson

Xie, F., De-shan, T., & Ying-de, L. (2009). Study on the utilization model of infrastructure in declined mine based on the circular economy. In *1st International Conference on Sustainable Power Generation and Supply, SUPERGEN '09*.

Xu, D. M., Zhan, C. L., Liu, H. X., & Lin, H. Z. (2019). A critical review on environmental implications, recycling strategies, and ecological remediation for mine tailings. *Environmental Science and Pollution Research, 26*, 35657–35669. https://doi.org/10.1007/s11356-019-06555-3

Xu, Y., & Zhang, T. (2006). Application of physical input-output table to material flow analysis in Yima City. *Zhongguo Huanjing Kexue/China Environmental Science, 26*(6), 756–760.

Yakovleva, N., Chiwona, A. G., Manning, D. A. C., & Heidrich, O. (2021). Circular economy and six approaches to improve potassium life cycle for global crop production. *Resources Policy, 74*, 102426. https://doi.org/10.1016/j.resourpol.2021.102426

Yu, X. (2017). Coal mining and environmental development in southwest China. *Environmental Development, 21*, 77–86. https://doi.org/10.1016/j.envdev.2016.12.001

Zeng, S., Hu, H., & Wang, W. (2009). Circular economy assessment for coal-fired power plants based on supper-efficiency DEA model. In *2009 International Conference on Energy and Environment Technology*. https://doi.org/10.1109/iceet.2009.19

Zeng, X. (2023). Win-win: Anthropogenic circularity for metal criticality and carbon neutrality. *Frontiers of Environmental Science & Engineering, 17*(2), 23. https://doi.org/10.1007/s11783-023-1623-2

Zeng, X., & Li, J. (2021). Emerging anthropogenic circularity science: Principles, practices, and challenges. *iScience, 24*(3), 102237. https://doi.org/10.1016/j.isci.2021.102237

Zhao, Y., Zang, L., Li, Z., & Qin, J. (2012). Discussion on the model of mining circular economy. *Energy Procedia, 16*, 438–443. https://doi.org/10.1016/j.egypro.2012.01.071

Chapter 5
Temple Waste and Its Mitigation Using Vermitransformation Technology for Green Future

Rajesh Kumar, Ajay Kumar, Younis Ahmad Hajam, Aabid Amin Hajam, and Neelam

Abstract The management of temple waste poses a significant challenge in the pursuit of a sustainable and green future. Temples, being central to religious and cultural practices, generate substantial amounts of organic waste from offerings, floral decorations, and daily rituals. Conventional waste disposal methods often lead to environmental pollution and resource wastage, exacerbating ecological concerns. This review article explores the potential of Vermitransformation Technology (VTT) as an ecofriendly and efficient solution for managing temple waste and promoting a greener future. Vermitransformation Technology is a novel approach that utilizes earthworms to transform organic waste into nutrient-rich vermicompost. By harnessing the natural abilities of earthworms, VTT accelerates the decomposition process, leading to a reduction in waste volume and odor while enriching the resulting compost with essential nutrients. The produced vermicompost offers a valuable resource for soil enhancement and sustainable agriculture, promoting the adoption of circular economy principles. In this review, we present an in-depth analysis of the challenges associated with temple waste management and the environmental implications of conventional disposal methods. Furthermore, we discuss the various applications of VTT, its advantages over traditional composting techniques, and its potential to create a positive impact on the environment. In conclusion, this review article underscores the urgent need for sustainable waste management solutions in religious establishments. Vermitransformation Technology emerges as a promising and ecofriendly approach to address temple waste challenges while promoting a greener future. By adopting VTT, temples can play a pivotal role in inspiring broader

R. Kumar · A. Kumar
Department of Biosciences, Himachal Pradesh University, Shimla, Himachal Pradesh, India

Y. A. Hajam (✉) · Neelam
Department of Life Sciences and Allied Health Sciences, Sant Baba Bhag Singh University, Jalandhar, Punjab, India
e-mail: younismajeeed64@mail.com

A. A. Hajam
Department of Allied Health Sciences, Golden Institute of Management and Technology, Gurdaspur, Punjab, India

R. A. Bhat et al. (eds.), *Zero Waste Management Technologies*,
https://doi.org/10.1007/978-3-031-57275-3_5

societal shifts toward sustainability, fostering an environment where tradition and innovation converge to protect our planet for generations to come.

Keywords Temple waste · Mitigation · Vermitransformation Technology · Green future · Sustainable waste management

1 Introduction

Waste is any unusable substance that has been discarded by people and includes trash, waste, junk, and litter. Medical waste and hazardous waste developed from well-known origins and are managed by the administrative unit. This rapid accumulation of waste occurs every minute of every day as a result of the world's overpopulation and makes it difficult for administrative bodies to collect and handle it. However, incorrect organic waste disposal is the main reason for environmental degradation (Hameed et al., 2021; Siddiqua et al., 2022). Since its improper disposal can have numerous negative effects on the environment, including offensive odors, ground-water contamination, communicable diseases (such as malaria, tuberculosis (TB), cholera), and vector-borne diseases (such as dengue, typhus), managing organic waste has become a top concern for the entire world (Cattunar et al., 2023). There are numerous physical, chemical, and biological methods of disposal, but they take a long time and cost a lot of money (Hajam et al., 2023; Roy et al., 2022). Domestic garbage, non-edible oil cakes, and agricultural wastes are all abundant in carbon (Bisht et al., 2022; Gupta et al., 2023a). For the disposal of organic waste, the land lifting remediation strategy is typically used.

2 Floral Waste

One of the main issues is floral waste, which is used in many other industries, including the textile, culinary, cosmetics, and perfume industries. Flowers left in bodies of water (such as rivers, oceans) can have an impact on the aquatic life there, which is a major contributor to water pollution. Flowers are used in India for many purposes and in various locations (religious places of worship, making bouquets and garlands, rituals, etc.). Every year, numerous festivals are observed. Flowers are offered in all sacred locations, but they are afterward abandoned and become garbage (Gupta et al., 2023b). Different representations of Durga are worshipped by the populace during Navratri. Various types of flowers are used to decorate temples and are donated to Goddess Durga during puja (Mukherjee & Dasgupta, 2023). Flowers are said to be spiritual beings that followers of God can provide. Flowers become useless after being offered to God, and they are later transformed into volatile solid waste. All of the flower trash that was thrown is scattered across mosques, gurudwaras, and temples (Gupta et al., 2023b). In terms of flower production in

India, Andhra Pradesh, Karnataka, Tamil Nadu, and West Bengal come in fourth. In temples, dargahs, gurudwaras, and other religious buildings, various floral offerings are made. In gurudwaras, marigold flowers are typically delivered; in dargahs, usually jasmine flowers. Numerous types of flowers, including marigolds, lotuses, roses, and many others, are brought to God in various temples (Elango & Govindasamy, 2018). Different important components are present in these flowers. Essential oils, flavonoids patuletin, quercetagetin, and quercetin are found in the flowers of the marigold (*Tagetes erecta*). Flavonoids, volatiles, myricetin, and quercetin are found in chrysanthemum flowers (Wu et al., 2023), whereas flavonoids, phenolics, saponins, and steroids are abundant in jasmine flowers (Kunjachan et al., 2012). In addition to essential oils, flavonoids, tannins, quinines, phenols, alkaloids, protein, carbs, and steroids, rose petals are the main source of riboflavin, sugar, tannins, pectin, mineral salts, and salt of tartaric acid (Thakare et al., 2017). If managed effectively, the proper use of floral waste could be advantageous (Table 1).

Compost, biofuels, biogas, bioethanol, organic acids, pigments, and dyes, as well as the manufacture of polyhydroxybutyrate-co-hydroxyvalerate, food items, biosurfactants, incense sticks, and other value-added goods, are all produced as a result of managing floral waste through solid-state fermentation (Atal, 2022). The residue from flower production is also used to make handmade paper. These goods will be used in a variety of ways, including promoting plant growth, producing biogas (which may be used to generate power), and adding nutrients and other ingredients to food products. Additionally, diverse textile sectors make dyes and pigments from flowers and biofuels, and bioethanol aids in the resolution of energy issues. As a result, waste is transformed into wealth and a product with value. MSW is combined with flower trash for additional treatment. One of the major components of municipal solid trash is floral waste, which is produced in quantities of roughly 300 MT daily. The majority of the waste dumped into waterbodies is flower waste. Waste from flowers is used for biosorption, which is useful for treating wastewater and industrial waste. The challenges with floral waste disposal will be resolved, which will ultimately lead to less water and environmental damage (Table 2).

Table 1 Flowers and their chemical ingredients

S. No.	Flowers	Essential components
1.	Chrysanthemum flowers	Flavonoids, volatiles, myricetin and quercetin
2.	Jasmine flowers	Essential oils, flavonoids, phenolic, saponins, and steroids
3.	Rose flower	Riboflavin, sugar, tannins, pectin, mineral salts, salt of tartaric acid
4.	*Hibiscus rosa-sinensis*	Essential oils, flavonoids, phenols, tannins, quinines, alkaloids, protein, carbohydrates, reducing sugars and steroids
5.	Marigold	Essential oils, flavonoids patuletin, quercetagetin, and quercetin

Table 2 Different studies reviewed by different researchers

S. No.	Different studies	References
1.	Disposal of flower waste	Gupta et al., (2023b)
2.	Generation of flower waste every day (*about 300 MT of flower waste*)	Singh et al. (2013)
3.	Study of production of biogas and analysis of nutrient reduction potential in flower	Kumar and Swapnavahini (2012)
4.	Flower waste produced in vast amounts from temples in India that further utilized to forms dies	Bhati et al. (2021)
5.	Trichoderma utilization as a biocontrol agent (BCAs)	Hidayah et al. (2022)
6.	Role of fungi in energy recycling and increase plant community	Kržišnik and Gonçalves (2023)
7.	Trichoderma utilized in "bioremediation"	Racić et al. (2023)
8.	The role of earthworms in different wastes	Hajam et al., (2023)
9.	Trichoderma fungal species origin	Cao et al. (2022)

3 Nutrient Assessment of Floral Waste

The organic content and nutrients included in flower debris are good for the growth of plants. Currently, the majority of flower waste is combined with animal and human waste during vermicomposting to create organic manures, which aids in accelerating the decomposition process (Raza et al., 2022; Shouche et al., 2011). There are many fruits, leaves, and flower petals in it. These flowering plant parts are entirely organic in origin and can be utilized widely in residential and business settings (Singh, 2023). In their 2015 study, Makhania and Upadhyay looked at the physical and chemical characteristics of manure made from floral debris. Makhania and Upadhyay looked at a variety of physical and chemical characteristics, including temperature changes, hydrogen potential, mechanical strength, humidity levels, and unstable solids. They computed a rise in temperature on the fourth day that demonstrates the effectiveness of the "zero-waste process" for floral trash as well as the benefits of composting. Agarbatti and Gulab-Jal (rose water) are manufactured from floral Tagetes and roses, and some organic items like colors and strains (dye) are made from flower waste. Therefore, diverse procedures (methods) like the application and enhancement of floral waste in various places have been investigated in the current paper.

4 Role of Earthworms in Waste Management

Earthworms have intrigued philosophers and naturalists around the world since prehistoric times since they have been shown to be beneficial to agriculture and human health. Aristotle referred to earthworms as the "Intestine of Earth" in 384–322 B.C. They support the preservation and improvement of soil fertility. Robert Templeton published an in-depth study on earthworms in India in 1844, tracing the origin of earthworms to the Indian subcontinent. Using zoological nomenclature for earthworms, Carolus Linnaeus wrote in his book "Systema Naturae" in 1758 that earthworms increase soil fertility through their feeding and burrowing activities. From Paris, Savigny (1826) described the variety of earthworms and around 20 species of Lumbricidae. Dr. Thomas J. Barrett (1959) was greatly pleased by Charles Darwin's studies on earthworms, who acknowledged their remarkable job activities in the waste decomposition process in 1881. Dr. Thomas J. Barrett tried to use them to create fertile, lush fields on arid, dry soil. Additionally, he advocated for the introduction of particular earthworm species to the barren agricultural area since they are known as "Earth master Systems" and because they are beneficial to soil quality. The "Humus Factories of Manure" are also known as earthworms. Dr. Thomas J. Barrett published a book in 1959 titled "Harnessing the Earthworm" that upholds the potential of earthworms in compost (forming healthy soil with a deep color) and is referred to be the "Greatest Servant of Nature". The earthworms have great importance in waste management, in organic farming and sustainable agriculture reported by different workers (Hajam et al., 2023; Singh & Sinha, 2022; Vats et al., 2023); Vermitechnology, helps to regulate the sustainable development in human society. Earthworms: Cinderella of Organic Farming, written by Professor Radha D. Kale, widely known as "Earthworm Lady", was released in 1998. A subsequent edition of the book, Vermicompost—The Crown Jewel of Organic Farming, was published in 2006.

The Lumbricidae family of naturally occurring invertebrates includes earthworms, which are mainly found in tropical and temperate soils. Natural hermaphrodites include earthworms. Self-fertilization does not, however, typically take place in them. Under ideal circumstances, an earthworm can produce 100 cocoons in the six weeks to six months that it takes to reach adulthood (Edwards & Arancon, 2022a, 2022b). The incubation period for cocoons lasts between three and five weeks on average; however it varies by region. It takes 3–30 weeks for temperate earthworms. But in tropical earthworms, 1–8 weeks. The ability to repair missing segments is possessed by earthworms. The earthworm life cycle is well understood, and a thorough examination of the biological potential of earthworm species is necessary to comprehend the impact of earthworms on various organic wastes. The lifecycle of these organisms in many species, including temperate species, Indian species, and tropical species, has been the subject of numerous research (Mendesil et al., 2023). The biology of various earthworm species. Many researchers have looked into the applications of biological decomposition agents, which also include the use of earthworms to dispose of organic waste (Ducasse et al., 2022). Epigeic earthworms, endogeic earthworms, and anecic earthworms are the three main forms of earthworms (Edwards & Arancon, 2022a, 2022b). Epigeic worms are surface-dwelling in these species. In soil minerals, endogeic worms can be found, and the anecic earthworm species *Lumbricus terrestris* digs vertical burrows. Epigeic and anecic earthworms are two of the species employed in vermicomposting. *Eisenia fetida*, *Eudrilus eugeniae*, and *Perionyx excavatus* are said to digest wastes (Jayappa et al., 2022).

One of these is the earthworm species known as *E. fetida*, also known as the brandling worm (He & Liu, 2023). They can adjust to cultural norms that are close to nature. Other earthworm species, on the other hand, are seasonal breeders with modest dietary needs, metabolic rates, reproductive potential, and population densities. As a result, they cannot be used for composting or trash management. During the absorption of these biodegradable materials, organic waste can be transformed into vermicompost (Hajam et al., 2023), and five to ten percent of this material is responsible for the growth and development of the earthworm material is expelled as cast. The mucus secreted by the earthworm's stomach wall and some bacteria are combined with the cast. Even after the cast has been removed, the breakdown process continues. Compost's nutritional content is influenced by the organic waste that is utilized as earthworm feeding.

Earthworms consume high-quality fungus, which contains calcium and has greater sugar content. Earthworms are heterothermic organisms, meaning that their metabolism and other essential functions depend on the surrounding temperature. *P. excavatus*, a native species, is adapted to the typical environment there. The concept of an in-situ soil community made up of epigeic and anecic varieties for the management of soil and garbage in tandem. Earthworms were examined by Charles Darwin (1809–1882) for 40 years. He calculated that approximately 50,000 worms in British agriculture produced 18 tons of worm cast annually.

Workers have shown the importance of earthworms in recycling organic materials (leaf litter) (Edwards & Arancon, 2022a, 2022b). Earthworms are one of the many soil creatures that can handle large volumes of waste. *E. eugeniae*, *Eisenia species*, and *P. excavatus*, three earthworm species are capable of degrading vegetable waste. *E. fetida* was shown to be more tolerant to high temperatures than *E. eugeniae* and *P. excavatus* (Tagliabue et al., 2023). It can withstand temperatures as low as 42 and 500 °C in the soil. Food quality and quantity have an impact on the growth and development of earthworms (Ahmed & Al-Mutairi, 2022). Earthworms play a vital part in boosting plant yield, soil fertility, and microorganism proliferation (Jing et al., 2023a, 2023b).

Because of its capacity to withstand extreme environmental conditions (pH, temperature, and moisture content), *E. fetida* is a common earthworm species utilized in vermicomposting for vegetable waste (Edwards & Arancon, 2022a, 2022b). *E. fetida* is additionally used in the control of sludge (Ameen & Al-Homaidan, 2022). Partially digested waste has good quality and a humus composition with a high phosphorus (P) and potassium (K) content (Anielak & Kłeczek, 2022). *E. fetida* reportedly accelerates mineralization and the degree of humification.

Jing et al., (2023a, 2023b) reported comparative study of *E. fetida* (exotic) and *P. excavatus* (local) species of earthworms in vermicomposting of municipal solid waste

(MSW), Role of bio-inoculants, i.e., "*Pleurotus sajor-caju*, *Trichoderma harzianum*, *Aspergillus niger* and *Azotobacter chroococcum*", in pre-decomposition of a mixed solid waste, i.e., municipal solid waste (MSW) and horticulture waste (70:30) vermicomposting then observed that it improves its quality of compost and reduces its stability time period.

Lufkin, USA starts the earthworms use in liquid waste and about 4500 kg earthworms consume 900 kg waste in a day and increase in population. Ahmed and Deka (2022) studied that *E. fetida* has ability to decompose different types of wastes or mixture of wastes in one month. The population growth of *E. fetida* and *E. eugeniae* was found to depend on the quality and quantity of the available food, culture media, temperature and moisture levels (Nsiah-Gyambibi et al., 2022). Aoka Sanagyo Co. Ltd. of Japan has a capacity to handle 3000 tons of garbage by converting it by vermicomposting per month (including the wastes from paper pulp and food industry), that leads to production of 400 tons casting along with 10 tons of earthworms.

5 Vermicomposting

Vermicomposting is one of the best methods to produce high-quality compost with the help of earthworms (Devi & Prakash, 2015). It is a biotechnical process, used in floral waste treatment that adopt new concept of "ecological design" with earthworms (Arora et al., 2014). It is also known as "bio-oxidative process" in which both earthworms and other microorganisms play an important role to form a stabilized vermicompost. It is a best method for the proper management of floral waste (Singh et al., 2013). However, lignin and cellulose (phenolic compounds) having complex structure may delay the process of vermicomposting (Greff et al., 2022). Hence, some other methods are required or in high demand to forms value-added products by disposing flower waste (Singh et al., 2018). For the bio-chemical degradation, microbes play an important role and earthworm is also helpful to increase biological activities. Composition plays an important role for maintaining better quality of compost. Different organic waste such as vegetable waste, cow dung, water hyacinth, municipal waste are used in vermicomposting to produce nutrition rich compost (Patra et al., 2022). It can be practiced and operated in small medium and large-scale for the management of garbage. It is also helpful to produce high-quality compost (fertilizer) by converting organic waste, i.e., agro-waste and animal waste into manure useful for plants and soil (Chojnacka et al., 2022). Vermicomposting is divided peat-like material to have excellent structure, proper aeration, and proper moisture-holding capacity (Edwards et al., 2011). Vermicompost various nutrients such as nitrogen, potassium, sodium, magnesium, phosphorous, iron etc. are essential for the maintenance of fertility of soil. Among all of these nitrogen (N_2), phosphorous (P), and potassium (K) are most important nutrients. Vermicomposting gets evolved in less carbon–nitrogen ratio, phyto-toxicities, and increases in microbial activities and its population, as compared to ordinary composting. Nitrogen fixing bacteria are commonly found in the gut of earthworms as together with other organisms (Ghosh et al., 2022). Concomitant increase in present nutrients, i.e., nitrogen, phosphorus and cellulose activity and reduction of microbial biomass carbon in waste. Vermicomposting method involves earthworms as versatile natural bioreactors and play an important role in decomposition of organic waste, management of soil fertility that enhance the plant growth. Vermicompost have many other benefits to enhance different activities simultaneously such as earthworm can be utilized in medicines as a protein rich animal, they can't grow in polluted wastes. They are utilized as anti-soil pollutants. Various studies have been conducted on decomposition of flower waste through vermicomposting. Gaurav and Pathade (2011) reported that Ganesh temple, Sangli, Maharashtra are the source of temple waste for vermicomposting. Organic waste vermicomposting plant was beginning in 1970 in Hollands Landing, Ontario (Canada) and was used to process about 75 tons waste in a week. Second vermicomposting plant was started in California in 1975. In 1978–79, a vermicomposting farm was started in which about 500 MT of vermicompost disposed per month. In USA, Philippines, Japan, UK, Cuba, France, and India are the most popular countries having subsequently vermicomposting practice. Among these Cuba is an ideal example of vermicomposting provided by nation-wide experiment. In various countries such as Japan, Canada and USA vermicomposting facilities are reported in commercial operations, also being efficiently practiced in Asia and in Philippines. Gaurav and Pathade used the waste of biogas plant mixed with flower waste and cow dung at 30 °C for 30 days. Singh et al. (2013) demonstrate that flower waste managed through vermicomposting has been related with kitchen waste (KW) and green waste compost. Recently, vermicomposting and composting has been beneficial for the disposal of organic waste to form nutrient-rich compost. In India, food production is enhanced 50 MT–170 MT due to high yielding verities, chemical fertilizers and improvement of irrigation facilities. Nowadays about 35 million tons buffer stock has of food grains in country. According to Dr. Kalam's estimation, about 14 million tons food grains was imported by India since 2010 and the rate will increased day by day with 2% per year due to stability and saturation stages in food grain production. In agricultural practices, there is a need of change. A fine organic waste can be used in fields instead of chemical fertilizer which increases the soil productivities. Scientific research has also recognized the anti-oxidant property of various flowers that have some essential components like carotenoids that also used to treat many of skin related problems (Adhikari et al., 2019).

6 Properties of Floral Waste

Different flowers have different properties such as anti-oxidant and anti-bacterial, and many more. These flowers are utilized in food industries and cosmetic industries in some places due to their anti-oxidant properties observed in many studies (Ulbricht et al., 2011). In vermicompost, various nutrients occur including nitrogen (N), phosphorus (P), potassium (K), calcium (Ca), magnesium (Mg), sulfur (S), iron (Fe), manganese (Mn), zinc (Zn), copper (Cu) and boron (B) show positive effects on plant nutrition, photosynthesis, the content of chlorophyll and nutrients in roots, shoots and fruits (Yadav et al., 2022). Frequently, researchers are being attracted to vermicomposting and literature is showing an enhancing trend. But enough and significant involvement of farmers to be lacking. Therefore, in addition to research work, promotional activities must also be conducted.

Various studies have been conducted to utilize floral waste in vermicomposting. Temple waste recycling mainly comprises the flower offerings carried out by vermicomposting process by using *E. fetida*. Kumar and Swapnavahini (2012) conducted a study in which the production of biogas and analysis of nutrient reduction potential of rose residue was analyzed through anaerobic digestion in a batch reactor. About 2.5L batch reactor was used and allowed for digestion for 30 days at room temperature. They observe different parameters like oil yield, color, and Physico-chemical properties of *Rosa damascena* and *Rosa centifolia*. They kept that there are differences in their qualitative and quantitative analysis (chemical composition, aroma constituents) in these species. Flower waste produced in vast amounts from temples in India that further utilized to forms dies for cotton, wool and silk. *T. erecta* (marigold) petals were used in dyeing industries and isolated, consisting of carotenoids-lutein and flavonoid-patuletin. For the breakdown of organic matter, a symbiotic relationship exists between the earthworms and microorganisms that help realize minerals and nutrients in the soil.

Due to this, the cast has more microbial flora than the surrounding soil. Some microorganisms are dead when passed through the gut, and then others proliferate. It is widely believed that Organic fertilizers provide nutrient-rich substrate and an increase in earthworm species population when they feed upon organic matter or microorganisms directly and it colonizes organic materials. In addition to this inorganic fertilizer also contributes indirectly to the quality of crop waste returned in soil. Further, these inorganic fertilizers cause disease in the earthworm, mainly if it is ammonia-based.

Vermibiotechnology: From Soil Health to Human Health, which indicates the importance of term technology for humans. It is used for the management and decreasing of organic waste material. It is a very challenging field for scientists. Continuous efforts are required to bridge the gap between the laboratory and the area. Establishment of vermicomposting units can encourage farmers, animal keepers, industrialists, and entrepreneurs to follow develop the vermicomposting unit at small as well as large scale. This technology could be beneficial to human society and sustainable development in various fields such as agriculture, environment, socio-economic sector and science and technology.

7 Role of Microbes in the Recycling of Biodegradable Waste

Fungi, a remarkably diverse group and its diversity can be classified taxonomically and by their roles. Rajesh et al. (2016) reported the classification of *Trichoderma* spp., i.e., division-Ascomycota, subdivision-Pezizomycotina, class-Sordariomycetes, order-Hypocreals, family-Hypocreaceae. Since 1920 it acts as a biocontrol agent (BCA) against the pathogen. It has recently included six phyla and four subphyla with five groups: Chytridiomycota, Zygomycota, Ascomycota, Basidiomycota and Glomeromycota. The fungi functional role can cross the taxonomic limits. It plays significant role in energy recycling and increases plants' community by symbiotic relationship. Sporocarps, the fungal fruiting body, are engulfed by different vertebrates and invertebrates. There is an important distinction, i.e., the difference among saprotrophic and mycorrhizal fungi. Saprotrophs are organisms that utilize energy and many nutrients from the waste, and Mycorrhizal fungi depend upon the living organism (plants), usually for their food. The major difference between macro fungi and macro fungi is their fungal fruiting structures. As a difference in classification, fungi show common features and being heterotrophic. They are highly amalgamated with the environment to forms communities. The *Trichoderma* fungal species reported in ancient period. It was proposed only one species, i.e., *Trichoderma viride*. In 1969, a particular protocol was developed for species identification. For 70 years, the *Trichoderma* spp. was identified. It can produce antibiotics that inhibit pathogen growth and development. It is also used as a biocontrol agent (BCAs). This genus (*Trichoderma*) firstly introduced based on its microscopic similarities. Four species are categorized in it, i.e., *T. viride*, *Trichoderma nigrscens*, *Trichoderma aureum* and *Trichoderma roseum*, collected in Germany. They evaluate that it is challenging to clear out the difference in *Trichoderma species* morphologically. However, its phylogenetic classification reached 100 constantly and expected to continue its increase rapidly. There are some modern methods such as "Genealogical concordance phylogenetic species recognition" (GCPSR) and "DNA-barcode system" to identifying the *Trichoderma* spp. But there are limitations due to this promising investigation occurs in future.

Trichoderma spp. has widespread application as biocontrol agents have been exploited. It is reported that it acts against some soil-borne phytopathogenic fungi. Similarly, to most BCAs, *Trichoderma* spp. can be systematically utilized as spores (conidia) that can tolerate adverse conditions during the production of the product. These fungi act as an adverse (enemy) against various soil-phytopathogens like fungi, invertebrates and bacteria. Their BCA activity is accountable to different antimicrobial compounds having aggressive modes of growth and physiology. It is also utilized in "bioremediation" due to the formation of metabolites that have laccase and peroxidases enzyme in massive amounts. *Trichoderma* spp. also used as a bio fungicide agent; thus, it acts as a bio fungicide. *Trichoderma* spp. is a mycoparasitic fungus that studied broadly. Different *Trichoderma species* are used not only for plant disease but with the ability to increase development, growth, the reproduction rate of the plant also having the ability to grow in harsh conditions, high aggressiveness against phytopathogenic fungi and to enhance defense system (a mechanism) (Daguerre et al., 2014; Keswani et al., 2014; Pandya et al., 2011; Schuster & Schmoll, 2010; Tripathi et al., 2013). However, due to these features, the *Trichoderma* genus can grow in broad habitats with high population densities (Chaverri et al., 2011). Therefore, the present study has been designed to explore the biodegradable role of *T. viride* in the decomposition of flower waste.

8 Anthropogenic Influence on Waste Generation and Management

Human beings are dependent on climate; however, overexploitation of resources causes environmental degradation. The primary cause of ecological degradation is overpopulation, pollution and generation of vast quantities of wastes (Bhat et al., 2021). As a developing country, India produces enormous waste, including industrial waste, municipal waste, etc. However, management processes are lacking. If these wastes are recycled and reused correctly, then they can become alternate sources for producing various products. In the nineteenth century, the first waste to energy (WtE) plant used to dispose of different waste types. Waste to energy is the recycling of waste to produce energy like electricity, fuel, etc. but not properly forming the benefits to the environment and as well as economically.

Generation of waste is unavoidable and the extra (remaining) waste should be treated to minimize its effect on environment (Bhat et al., 2012, 2013). Waste to energy (WtE) recognized good source for the management of waste. There are about 800 thermal wastes to energy plants in forty countries. About 11% municipal solid waste and about 429 TWh power is produced broadly (Martin, 2012). In Japan, Germany, Sweden, Denmark and UK other alternate sources implanted at large scales. About 80% municipal waste dispose of in Japan and 10% of which utilized to generates power (Tanaka, 2014). The drawbacks of waste disposal are well recognized. In Europe, landfill disposal is the only way regulated by the Council Directive on the Landfill of Waste (1999). There are many other waste like hazardous waste that were not treated properly with WtE treatment. Different countries have high demand of energy, and WtE is a very good and efficient source of it. In contrast, only electricity is generated in warm regions. At the beginning of the twentieth century, various strict rules were announced, such as in 1986, as per procedures for setting emission standards or discharge of environmental pollutants formulated to manage environmental issues. In 1989, Hazardous Waste (Management and Handling) aimed to control the generation, collection, treatment, import, storage, and handling of hazardous waste.

The Municipal Solid Wastes Management and Handling ruled that municipal authority was responsible for collecting, segregation, storage, transportation, processing, and disposal of municipal solid wastes (Bhat et al., 2014). The old disposal processes include throwing debris in water bodies, empty spaces, barren lands, small pits, etc. However, these disposal processes are unhygienic for living beings. Traditionally wastes are assimilated as extraneous for production and are managed when they create any unfeasible environmental problem (Seadon, 2010). The significant sources include rice, wheat, sugarcane residues whose burning causes environmental pollution like smog. Therefore, waste management has become an enormous problem that needs ken attention and the design of scientific disposal techniques. Development in science and technology has changed the lifestyle of humankind by manufacturing advanced tools, instruments, and practices carry to fulfill essential amenities. To manage various types of waste, i.e., agricultural waste, commercial wastes, domestic waste, industrial waste, modern decomposition or degradation methods (aerobic and anaerobic). However, the decomposition method for temple wastes has not explored yet. A massive quantity of municipal solid waste (MSW), agro-waste, food waste, vegetable waste etc. are being generated due to rapid civilization, green revolution and mechanization (Gabhane et al., 2016). The decomposition of these all types of wastes, composting is best option among the decomposition processes for managing such colossal biomass. It produces nutrient-rich compost that helps to increase fertility of soil. Composting is the biotransformation process of complex organic material to stable and straightforward material that supplies essential soil minerals for the plant with implementing soil conditioning properties to reduce toxicity and pathogenic organisms. It is a good and effective method utilized for the recycling of the waste (animal manure, industrial waste, crop straws, municipal solid waste (MSW)) in a better way all over the world (Odejobi et al., 2022). However, composting is a slow process and not favorable for future prospects of entrepreneurship. Thus, rapid composting is the "need of the hour". It reduces the time of composting and giving highly stable and nutrient-rich compost. Composting is an essential step in the reduction and preparation of a naturally beneficial soil additive. It is the most economical and sustainable option for organic waste management. It is easy to operate and manage appropriately. It is a natural process of organic waste treatment operated with different technological modifications.

9 Factors Affecting the Recycling of Waste

The optimum temperature in composting process is minimum of 45 °C (thermophilic conditions), in which proper aeration and moisture level play a significant role in the maintenance of ambient conditions. During composting, the benefit-cost ratio (BMR) constantly increases under all conditions. This animal (livestock) culture is easy to handle by a professional and an unprofessional human being. Manure reduced by the decomposition of cattle dung, farm and forest waste, as well as MSW (municipal solid waste), enhanced the growth of plants and called a "**miracle**". Different methods were used to detect some carbon and nitrogen sources. The microflora of compost with the bulk agent holds maximum moisture and provides good aeration and a hydrolytic enzyme's surfactant production (Gabhane et al., 2012). The effect of commercial additives containing iron, magnesium, manganese, and calcium hydroxide on composting and reported mineral addition improving the nutritional value of compost. Compost activators are usually mixtures of various microorganisms, nutrients, enzymes, and pH-balancing compounds that helps to increase microbial action and also enhance the stability of compost (Gabhane et al., 2012). Commercially available activators are sugar solutions congaing fungal spores. Which initially boosts microbes but fail to produce enzymes. Thus, a compost activator is needed to promote the microbial population and increase plant dry matter production, degrading enzymes such as cellulose, dietary fiber, versatile enzyme (laccase), amylase enzyme, etc. Various researchers will develop a commercial compost activator, but the effective compost activator not developed yet.

10 Earthworms

Earthworms have essential role in solid waste management and are the primary component of soil fauna communities in most ecosystems. Earthworms are not only the "**intestine of the earth**" but also an essential source of minerals and trace elements (Ayoola & Olayiwola, 2014). More than 4000 earthworm species are classified in three major groups such as epigeic, anecic and endogeic. Epigeic earthworms are surface feeder. Endogenic earthworms are living in horizontal barrows in soil, and are not permanent. Anecic earthworms are bottom feeder with permanent vertical holes. Several researchers have investigated the vermicomposting potential of earthworms from all three groups (Fig. 1). Among all the earthworms, selective earthworms are used to produce effective and nutrient-rich compost through vermicompost process.

Some epigeic species *E. fetida*, *Eisenia andrei*, *E. eugeniae* and *P. excavatus* are shows effective role in vermicomposting. They have high fecundity rate, high tolerance power and fast conversion of waste into compost. Earthworms are natural decomposers because they can degrade all types of trash. The proper selection of earthworm species can affect the stabilization of various physical and chemical

Fig. 1 Earthworms in vermicompost

parameters in vermicomposting in which many species of earthworm used in it for better results. They tend to tolerate unfavorable conditions. The productivity rate increases and cocoons have less hatching period, fast maturation rate, suitable for the vermicompost process. They survive under aerobic conditions and convert the mixture of wastes into renewable energy sources. The partially stabilized material is vermicast utilized as "bio-compost". The quantity of organic matter utilized by earthworms depends on a variety of factors, including the type of earthworm species and waste material, the composition of waste stuff, particularly C/N ratio, particle size of waste, moisture and pH of the bedding, organizing system to favor proper aeration, protection from direct sunlight, extremes of weather and enemies, etc. If the physical, chemical and biological (activity of bacteria) conditions are suitable, than they increase their population as long as food available for them. Mostly the juvenile feed upon the small sized organic waste produced by adult earthworms. However, multiple species used rather than single specie for quick decomposition of waste and that compost contain high microbial biomass and proliferation. And it was analyzed that multispecies more beneficial that use of single species. Bacteria and earthworm combination increasing the decomposition rate that were reported previously. *E. fetida*, an earthworm species, is mainly used to produce compost by vermicomposting because it can tolerate a high level of atmospheric conditions such as climate change. It reported that *E. fetida* is best suitable for compositing of flower waste and cow dung having a higher maturity rate and high micronutrient and macronutrient concentration during the different stages of the composting process.

11 Vermitransformation and Vermicompost

Vermicompost is an effective and alternative composting method that allowed for minimum composting space. It is a biological process, earthworms and microorganisms are interacting with each other and helps to produce a nutrient-rich manure (Amouei et al., 2017; Das et al., 2019; Pigatin et al., 2016; Soobhany, 2019). This process evolved in mid of twentieth century. In the vermicomposting process, cattle dung is used with agricultural waste and other wastes like vegetable waste, kitchen waste, garden waste or others and has less intentioned by vermicomposter. The production of quality vermicompost from raw waste requires much knowledge of the process. Vermicompost contributes significant role to maintain soil fertility. Vermicompost contains high-quantity nutrients, and it is less toxic in nature. It is a simple process, cost effectiveness and has ability to decompose various all type of organic waste. Vermicompost is good quality manure containing various essential nutrients required by crops such as nitrogen (N), phosphorus (P), potassium (K), calcium (Ca), magnesium (Mg) and micronutrients, i.e., iron (Fe), zinc (Zn), copper (Cu) and manganese (Mn) in adequate quantities to increase the quality and quantity of crops. It improves the biological, chemical and physical characteristics of the soil. It also improves the soil structure. Due to that, soil permeability and porosity increased. However, it also helps in decomposition of flower waste in high-quality organic manure (Shouche et al., 2011). In Nirmalya, a temple in South Mumbai was pre-composted at 30 °C and used as a substrate for vermicomposting by earthworm species *E. fetida* for 90 days. Vermicompost contains an excessive amount of antibiotics, vitamins, hormones, enzymes and amino acids that are helpful in the growth and development of plants. It provides resistance and nutrient to plants and decreases the cost of production and overall cultivation by reducing chemical fertilizer application. Application of bio-manure could be a better option to maintain sustained agricultural practices. In bio-manure, micro and macronutrients such as nitrogen, phosphorus and potash contain ten 1bs, 51bs and ten 1bs respectively in 1-ton compost.

12 Contamination of Water Due to Disposal of Waste

About 95% of human beings are dependent upon groundwater. Still, water becomes polluted because of the waste's misuse by putting the waste into it without any treatment. About 70% of water is contaminated badly due to the industrial and sewage waste into natural water bodies. Out of all the organic waste, floral waste management is yet to explore. Flower waste is also a semi-solid waste produced from temples and rituals like marriages, etc.

It is a religious matter offered by devotees to the gods and discarded in the environment, which converted into solid waste. Pilgrims (devotees) offers many things to gods like fruits, coconut, flowers, etc. and the things that are edible and further distributed in pilgrims as a Prasad and non-edible things were thrown away. Annually, in India about 8,000,000 tons of floral waste thrown in waterbodies mainly in rivers. Mostly marigold flower is used in abundance by the worshipers. The degradation of floral waste is a prolonged process compared to kitchen waste degradation. So, the proper management of floral waste through ecofriendly process is needed. In few studies flower waste management and use was carried out.

13 Generation of Waste in Temples

About 2 million temples in our country in different locations that includes Haridwar, Kedarnath, Tirupati, Patna Gaya, etc. (Ramachandra, 2012). Among all the temples situated in India, in Shraavana month Kashi Viswanathan temple has maximum pilgrims visit in whole year than others. In this temple, the proper management of disposing the flower waste is coordinated accordingly and converted into compost. Many methods can be used to treat this waste, but they are slow and very costly, and the natural processes are slightly tough to manage. Many festivals and functions organized by the people in which flowers play an important role. But after using them, they are thrown away as they become waste. Quantity and quality of flower waste differ in different temples, and every God has their own worshipped flower as considered in Vedas. In Lord Shiva temple maximum of flowers are offered on

Monday and Saturday, then that on other days and the occasion of Shivratri (Dwivedi et al., 2019). There are many practical examples where we can see the generation of floral waste near or in vicinity places or holy places, Chitrakoot is one of them. Kamtanath Mandir and Bharat Koop are the places in Chitrakoot where flowers are sold and situated in the Ramghat on Mandakini River and another on the platform of Lord Kamtanathji respectively. Total floral waste estimated in both places generated is 5.48 tons per year. The studies found that about 10 tons flower waste generated in city that estimated by Varanasi Nagar Nigam. Kaur and Joshi (2002) reported that large amounts of flowers are offered in temples of Jaipur city (India), that leads to solid waste, and forms different health issues and environmental pollution. About 23,000 temples situated in Varanasi only situated at river Ganga where almost all floral waste dipped into it. It leads to a very odor smell that destroys its ecology (Wani & Rao, 2013).

To evade effects caused by the disposal of floral-related offerings, they can use to make some valuables. Various value-added products, i.e., incense sticks are makes from flowers like *Tagetes* spp. and gulab-jal (rose water) from rose spp. Instead of all these, flowers are also used to make some organic products like colors, dyes etc. We can also recycle this floral waste by reusing it for various purposes. In Ajmer, floral waste management has yielded good payoffs like that of Ajmer Sharif Dargah of Khawaja Moinuddin Chishti, where about 15–18Q of flowers were offered every day and dumped in a well. Tiwari (2014) researched with the aspire to utilize and manage floral waste obtained from ten popular temples of Jaipur city. Makhania and Upadhyay did another experiment (2015), studied various physical and chemical parameters of flowers collected from temples of Surat. They examined the parameters like temperature, pH, electrical conductivity (EC), moisture content, and different volatile solid (VS) samples. They measured the highest temperature on the 4th day of heap composting and proved that composting is an effective "**zero-waste**" process for the flower waste treatment. Kashi Viswanathan temple which draws the attention of many devotees round the year especially in Shrawan. The flower waste thrown into the water bodies (rivers, oceans, ponds etc.), which leads to water pollution and harms aquatic organisms living in the water. Every year about 50 MT of municipal waste (MW) is produced in India (Bhat et al., 2018b; Hajam et al., 2023). An increase in the amount of the waste material causes various discarded related issues. The mismanagement of this waste creates many health problems and environmental issues because about 90% of waste managed through unscientific manners (Bhat et al., 2018a). Recently, many natural processes treat waste with no toxicity and change it into useful resources like organic compost. Organic waste utilized as the production of important sources like the formation of energy and nutrition in agriculture.

Floral waste gets decomposed aerobically and anaerobically and produces bad smells, toxic gases, and solid–liquid waste, leading to health problems in humans. It already reported in traditional and historical evidence from the different places of India such as Andhra radish, Maharashtra, and Chhattisgarh are occupied cultivation or mahua flower and harvesting for alcohol industries through traditional methods. Mahua flower exploitation as a substrate to prepare ethanol is an immense advantage. It observed that flower waste also used to produce biogas by using it as raw material.

Also, it reported that biogas production from rose through anaerobic digestion in a batch reactor. Biogas produces from flower waste can be utilized as fuel and electricity generation. The scheme represents the flower as a substrate for "**bio-fuel**" production in fermentation. Ranjitha et al., (2014) reported a huge quantity of biogas kg^{-1} substrate from floral waste in Kenya. In India commonly, sugarcane is a chief raw material for oxalic acid. In the mahua flower, there is more sugar than others and is considered an alternate potential in the production of oxalic acid (Das, 2010). It is mostly utilized as a preservative and as a chelating agent in different ways. Marigold and Rose's flowers were collected from the Aurangabad temples to utilize dyes preparation. Recently, biodegradable dyes emerged as an important substitute the synthetic dyes. It reported in previous studies that Saffron flower used to extract dye for Pashmina shawl (Raja et al., 2012). Also, Hibiscus flowers used as a natural dye in industries (Teli et al., 2013). Mostly, they are mixed with municipal solid waste (MSW). Most of the time, flowers dipped into water bodies like rivers, lakes, and ponds, which cause a major issue of environmental pollution and water pollution and also causes harmful effects on human life. Leaves are solid waste having many essential nutrients and minerals. A major cause of solid waste generation in India is rapid urbanization, industrialization, and population. Yadav and Garg (2011) reported that in India, about 133,760 tons of waste generated in a single day. Most of the time, the municipal solid waste contains about 70% organic waste, including vegetables waste, flower waste, leaves, kitchen waste (KW), and fruits (Potdar et al., 2016).

Generally, about 300 tons of flower waste segregated per day. About 1.5 tons of flower waste generated from Surat city per day (separates at flower waste (FW) collecting center). In the present study, temple waste, farmyard and kitchen waste utilized as feedstock. However, *Aegle marmelos* leaves, *Datura stramonium*, *T. erecta* and *H. rosa-sinensis* flower mainly considered in temple waste. Flowers collected from different city temples, while most of the offerings collected from two temples, i.e., "Viswanathan temple" and "Sankatmochan temple". Mostly flowers, leaves, coconut etc., constitutes temple waste, resulting in various pollution and hygiene problems without proper management. Flowers used as spiritual entities offered by pilgrims to gods daily and dumped into water bodies or in open land converted into solid waste. Sometimes flower waste accumulated at temples, Gurudwaras, and Mosques land sites.

In India, Andhra Pradesh, Karnataka, West Bengal and Tamil Nadu promote flowers, in which West Bengal leads the fourth position. India is a city of temples, visited by many devotees and various things like fruits, floral garlands, coconut and sweets etc. offered to gods. And all things are distributed by "Pandits" among the pilgrims as parshad. Among these, floral garlands non-consumable materials thus get discarded. Various practical examples are there, where lots of flower waste can see near holy places (Temple and Gurudwaras etc.). One of the famous temples is situated in Chitrakoot, where many pilgrims offer flowers. As research is concerned, huge work done on the utilization of flower waste in vermicomposting. Gaurav and Pathade (2011) worked on temple waste produces from Ganesh temple, Sangli, Maharashtra, through vermicomposting. Micronutrients and macronutrients such as Nitrogen (N), phosphorous (P), potassium (K), calcium (Ca), iron (Fe), manganese (Mn), zinc (Zn) and copper (Cu) present in compost used for agricultural purposes (Dhaliwal et al., 2021). These two methods, i.e., composting and vermicomposting are ecofriendly and the best methods for treating hazardous waste (Dominguez & Edwards, 2010). Recently, a new strategy developed for sustainable economic growth, which European Commission adopted. Yeboah (2011) reported that temple floral waste management could provide a sustainable source for handmade paper production. This method is useful to discard waste produced from temples and also to recycle and reuse it. This handmade paper has various applications such as 100% chemicals free, no harmful products produced at the time of production.

Flowers are also having medicinal properties because of the presence of many metabolites in them. Various papers show calendula oil as a value-added product from floral waste used as high-quality massage oil after pretreatment, olive oil gets infused with dried calendula flowers. "Passion flower" is helpful to decrease anxiety stress and insomnia. Lily flower is used to prevent jaundice, respiratory disorders and gastrointestinal disorders. Rhododendron flower juice can help to prevent high BP (blood pressure). Chamomile oil prevents skin disorders such as eczema and relieves muscular cramps (Waghmode et al., 2016). In horticulture practices, flowers cultivated in large amounts, and flower waste generated in high amounts. Many pathogens attracted toward flower waste that causes multiple health problems. If proper management of flowers done, then many value-added products are produced. Bacteria consortium is a bacteria having an effective process in the flower degradation process. Composting is a better option for the degradation of flower waste that forms nutrient-rich compost, which further increases soil fertility. This organic compost increases the water holding capacity that helps to prevent soil erosion. To solve all these problems, Government provides various techniques such as composting and vermicomposting. Vermibiotechnology is the most common and traditional method for the management and decomposition of various wastes. It is a very ecofriendly, viable and very cost-effective process that finally converted into "**black gold**".

In India, Bhawalkar Earthworms Research Institute (BERI) is a vermiculture institute in Pune (Griffin et al., 2018). Vermicompost is very useful and highly nutritious for farming than chemical fertilizers. It contains micronutrients and macronutrients, including humic acid, growth hormones utilized primarily (Dominguez & Edwards, 2011). Vermicomposting is an ecofriendly method to process waste produced by rural and urban populations (Bhat et al., 2018c; Boruah et al., 2019; Paul et al., 2018). Bio-oxidative process in which earthworms play a vital role to decompose waste and stabilize it by changes its physicochemical activities. Both processes (composting and vermicomposting) are different in different ways. Vermicomposting is a mesophilic process in which earthworm lives at moderate temperature, i.e., 10–32 °C. The vermicomposting process is faster than that of composting because earthworms used in it. Microbial activity is best in it and also promotes plant growth. It is a process in which the conversion of waste to "**gold**" occurs with the help of worms. In the city and villages, flowers are mainly used and seen throughout there and by keeping it in mind various biogas plants established at suitable places. In industrialized countries, an increase in flower waste quantities production and various dumpsites are enlarges their limits of capability. About half by mass floral waste made up of biodegradable waste. In India, religion is a path of life due to cultural diversity, hence known as the "**land of God**". Our religious belief is that flower are a holy entity so avoid throwing flowers into dustbins. Generally, flowers are packed in plastic bags and thrown directly in the water bodies or in open land. Also, about 960 million tons of solid waste are produced in India per year from different industrial, agricultural, and many more. About 350 MT of waste produced from agriculture practices.

Nowadays, to enhance the productivity and quality of crops, various chemical fertilizers are used that causes environmental pollution. Bio-fertilizers and bio-pesticides are alternative sources of sizeable sustainable production having a less ecological impact. Various soil-borne bacteria and fungi can colonize the roots of plants, having positive effects. Different bacteria, fungi and plant-growth-promoting rhizobacteria (PGPR) in which fungi genus *Trichoderma* spp. and *Piriformospora indica* can stimulate plant growth. *Trichoderma* is a fungal genus present in a different ecosystem. Some strains can decrease the plant disease severity by inhibiting pathogen present by their high antagonistic and mycoparasitic potential (Viterbo & Horwitz, 2010). *Trichoderma* is an asexual spore-producing filamentous ascomycetic fungi. It belongs to the Deuteromycetes family. *T. viride* species of fungus present in all types of soil. It is an important prevalent and culturable fungi (Leta et al., 2013; Martínez-Medina et al., 2011). It also used as a biocontrol in industries. It is host-specific and colonized. *Trichoderma* studied worldwide utilized as a biological agent in agriculture. The products form from them, and they used as an alternative to synthetic agro-chemical. Among this *Trichoderma* spp. They used most abundantly against plant pathogens. Many species of *Trichoderma* can produce both volatile and non-volatile antibiotics. However, *Trichoderma* grows more rapidly than others. It forms large colonies to inhibit growth and reduce colony counts. It promotes fungus in plants, which helps activate change and hinders plant diseases.

These microorganisms show endophytic associations that interact in the rhizosphere with other microbes. It is a growth-promoting fungus, increasing nutrients availability of iron (Fe), nitrogen (N) and phosphorous (P) and also phytohormones productions, i.e., 1-aminocyclopropane-1-carboxylate deaminase, 3-indoleacetic acid, indol-3-acetaldehyde, indole-3-ethanol and indole-3-carboxaldehyde (indoles compounds) (Martínez-Medina et al., 2011). There are limited studies that demonstrate the efficiency of active fungal strains in plant rhizosphere and pre-planting inoculation. Therefore, the development of multifunctional biological preparations occurs with efficiency, including the *Trichoderma* preparations. It helps to increases active fungal colonization of the plant rhizosphere. The amount of organic matter turned over by earthworms depends on a variety of factors, including the type of earthworm species and waste material, the composition of waste stuff, particularly C/N ratio, particle size of waste, moisture and pH of the bedding, organizing system to favor proper aeration, protection from direct sunlight, extremes of weather and enemies, etc. Therefore, current study having designed to evaluate the physio-chemical characteristics of flower waste at pre-decomposed and post-decomposed vermicompost (Fig. 2).

Fig. 2 Different waste product forming nutrient-rich compost

References

Adhikari, S., Gascó, G., Méndez, A., Surapaneni, A., Jegatheesan, V., Shah, K., & Paz-Ferreiro, J. (2019). Influence of pyrolysis parameters on phosphorus fractions of biosolids derived biochar. *Science of the Total Environment, 695*, 133846.

Ahmed, N., & Al-Mutairi, K. A. (2022). Earthworms effect on microbial population and soil fertility as well as their interaction with agriculture practices. *Sustainability, 14*(13), 7803.

Ahmed, R., & Deka, H. (2022). Vermicomposting of patchouli bagasse—A byproduct of essential oil industries employing *Eisenia fetida*. *Environmental Technology & Innovation, 25*, 102232.

Ameen, F., & Al-Homaidan, A. A. (2022). Treatment of heavy metal–polluted sewage sludge using biochar amendments and vermistabilization. *Environmental Monitoring and Assessment, 194*(12), 861.

Amouei, A. I., Yousefi, Z., & Khosravi, T. (2017). Comparison of vermicompost characteristics produced from sewage sludge of wood and paper industry and household solid wastes. *Journal of Environmental Health Science and Engineering, 15*(1), 1–6.

Anielak, A. M., & Kłeczek, A. (2022). Humus acids in the digested sludge and their properties. *Materials, 15*(4), 1475.

Arora, S., Rajpal, A., Bhargava, R., Pruthi, V., Bhatia, A., & Kazmi, A. A. (2014). Antibacterial and enzymatic activity of microbial community during wastewater treatment by pilot scale vermifiltration system. *Bioresource Technology, 166*, 132–141.

Atal, H. L. (2022). Sustainable management of floral waste to produce bioenergy and valuable products. *Journal of Indian Association for Environmental Management (JIAEM), 42*(2), 26–34.

Ayoola, P. B., & Olayiwola, A. O. (2014). Trace elements and major minerals evaluation of earthworm casts from a selected side in Southwestern Nigeria. *Journal of Agricultural and Biological Science, 9*(6), 216–218.

Bhat, R. A., Dar, G. A., Jehangir, A., Bhat, B. A., & Yousuf, A. R. (2012). Municipal solid waste generation and present scenario of waste management during Yatra season in Pahalgam: A tourist health resort of Kashmir valley. *International Journal of Current Research, 4*, 004–009.

Bhat, R. A., Dar, S. A., Dar, D. A., & Dar, G. H. (2018b). Municipal solid waste generation and current scenario of its management in India. *International Journal of Advance Research in Science and Engineering, 7*(2), 419–431.

Bhat, R. A., Dervash, M. A., Mehmood, M. A., & Hakeem, K. R. (2018a). Municipal solid waste generation and its management, a growing threat to fragile ecosystem in Kashmir Himalaya. *American Journal of Environmental Sciences*. https://doi.org/10.3844/ajessp.2018

Bhat, R. A., Kamili, A. N., & Bandh, S. A. (2013). Characterization and composition of municipal solid waste (MSW) generated in Yusmarg: A health resort of Kashmir valley. A glance at the world. *Waste Management, 33*, 774–777.

Bhat, R. A., Nazir, R., Ashraf, S., Ali, M., Bandh, S. A., & Kamili, A. N. (2014). Municipal solid waste generation rates and its management at Yusmarg forest ecosystem, a tourist resort in Kashmir. *Waste Management and Research, 32*, 165–169.

Bhat, R. A., Singh, D. V., Qadri, H., Dar, G. H., Dervash, M. A., Bhat, S. A., Unal, B. T., Ozturk, M., Hakeem, K. R., & Yousaf, B. (2021). Vulnerability of municipal solid waste: An emerging threat to aquatic ecosystems. *Chemosphere, 287*, 132223.https://doi.org/10.1016/j.chemosphere.2021.132223

Bhat, S. A., Singh, S., Singh, J., Kumar, S., & Vig, A. P. (2018c). Bioremediation and detoxification of industrial wastes by earthworms: Vermicompost as powerful crop nutrient in sustainable agriculture. *Bioresource Technology, 252*, 172–179.

Bhati, M. I., Dubey, R. K., & Singh, S. (2021). Review on approaches to floral waste management for sustainable environment. *Climate Change and Environmental Sustainability, 9*(2), 110–116.

Bisht, A., Kamboj, N., Bisht, A., Kamboj, V., & Bharti, M. (2022). An intensive approach to the renewable energy recovery from agro waste—A review. In *Environmental pollution and natural resource management* (pp. 19–38).

Boruah, T., Barman, A., Kalita, P., Lahkar, J., & Deka, H. (2019). Vermicomposting of citronella bagasse and paper mill sludge mixture employing *Eisenia fetida*. *Bioresource Technology, 294*, 122147.

Cao, Z. J., Qin, W. T., Zhao, J., Liu, Y., Wang, S. X., & Zheng, S. Y. (2022). Three new *Trichoderma* species in *harzianum* clade associated with the contaminated substrates of edible fungi. *Journal of Fungi, 8*(11), 1154.

Cattunar, A., Marijančić, V., Rončević, D., & Mićović, V. (2023). Natural health remedies.

Chaverri, P., Gazis, R., & Samuels, G. J. (2011). *Trichoderma amazonicum*, a new endophytic species on *Hevea brasiliensis* and *H. guianensis* from the Amazon basin. *Mycologia, 103*, 139–151.

Chojnacka, K., Mikula, K., Skrzypczak, D., Izydorczyk, G., Gorazda, K., Kulczycka, J., Kominko, H., Moustakas, K., & Witek-Krowiak, A. (2022). Practical aspects of biowastes conversion to fertilizers. *Biomass Conversion and Biorefinery*, 1–19.

Daguerre, Y., Siegel, K., Edel-Hermann, V., & Steinberg, C. (2014). Fungal proteins and genes associated with biocontrol mechanisms of soil borne pathogens: A review. *Fungal Biological Revolution, 28*, 97–125.

Das, B. (2010). Identification of abiotic and biotic factors causing deterioration during storage and development of storage techniques for mahua (*Madhuca indica* Syn. *Bassia latifolia*) flowers. *Agriculturae Conspectus Scientificus, 75*, 119–125.

Das, S., Lee, S. H., Kumar, P., Kim, K. H., Lee, S. S., & Bhattacharya, S. S. (2019). Solid waste management: Scope and the challenge of sustainability. *Journal of Cleaner Production, 228*, 658–678.

Devi, J., & Prakash, M. (2015). Microbial population dynamics during vermicomposting of three different substrates amended with cow dung. *International Journal of Current Microbiology and Applied Science, 4*(2), 1086–1092.

Dhaliwal, S. S., Sharma, V., Mandal, A., Naresh, R. K., & Verma, G. (2021). Improving soil micronutrient availability under organic farming. In *Advances in organic farming* (pp. 93–114). Woodhead Publishing.

Dominguez, J., & Edwards, C. A. (2010). *Biology and ecology of earthworm species used for vermicomposting* (pp. 28–38). CRC Press Taylor & Francis Group.

Dominguez, J., & Edwards, C. A. (2011). Biology and ecology of earthworm species used for vermicomposting. In *Vermiculture technology: Earthworms, organic wastes, and environmental management* (pp. 27–40). CRC Press.

Ducasse, V., Capowiez, Y., & Peigné, J. (2022). Vermicomposting of municipal solid waste as a possible lever for the development of sustainable agriculture. A review. *Agronomy for Sustainable Development, 42*(5), 89.

Dwivedi, A., Dubey, R., Singh, P. K., & Ohri, A. (2019). Scientific management of municipal solid waste in an academic campus—A case study of IIT (BHU). *Journal of Materials and Environmental Sciences, 10*(10), 1–9.

Edwards, C. A., & Arancon, N. Q. (2022a). The influence of environmental factors on earthworms. In *Biology and ecology of earthworms* (pp. 191–232). Springer US.

Edwards, C. A., & Arancon, N. Q. (2022b). The role of earthworms in organic matter and nutrient cycles. In *Biology and ecology of earthworms* (pp. 233–274). Springer US.

Edwards, C. A., Arancon, N. Q., & Sherman, R. (2011). *Vermiculture technology: Earthworms, organic wastes, and environmental management*. CRC Press.

Elango, G., & Govindasamy, R. (2018). Analysis and utilization of temple waste flowers in Coimbatore District. *Environmental Science and Pollution Research, 25*(11), 10688–10700.

Gabhane, J., Tripathi, A., Athar, S., William, P., Vaidya, A. N., & Wate, S. R. (2016). Assessment of bioenergy potential of agricultural waste: A case study cum template. *Journal of Biofuels & Bioenergy*.

Gabhane, J., William, P., Bidhyadhar, R., Bhilawe, R., Duraisamy, A., Vaidya, A. N., & Wate, S. R. (2012). Additive aided composting of green waste: Effects on organic matter degradation, compost maturity and quality of finished product. *Bioresource Technology, 114*, 382–388.

Gaurav, M. V., & Pathade, G. R. (2011). Production of vermicompost from temple waste (Nirmalya): A case study. *Universal Journal of Environment and Research Technology, 1*(2), 182–192.

Ghosh, S., Sarkar Paria, D., & Chatterjee, S. (2022). Comparative study on bacterial population dynamics of foregut, midgut, and hindgut content of *Perionyx excavatus* (Perrier) isolated from eco-friendly, non-hazardous vermicompost. *Applied Biochemistry and Biotechnology, 194*(12), 6126–6139.

Greff, B., Szigeti, J., Nagy, Á., Lakatos, E., & Varga, L. (2022). Influence of microbial inoculants on co-composting of lignocellulosic crop residues with farm animal manure: A review. *Journal of Environmental Management, 302*, 114088.

Griffin, S., Sarfraz, M., Farida, V., & Jawad, M. (2018). No time to waste organic waste: Nanosizing converts remains of food processing into refined materials. *Journal of Environmental Management, 210*, 114–121.

Gupta, A. P., Upadhyay, P., Sen, T., & Dutta, J. (2023a). Agricultural waste as a resource: The lesser travelled road to sustainability. In *Agriculture waste management and bioresource: The circular economy perspective* (pp. 1–20).

Gupta, S., Tewari, S. K., & Pathak, S. (2023b). Temple floral waste for various bio-products in India. In *Recent trends in solid waste management* (pp. 293–307). Elsevier.

Hajam, Y. A., Kumar, R., & Kumar, A. (2023). Environmental waste management strategies and vermi transformation for sustainable development. *Environmental Challenges*, 100747.

Hameed, M., Bhat, R. A., Pandit, B. A., Ramzan, S., Dijoo, Z. K., & Wani, M. A. (2021). Qualitative assessment of compost engendered from municipal solid waste and green waste by indexing method. *Journal of the Air & Waste Management Association*. https://doi.org/10.1080/10962247.2021.1959466

He, F., & Liu, R. (2023). Mechanistic insights into phenanthrene-triggered oxidative stress-associated neurotoxicity, genotoxicity, and behavioral disturbances toward the brandling worm

(*Eisenia fetida*) brain: The need for an ecotoxicological evaluation. *Journal of Hazardous Materials, 450*, 131072.

Hidayah, B. N., Khangura, R., & Dell, B. (2022). Biological control potential of Trichoderma species and bacterial antagonists against *Sclerotinia sclerotiorum* on canola in Western Australia. *International Journal of Agriculture and Biology, 27*(3), 215–227.

Jayappa, V. S., Shivanand, K., & Mahadevappa, P. (2022). Bioinoculants for rapid production of vermicompost. In *Enzymes for pollutant degradation* (pp. 309–325). Springer Nature Singapore.

Jing, L., Kakati, L. N., Ao, B., & Kiewhuo, P. (2023a). Augmentation of plant biomass productivity using epigeic earthworm *Perionyx excavatus* and *Eisenia fetida* as soil nutrient facilitators. *Scientific Reports, 13*(1), 18648.

Jing, L., Kiewhuo, P., Ao, B., & Kakati, L. N. (2023b). Nutrient stabilization and heavy metal reduction in organic wastes using *Eisenia fetida* (Savigny) and *Perionyx excavatus* (Perrier). *Environment, Development and Sustainability*, 1–20.

Kaur, S., & Joshi, N. (2002). The solid waste generation during normal days in Haridwar city. *Himalayan Journal of Environmental Zoology, 16*(2), 266–270.

Keswani, C., Mishra, S., Sarma, B., Singh, S., & Singh, H. (2014). Unraveling the efficient applications of secondary metabolites of various *Trichoderma* spp. *Applied Microbiology and Biotechnology, 98*, 533–544.

Kržišnik, D., & Gonçalves, J. (2023). Environmentally conscious technologies using fungi in a climate-changing world. *Earth, 4*(1), 69–77.

Kumar, M. S., & Swapnavahini, K. (2012). Nutrient reduction and biogas production of rose residues by anaerobic digestion in a batch reactor. *International Journal of Advanced Research in Science and Technology, 1*(2), 125–129.

Kunjachan, S., Błauż, A., Möckel, D., Theek, B., Kiessling, F., Etrych, T., Ulbrich, K., van Bloois, L., Storm, G., Bartosz, G., & Rychlik, B. (2012). Overcoming cellular multidrug resistance using classical nanomedicine formulations. *European Journal of Pharmaceutical Sciences, 45*(4), 421–428.

Leta, A., Selvaraj, T., & Ambo, P. O. (2013). Evaluation of arbuscular mycorrhizal fungi and Trichoderma species for the control of onion white rot (*Sclerotium cepivorum* Berk). *Journal of Plant Pathology & Microbiology*.

Makhania, M., & Upadhyay, A. (2015). Study of flower waste composting to generate organic nutrient. *International Journal of Innovative and Emerging Research in Engineering, 2*(1), 145–149.

Martin, R. (2012, March 23). Global waste-to-energy market to reach $29.2 billion by 2022.

Martínez-Medina, A., Roldán, A., Albacete, A., & Pascual, J. A. (2011). The interaction with arbuscular mycorrhizal fungi or *Trichoderma harzianum* alters the shoot hormonal profile in melon plants. *Phytochemistry, 72*(2–3), 223–229.

Mendesil, E., Tefera, T., Blanco, C. A., Paula-Moraes, S. V., Huang, F., Viteri, D. M., & Hutchison, W. D. (2023). The invasive fall armyworm, *Spodoptera frugiperda*, in Africa and Asia: Responding to the food security challenge, with priorities for integrated pest management research. *Journal of Plant Diseases and Protection, 130*(6), 1175–1206.

Mukherjee, T., & Dasgupta, N. (Eds.). (2023). *Religion, landscape and material culture in pre-modern South Asia*. Taylor & Francis.

Nsiah-Gyambibi, R., Essandoh, H. M. K., Asiedu, N. Y., & Fei-Baffoe, B. (2022). Vermistabilization of fecal sludge with organic enriched substrates for compost production using two earthworm species (*Eisenia foetida* and *Eudrilus eugeniae*). *Environmental Quality Management, 32*(2), 357–365.

Odejobi, O. J., Ajala, O. O., & Osuolale, F. N. (2022). Review on potential of using agricultural, municipal solid and industrial wastes as substrates for biogas production in Nigeria. *Biomass Conversion and Biorefinery*, 1–13.

Pandya, J. R., Sabalpara, A. N., & Chawda, S. K. (2011). Trichoderma: A particular weapon for biological control of phytopathogens. *Journal of Agriculture and Technology, 7*, 1187–1191.

Patra, R. K., Behera, D., Mohapatra, K. K., Sethi, D., Mandal, M., Patra, A. K., & Ravindran, B. (2022). Juxtaposing the quality of compost and vermicompost produced from organic wastes amended with cow dung. *Environmental Research, 214*, 114119.

Paul, S., Das, S., Raul, P., & Bhattacharya, S. S. (2018). Vermi-sanitization of toxic silk industry waste employing *Eisenia fetida* and *Eudrilus eugeniae*: Substrate compatibility, nutrient enrichment and metal accumulation dynamics. *Bioresource Technology, 266*, 267–274.

Pigatin, L. B. F., Atoloye, I. A., Obikoya, O. A., Borsato, A. V., & Rezende, M. O. O. (2016). Chemical study of vermicomposted agro industrial wastes. *International Journal of Recycling of Organic Waste in Agriculture, 5*(1), 55–63.

Potdar, A., Singh, A., Unnikrishnaa, S., Naik, N., Naik, M., & Nimkar, I. (2016). Innovation in solid waste management through the Clean Development Mechanism in India and other countries. *Process Safety and Environment Protection, 101*, 160–169.

Racić, G., Vukelić, I., Kordić, B., Radić, D., Lazović, M., Nešić, L., & Panković, D. (2023). Screening of native Trichoderma species for nickel and copper bioremediation potential determined by FTIR and XRF. *Microorganisms, 11*(3), 815.

Raja, A., Pareek, P., Shakyawar, D., Sarfaraz, A., NehviWani, F., & Sofi, A. (2012). Extraction of natural dye from saffron flower waste and its application on Pashmina fabric. *Advance Applied Science Resource, 3*, 156–161.

Rajesh, R. W., Rahul, M. S., & Ambalal, N. S. (2016). Trichoderma: A significant fungus for agriculture and environment. *African Journal of Agricultural Resources, 11*, 1952–1965.

Ramachandra, L. (2012). Natural dye yielding plants of Mizoram, Northeast India. *Indian Journal of Tropical Biodiversity, 20*(1), 89–91.

Ranjitha, J., Vijayalakshmi, S., Vijayakumar, P., & Ralph, N. (2014). Production of biogas from flowers and vegetable wastes using anaerobic digestion. *International Journal of Resource Engineering Technology, 3*, 279–283.

Raza, S. T., Wu, J., Rene, E. R., Ali, Z., & Chen, Z. (2022). Reuse of agricultural wastes, manure, and biochar as an organic amendment: A review on its implications for vermicomposting technology. *Journal of Cleaner Production, 360*, 132200.

Roy, H., Alam, S. R., Bin-Masud, R., Prantika, T. R., Pervez, M. N., Islam, M. S., & Naddeo, V. (2022). A review on characteristics, techniques, and waste-to-energy aspects of municipal solid waste management: Bangladesh perspective. *Sustainability, 14*(16), 10265.

Schuster, A., & Schmoll, M. (2010). Biology and biotechnology of Trichoderma. *Applied Microbiology and Biotechnology, 87*(3), 787–799.

Seadon, J. K. (2010). Sustainable waste management systems. *Journal of Cleaner Production, 18*(16–17), 1639–1651.

Shouche, S., Bhati, P., & Pandey, P. (2011). Study about the changes in physical parameters during vermicomposting of floral wastes. *Journal of Environmental Research and Development, 6*(1), 63–68.

Siddiqua, A., Hahladakis, J. N., & Al-Attiya, W. A. K. (2022). An overview of the environmental pollution and health effects associated with waste landfilling and open dumping. *Environmental Science and Pollution Research, 29*(39), 58514–58536.

Singh, A., Jain, A., Sarma, B. K., Abhilash, P. C., & Singh, H. B. (2013). Solid waste management of temple floral offerings by vermicomposting using Eisenia fetida. *Waste management, 33*(5), 1113–1118.

Singh, A. K. (2023). *Horticultural practices and post-harvest technology*. Academic Guru Publishing House.

Singh, G., Kumar, S., Sidhu, G. S., & Kaur, R. (2018). Effect of integrated nutrient management on yield of wheat (*Triticum aestivum* L.) under irrigated conditions. *International Journal of Chemical Study, 6*(2), 904–907.

Singh, J., & Kalamdhad, A. S. (2013). Effect of rotary drum on the speciation of heavy metals during water hyacinth composting. *Environmental Engineering Research, 18*(3), 177–189.

Singh, S., & Sinha, R. K. (2022). Vermicomposting of organic wastes by earthworms: Making wealth from waste by converting 'garbage into gold' for farmers. In *Advanced organic waste management* (pp. 93–120). Elsevier.

Soobhany, N. (2019). Insight into the recovery of nutrients from organic solid waste through biochemical conversion processes for fertilizer production: A review. *Journal of Cleaner Production, 241*, 118413.

Tagliabue, F., Marini, E., De Bernardi, A., Vischetti, C., & Casucci, C. (2023). A systematic review on earthworms in soil bioremediation. *Applied Sciences, 13*(18), 10239.

Tanaka, M. (2014). Municipal solid waste management in Japan. In A. Pariatamby & M. Tanaka (Eds.), *Municipal solid waste management in Asia and the Pacific Islands* (pp. 157–171). Springer Singapore.

Teli, M. D., Valia, S. P., & Kolambkar, D. (2013). Flower waste from temple for dyeing of cotton and cotton/silk. *Journal of the Textile Association,* 210–214.

Thakare, D., Zhang, J., Wing, R. A., Cotty, P. J., & Schmidt, M. A. (2017). Aflatoxin-free transgenic maize using host-induced gene silencing. *Science Advances, 3*(3), 1602382.

Tiwari, P. (2014). *Utilization and management of floral waste generated in popular temples of Jaipur city*. The IIS University.

Tripathi, P., Singh, P. C., Mishra, A., Chauhan, P. S., Dwivedi, S., Bais, R. T., & Tripathi, R. D. (2013). Trichoderma: A potential bioremediator for environmental cleanup. *Clean Technology Environment Policy, 15*, 541–550.

Ulbricht, C., Conquer, J., Costa, D., Hollands, W., Iannuzzi, C., Isaac, R., Jordan, J. K., Ledesma, N., Ostroff, C., Grimes Serrano, J. M., Shaffer, M. D., & Varghese, M. (2011). An evidence-based systematic review of saffron (*Crocus sativus*) by the natural standard research collaboration. *Journal of Dietary Supply, 8*, 58–114.

Vats, P., Hiranmai, R. Y., & Neeraj, A. (2023). Bioconversion of organic waste for solid waste management and sustainable agriculture—Emphasized impact of bioelectromagnetic energy. In *Sustainable and cleaner technologies for environmental remediation: avenues in nano and biotechnology* (pp. 193–220).

Viterbo, A., & Horwitz, B. A. (2010). Mycoparasitism. In *Cellular and molecular biology of filamentous fungi* (pp. 676–693).

Waghmode, M. S., Gunjal, A. B., Nawani, N. N., & Patil, N. N. (2016). Management of floral waste by conversion to value-added products and their other applications. *Waste and Biomass Valorization.*

Wani, K. A., & Rao, R. J. (2013). Bioconversion of garden waste, kitchen waste and cow dung into value-added products using earthworm *Eisenia fetida. Saudi Journal of Biological Sciences, 20*(2), 149–154.

Wu, D., Wu, J., Cheng, X., Qian, J., Wang, W., Zhang, Y., et al. (2023). A research of marigold flavonoids from marigold residues. *Industrial Crops and Products, 191*, 115898.

Yadav, A., & Garg, V. K. (2011). Industrial wastes and sludges management by vermicomposting. *Reviews in Environmental Science and Bio-Technology, 10*(3), 243–276.

Yadav, A. K., Gurnule, G. G., Gour, N. I., There, U., & Choudhar, V. C. (2022). Micronutrients and fertilizers for improving and maintaining crop value: A review. *International Journal of Environment, Agriculture and Biotechnology, 7*(1).

Yeboah, R. (2011) *Waste fabrics that can be used to produce handmade papers when combined with the paper mulberry bark*. PhD diss..

Chapter 6
Plastic Waste Management for Zero Waste to Landfills: Potential, Challenges, and Opportunities

Florin-Constantin Mihai, Christia Meidiana, Sherien Elagroudy, Simona-Roxana Ulman, Jutta Gutberlet, and Carolina Carvalho

Abstract Linear economy reliance on plastic goods and materials produced by plastic industries translates into plastic pollution of all environmental compartments. The plastic goods reach a global market including remote regions while waste management performances are still limited all over the world, but particularly in low- and middle-income countries. This work aims to examine the plastic waste diversion from landfill-based traditional waste management systems toward zero-waste approach. In the first phase, it points out the critical steps in waste management practices to reduce environmental contamination with plastics. The plastic waste diversion from landfills is examined at downstream level to improve the plastic waste capture into circular mechanism (deposit return schemes, waste banks, reuse and recycling practices, recycling technologies, energy recovery). The upstream level is examined through emphasizing the role of circular business models in plastic industries, policy incentives (e.g., extended producer responsibility), regulatory framework (e.g., single-use plastic ban and restrictions), all supported by innovation. Secondly, the best practices from five countries (i.e., Romania, Canada, Egypt, Indonesia, and

F.-C. Mihai (✉) · S.-R. Ulman
CERNESIM Environmental Research Center, Department of Exact Sciences and Natural Sciences, Institute of Interdisciplinary Research, "Alexandru Ioan Cuza" University of Iasi, 700506 Iasi, Romania
e-mail: florin.mihai@uaic.ro

C. Meidiana
Department of Regional and Urban Planning, Faculty of Engineering, Brawijaya University, Malang 65145, Indonesia

S. Elagroudy
Egypt Solid Waste Management Center of Excellence, Faculty of Engineering, Ain Shams University, Cairo 11566, Egypt

J. Gutberlet · C. Carvalho
Department of Geography, University of Victoria, P.O. Box 3060 STN CSC, Victoria, BC V8W 3R4, Canada

C. Carvalho
Institute for Energy and the Environment, Universidade de São Paulo, São Paulo, Brazil

R. A. Bhat et al. (eds.), *Zero Waste Management Technologies*,
https://doi.org/10.1007/978-3-031-57275-3_6

Brazil) in different geographical areas might be able to reveal the current progress and challenges in the waste management sector with a focus on plastic fraction. Finally, alternatives to the use of plastics, like plant-based and compostable materials, glass bottles and refilling systems, or cardboard mono-material, are further examined as possible solution to replace the plastic materials in the packaging industries. While having into attention both the theoretical and practical levels, this study could prove to be useful in the process of supporting the implementation of pro-environmental policies and good practices aimed at decreasing the level of plastic pollution to landfills across societies.

Keywords Plastic pollution · Nanoplastics · Environmental contamination · Zero waste · Management · Biocompatible · Microplastics

1 Introduction

Plastic waste management poses serious challenges in both developed and emerging economies around the globe. Macro- and microplastics invade every environmental compartment of Earth system with disruptive effects on wildlife. Recent studies reveal that different parts of human body are contaminated to micro- and nanoplastics with unknown repercussions in the long term (Casillas et al., 2023). The air-soil-water-biosphere nexus is affected by macro- and microplastics while production of plastic materials is expected to keep an increasing trend in the following years (OECD, 2022). Therefore, a shift from linear economy model toward circular economy is required and supported by the improvement of waste management sector and the adoption of zero-waste strategies. Low- and middle-income countries should focus to improve the current waste management services in order to divert plastic waste disposal from both terrestrial and aquatic environments and to eliminate the open burning practices. On the other hand, high-income countries must improve the domestic plastic recycling capabilities and reduce the transboundary movements of plastic waste to abroad. On this regard, for lowering the plastic pollution crisis, an international agreement of United Nations is discussed (UNEA, 2022).

Emerging technologies to stimulate better plastic recycling performances are another bet in the coming years. However, plastic waste management is complex as it includes a wide range of plastic waste streams in addition to packaging or municipal waste streams. This book chapter aims to reveal the critical steps in the both upstream and downstream sides of plastic waste management in line with circular economy principles with the ultimate goal to achieve zero plastic waste to landfill sites as shown in Fig. 1.

Furthermore, mini case studies related to best plastic waste management practices for five countries in different geographical regions around the globe (Romania, Canada, Egypt, Indonesia, and Brazil) are outlined, while the current status of plastic alternatives is examined in the last section.

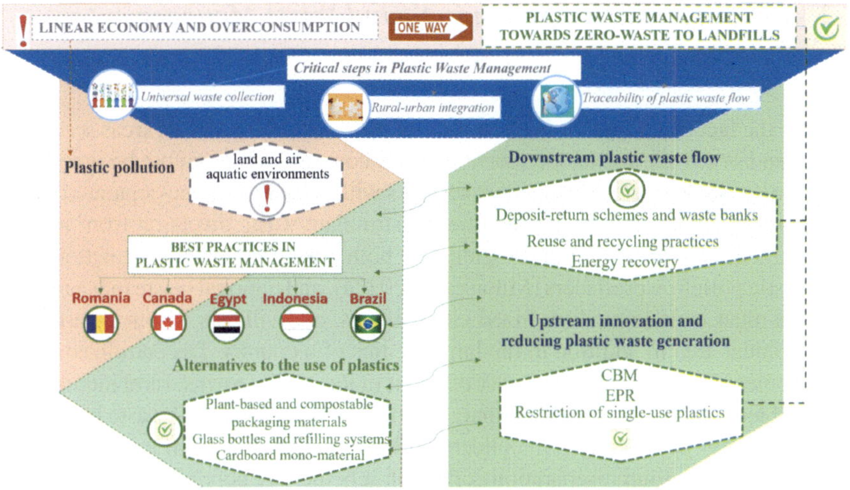

Fig. 1 Plastic waste management transition from linear economy toward zero-waste approach

2 Critical Steps in Waste Management Services

2.1 Universal Waste Collection Coverage

This is a basic step to reduce environmental contamination sources via open dumping and open burning of plastics across the globe. Around 1.9 billion people lack access to waste collection service in rural areas while the rudimentary infrastructure related to water and sanitation facilities is the norm (Mihai, 2017). To combat these poor statistics and to achieve the Sustainable Development Goals (SDGs), crucial investments in public utilities are required. The increasing access to plastic goods releases huge amounts of plastic waste into natural environment (OECD, 2022). The uncollected plastic waste flow is managed either at household level (burying, backyard burning) or through open dumping on surroundings (water bodies, roads, pastures, forest areas, etc.). Beside solid waste, wastewater is another key sector that releases huge amounts of plastic waste into natural environment without proper treatment. In emerging economies, informal settlements of larger urban areas are neglected by basic public utilities raising concerns in regard to public health and pollution risks. In addition, informal sector plays a key role in providing the primary waste collection services in marginalized communities through recovering of dry recyclables (metals, plastic, paper/cardboard, etc.) to feed the urban recycling facilities.

2.2 Rural-Urban Integration in Sound Waste Management Systems

Despite the fact that population tend to be concentrated in large urban areas served by public and/or private waste operators, the rural communities must not be ignored in providing reliable waste management services with access to source-separated waste collection schemes and sorting facilities to stimulate waste diversion from natural environment and landfills. Rural regions are both plastic pollution destination sites and plastic pollution producers (Mihai et al., 2022a), and regional waste management systems must include medium-sized cities, towns, and villages within a particular geographical area in addition to the largest cities of a country. Decentralized waste management systems seem to be cost-efficient in the case of towns, rural municipalities, and less populated regions (Mihai, 2023). Combined efforts of waste banks and informal sector have benefits in collecting plastic recyclables from urban and rural population where source-separation collection schemes are poor or non-existent like in the emerging economies of Asia (Meidiana et al., 2019). Therefore, rural-urban collaboration on regional scale is important to provide basic services in collection and treatment of municipal plastic waste flows.

In the EU countries, regional waste management system comprises several urban and rural municipalities at county level (equivalent to NUTS3 region level) where transfer stations serve smaller cities and surrounding rural communities. These transfer stations have material recovery facilities where plastic waste is treated and baled to be sent toward recycling facilities and plastic manufactures. However, the performance of such system depends on community awareness and implication in providing an effective and clean plastic source-separated waste. The waste management systems must be flexible and adaptable to local geographical realities to provide optimal services for both urban and rural municipalities. In addition, reliable waste statistics and field measurements regarding waste composition and waste generation rates are necessary to account and predict the plastic waste flows in particular areas.

2.3 Traceability of Plastic Waste Flow at International and National Levels

Beside the provision of regular waste collection schemes in all municipalities within a country, law enforcement and monitoring process of plastic waste flows are critical actions to divert plastic waste from natural environment toward proper management facilities. After the China ban on importing poor quality of plastic waste the world recycling market faced serious crisis (Yoshida, 2022). Overconsumption of plastic materials in well-developed countries surpasses the domestic recycling capacities; therefore, the surplus of plastic waste (often in poor quality) is transported to overseas with new destinations in countries from the South-east Asia or Africa that must manage the domestic gaps of waste management infrastructure. In addition to plastic packaging materials, other plastic materials and various polymers are present in other waste streams such as e-waste, textile waste, end-of-life vehicles, construction, demolition, or agricultural waste. E-waste is another emerging waste stream that is exposed to international trade and illegal traffic of e-waste where plastic materials are not valuable as metals, receiving less attention to proper treatment (Mihai et al., 2022b). Guidelines for monitoring of plastic waste flows should be formulated, and a global consensus is required among national authorities and plastic industries. The improvement of plastic waste statistics at different scales is crucial to monitor the import-export of plastic flows versus domestic plastic waste management flows to depict the geographies of plastic waste trades.

3 Plastic Waste Management and Environmental Pollution

3.1 Linear Economy and Overconsumption

The increasing of living standards fed by plastic goods in daily lives, nearby the industrial and agricultural applications, generates huge amounts of plastic waste at global level (Geyer et al., 2017) damaging marine and freshwater environments. Mixed waste collection and landfilling represent the baseline waste management practice of linear economy. The COVID-19 pandemic increased the plastic food packaging demand and caused the releasing of huge amounts of Personnel Protective Equipment in natural environment (Mahmoudnia et al., 2022). Moreover, the rapid development of food delivery systems creates additional plastic packaging flows in the domestic waste management systems. Africa is already a well-known destination for hazardous e-waste, and, in the latter years, textile waste importation and second-hand clothes are on rise due to fast fashion consumption model (BBC News, 2021).

Dumpsites or uncontrolled landfills are hotspots of critical environmental pollution in Africa, Asia, or Latin America. The closure of world's largest dumpsites is necessary due to public health concerns. In EU, non-compliant landfills should be closed and replaced with transfer stations with sorting capabilities and regional sanitary landfills. Therefore, the recyclable plastic materials should be diverted from landfills toward recycling facilities. However, the plastic production is expected to increase in both household and industrial demands. The secondary plastics could reclaim only 11–17% of the total plastic demand in 2025 taking into consideration a favorable scenario of high plastic waste collection rates (Klotz et al., 2022). The new circularity gap report reveals that circularity drops to 7.2% due to emerging of natural resources demands (Circle Economy, 2023). In terms of plastic waste, new assessment revealed that a larger amount of plastic waste is produced than the current statistics show (Material Economics, 2022). The multi-nodal statistical sources (authorities, waste operators, sorting stations, plastic recycling companies, and virgin plastic producers) in a multi-scale context (local-regional-national-EU-global) plus global plastic trades make very difficult the quantification of the plastic waste statistics.

The current studies reveal that plastic demand and plastic waste flow are on a rising trend, and without any effective circularity mechanisms, the plastic waste accumulation rate in the natural environment will increase despite some improvements in the plastic waste capture (OECD, 2022).

3.2 Plastic Pollution of Aquatic Environments

Macro- and microplastics are polluting the full spectrum of aquatic environments. In the first phase, plastic pollution research concentrated on marine environments (Allen et al., 2022) but the scale of plastic pollution reaches rivers from various geographical conditions, lakes, deltas, estuaries, ice caps, and mountain glaciers (Mihai et al., 2023). Rivers and tributaries are facing massive plastic pollution events during the seasonal floods. For example, Bistrita River catchment upstream to Bicaz Lake dam is located in a mountainous area (the Eastern Carpathians) in Romania (Mihai, 2018). This section of mountain catchment area is affected by plastic pollution scattered among tributaries network and villages due to the current waste management deficiencies. The plastic bottles (PET) are the most prevalent plastic type collected on the lake surface and riverbanks during the cleaning-up events (Mihai, 2018). This suggests that post-consumer plastic waste management is poorly developed despite the recycling potential as shown at national scale (Mihai & Ulman, 2022). In addition to PET bottles, HDPE packaging and plastic films are often seen during the field observations trapped by riparian vegetation areas along Bistrita River and its tributaries as shown in Fig. 2. Mountain regions could act like microplastic factories due to plastic fragmentation associated with local hydro-geomorphologies (Liro et al., 2023). Mismanagement plastic waste models reveal the scale of plastic pollution in world's rivers and the role of floods in the plastic mobilization in downstream areas or to marine environments (Roebroek et al., 2021). Over 1000 rivers contribute to most plastic pollution but every catchment level has its own geographical features related to landscape and socio-economic characteristics with particular plastic flows (Meijer et al., 2021).

The lack or underdeveloped wastewater treatment facilities lead to plastic contamination of rivers. The advanced tertiary-stage wastewater treatment facilities retained most of microplastics (Chen et al., 2022). However, these facilities are not available in all countries or less developed regions. As outlined in Sect. 2.1, universal waste collection coverage in line with sanitation facilities using centralized and flexible systems for rural and remote localities is critical steps to reduce plastic pollution of natural environment.

Fig. 2 PET bottles on river course (Bolatau-left, Sept 2022) and plastic foils films captured by riparian vegetation (right) on Bicaz river (Sept 2022), Romania. *Photos* FCM

3.3 Land and Air Pollution with Macro- and Microplastics

Plastic waste disposal in uncontrolled disposal sites is susceptible in areas with poor waste collection services. Informal settlements, small towns, and rural regions are exposed to illegal dumping practices. These wild dumps are found along the roadsides, pastures, forest areas or in the proximity, or freshwater bodies (riverbanks, floodplains). Illegal dumpsites contain plastic from household waste or from business sector (packaging, construction and demolition waste, e-waste items). Even conventional landfills pose risks to plastic contamination of soil-groundwater systems or freshwater bodies in the proximity (Yadav et al., 2022).

Open burning of plastics raises public health and environmental concerns particularly in the Global South (Cook & Velis, 2021). The hazardous waste streams such as e-waste and ELVs are processed through burning in rudimentary conditions to recover the metals with demands for recycling market. The plastic materials are often burned, buried, or uncontrollably disposal, polluting the surrounding land.

Agricultural practices such as plastic mulch contaminates the soil with macroplastics and microplastics (Mihai et al., 2022a). The soil-groundwater systems are exposed to micro- and nanoplastics contamination that poses further risks to freshwater supply for human settlements. Microplastic contamination routes of terrestrial environments including agricultural activities require more research in the coming years for filling the current knowledge gaps. Other economic sectors such as transportation contribute to soil and air microplastic contamination. Airborne microplastics contribute to overall particle pollution, but the impact on human health (e.g., workers in plastic industries), interaction of outdoor-indoor activities, effect on aerial microbiota, and tree canopies are largely unknown (Sridharan et al., 2021).

In addition, a recent study found that mechanical recycling plant is a source of microplastics for environment, and further investigations are required in the future since expanding of recycling capabilities around the globe is expected (Suzuki et al., 2022).

4 Plastic Waste Management Toward Zero Waste to Landfill

4.1 Upstream Innovation and Reducing Plastic Waste Generation

4.1.1 Circular Business Models in Plastic Industries

As it is commonly known, plastic is used almost everywhere, being met in almost each place or activity and even in almost every industry. As a consequence, this material is directly or indirectly linked to a wide range of business models (Dijkstra et al., 2020). This is why the analyses of the circular business models in the context of plastics industry, besides polymer producers, plastic components, plastic converters, plastic goods distributors and users, or plastic waste operators (Paletta et al., 2019), also include, for example, second-hand car sales, repairs of toys, electronics, or textiles (see Retamal et al., 2021). There is an intense work at different levels across societies toward a zero-emission economy, and the companies are playing an important role in this process. Considering that plastic still remains one of the most stringent environmental concerns, the discussion about circular business models (CBMs) in the plastic industry is relevant in this goal of decreasing global pollution. The study of Geissdoerfer et al. (2020) offers a distinct approach, categorizing CBMs into four groups as follows: cycling (reuse, repair, remanufacturing/refurbishing, recycling, design for modularity, reverse logistics), extending (long-lasting products, upgradability, timeless design, marketing/consumer education for long product life, maintenance/product support), intensifying (sharing models, rental/leasing models, user cooperatives, creative commons, pooling models), and dematerializing (software instead of hardware, service instead of product, consumer education, rationalizing demand). Particularizing the discussion regarding these principles, in the case of plastic industry, across the academic studies, there is a general consensus on the fact that recycling is the most frequently used practice (Barford & Ahmad, 2022; Dijkstra et al., 2020). Still, if considering that it requires more processing, labor, energy, and new materials than other circular processes, it becomes obvious that it is not the most preferred or eco-friendly loop from the technical cycle of material reuse (EMF, 2015). In the same way, plastics are unchangeably perceived especially as a waste and, consequently, as a problem that must be solved (de Jong & Mellquist, 2021; Dijkstra et al., 2020; Jørgensen & Pedersen, 2018; Kibria et al., 2023; Kruse et al., 2023; Paletta et al., 2019). Therefore, if also considering the concerns regarding the quality of waste material nearby other large challenges met in the reuse of plastics across the entire cycle, the current perspective seems to be pessimistic. However, one conclusion of the study of de Jong and Mellquist (2021) indicated that "there are opportunities for the plastic industry to be at the forefront of a circular initiative". Adapting their framework from Geissdoerfer et al. (2020), Retamal et al. (2021) proposed a circular business model typology for plastics and, based on it, offered some practical examples from India. The authors divided the business models for plastics into four groups: substituting (service instead of product, deliver functionality/performance, software instead of hardware, rationalizing demand, encourage sufficiency, light weighting, substitution with renewables and natural processes), extending (long-lasting products, upgradability, design for modularity, timeless design, marketing/consumer education for long product life, maintenance/product support, reuse, repair), intensifying (sharing models, rental/leasing/access models, user cooperatives, pooling models), and cycling (remanufacturing/refurbishing, recycling, reverse logistics, tech platforms to facilitate recycling, industrial symbiosis). Although these proposed models are intended to be specifically dedicated to plastics, it could be easily seen that they are not necessarily particular to this industry, but might be applied to different other areas, as they are following the theory of general circular business models and, largely, of circular economy, especially in regard to circular practices. This situation could be explained, firstly, by the fact that, as mentioned above, plastic is one material met everywhere and, secondly, by the theoretical development of circular business model in general, and, particularly, of the ones dedicated to plastic industry, that are on infant stages (Balaji & Liu, 2020). In the same time, the circular principles are incorporated into business models at different levels (Pieroni et al., 2019), three modes of integration being emphasized: downstream circular (linked to value capture and delivery), upstream circular (changing value creation systems), or fully circular (both upstream and downstream) (Urbinati et al., 2017).

Including the theory of change (Mattos et al., 2022) and considering the importance of different actors implied in the process of transformation from a linear business model to a circular one, the primary stakeholders like producers, designers, and municipalities, secondary stakeholders such as NGOs, journalists, academics, nearby consumers (Grodzińska-Jurczak et al., 2022) have to be integrated into such a discussion. Each of these groups of stakeholders plays a significant role, while adopting the circular business model approach represents a synergy among the perceptions, attitudes, and actions specific to each of them. Moving our discussion to plastic industry, in a work analyzing the levers for a corporate transition to a plastics circular economy, Barford and Ahmad (2022) emphasized the essential role of partnerships for attaining circularity in the context in which materials flow between companies, individuals, and other actors, also pointing on the fact that this idea was deduced from all the interviews applied for the study. Moreover, it was specified that one business' circular processes enable others to improve their own targeted or even implemented circular principles (Barford & Ahmad, 2022). In the same way, Balaji and Liu (2020) referred to the impossibility of adopting plastics circularity by a single stakeholder, while collaboration of all stakeholders involved in the product and supply chain represents a must in the context of aiming to achieve "a smooth progression from linear to circular economy". One example could be the one of recycled plastic demand, case in which, for achieving maximization, the establishment of a strong connection between recyclers and converters is a necessary condition, with needs to be optimized according to the prescription for various applications (Paletta et al., 2019).

Finally, some of the barriers toward the *activities* from the circular business model that were pointed out in the literature referred to specific environmental tradeoffs such as rebound effects, lower recyclability, and unsustainable raw material sourcing (Dijkstra et al., 2020), lack of proper technology and technical resources such as, for example, insufficient collection centers and recycling plants, lack of logistics and reverse logistics in some places, lack of information system to track recycled materials, lack of mechanical performance in recycled plastic-based products, incompatibility of recycled plastics with product manufacturing process (Di Bartolo et al., 2021; Ormazabal et al., 2018; Retamal et al., 2021), lack of support from public institutions (Di Bartolo et al., 2021; Dijkstra et al., 2020; Ormazabal et al., 2018), organizational behavioral barriers (prior support and commitment from management, lack of strategic planning, strong industrial focus on traditional business model of take-make-dispose), and lack of interdepartmental flexibility and coordination (Retamal et al., 2021).

4.1.2 Extended Producer Responsibility Policies in Plastic Industries

Extended producer responsibility (EPR) is becoming more commonly acknowledged as a successful waste management technique for increasing recycling and lowering product and material landfilling. It is essentially a policy instrument used to extend the producer's obligation to the end-of-life management stage of their products (OECD, 2014). Adopting an EPR strategy for plastic waste has two primary goals. Firstly, the policy allows producers and the government to collaborate on sharing waste management obligation, both physically and financially, relieving municipalities of the load. This enhances waste collection, management, reuse, and recycling rates. Secondly, it pushes manufacturers to create products that are eco-designed and have a minimal environmental effect. As a result, EPR assists governments in finally achieving plastic circularity by preventing the use of raw materials and using secondary raw materials to manufacture plastic products (WWF, 2019).

The EPR policy has different aspects that should be taken into consideration when designing the EPR scheme, (i) types of products, materials, and product consumers must be clearly defined; (ii) producers, that should be subject to EPR legislations, should be determined; (iii) requirements that producers should meet must be outlined; (iv) EPR scheme should be defined as either individual (individual producers responsibly) or collective (producer responsibility organization (PRO)); (v) establishing targets for collection, reduction, and recycling; (vi) establishment of a charge system and parameters such as product recyclability and recycled content percentage; (vii) fees should be set by PROs to cover the full amount of end of life of the products; (viii) governments and/or PROs must give enough information to consumers and stakeholders for the EPR plan to work well; and (ix) authorities and obliged producers should make information available publicly, monitor, and enforce EPR programs (Watkins & Gionfra, 2020; WWF, 2019).

Moreover, the deployment of EPR systems in some countries must be carefully tailored to the relevant local conditions of countries. The design of an EPR scheme must consider the country's waste management infrastructure, existing waste management policies, and social circumstances such as the local culture and readiness for implementation by key stakeholders and their ability to enforce requirements. It is equally important that the design of the EPR scheme results in funding that meets the three key criteria of being consistent, continuous, and adequate (Ellen MacArthur Foundation, n.d.).

The modulation of costs imposed to producers to participate in EPR programs offers a significant potential to the success of EPR system. Fees may be regulated based on a variety of product design characteristics, as reusability, toxicity, durability, reparability, and recyclability/composability, which have potential effects on the end-of-life phase and environmental consequences (Watkins et al., 2017; Zero Waste Europe, 2017).

According to Watkins et al. (2017), there are different EPR schemes that can be used in plastic industry. The most common of all is the product take-back requirement which assigns responsibility to producers in regard to the end-of-life management of their products by securing either mandatory or voluntary collection and recycling targets for specific products/materials. Governments often implement such rules by requiring the establishment of drop-off facilities for the return of end-of-life items, as well as by setting collection and/or recycling objectives. Collection targets are often expressed as a percentage of end-of-life products or a proportion of the quantity placed on the market (Dimitropoulos et al., 2021).

Another scheme used is the advanced disposal fees (ADF) which are fees that are assessed and imposed on selected products at the time of purchase based on expected collection and treatment expenses. The fees might be used to fund the products' end-of-life management.

Last but not least, the deposit-refund systems (DRSs) are the systems in which customers who purchase a product pay a nominal fee that will be refunded when they return the empty container to a collecting station. The objective of such systems is to encourage the return back of used products. Producers can implement this system voluntarily or as a part of a national legislation. This system has been implemented in many European countries for PET bottles.

Each of the previously suggested EPR schemes offers a distinct incentive to producers and consumers and/or contributes to waste management financing. Although take-back requirements compel producers to provide for the collection and appropriate disposal of end-of-life products, they do not give direct financial incentives for consumers to segregate their trash and return the relevant products to designated collection places. Advance fees, from the standpoint of producers, offer a direct means to fund the commitments defined undertake-back requirements. On the other hand, consumers see advance disposal/recycling fees and deposit-refund schemes as increasing the price to be paid at the time of purchase (Dimitropoulos et al., 2021).

According to Watkins et al. (2017), other EPR policy options are also believed to promote the success of EPR schemes. Firstly, harmonization of different EPR schemes could result in better results regarding collection, recycling, and eco-design rates of plastic products. Secondly, in order to assess the effectiveness of EPR schemes common terminologies and standards must be set and unified such as recyclability and composability. Thirdly, once EPR schemes proved successful to a certain plastic product, the EPR scheme should be tailored to include other plastic products to increase its scope. Fourthly, costs of collection, sorting, and treatment of plastic waste should all be considered when setting the cost of an EPR schemes. This would, in turn, assure the application of the polluter pays principle, aid in lowering the cost of waste management within municipal budgets, and minimize the quantity of packaging that manufacturers put out on the market. However, the full potential of EPR policy cannot be realized unless other instruments and policies are also in play. Pay-as-you-throw systems, prohibitions, a comprehensive legislative framework, trash taxes, information, and awareness efforts, and consumer desire to reduce, reuse, and recycle plastic waste will all contribute to the policy's success.

4.1.3 Restriction of Single-Use Plastics

According to the United Nations Environmental Programme (2018), single-use plastics include, but not limited to, plastic beverage bottles, plastic bags, food wrappers, plastic lids, straws, and foam takeaway containers. They are also commercially present in large quantities due to certain characteristics like durability, price, and resistance. Plastic bags and foamed plastic items, however, are the most prevalent examples of single-use plastics that governments find challenging and attempting to regulate because of how easily they may be seen in nature.

Mismanagement of single-use plastics has not only environmental and health consequences, but also economic effects due to high transportation costs, future removal/remediation costs, and output of businesses such as tourism and fishing.

According to the 73rd session of the General Assembly of the United Nations, two key aspects need to be considered to compact plastic pollution in the long term: (i) awareness and advocacy on a worldwide scale and (ii) effective policies and direction.

More and more waste avoidance measures have been put into place in countries all over the globe in recent years as people have become more conscious of the problem of plastics pollution. According to the United Nations Environment, 127 nations have banned or charged plastic bags as of July 2018, while 56 countries had prohibited or taxed other single-use plastic items (such as cutlery, bottles, food packaging) or certain polymers (Cornago & Borkey, 2021).

According to Cornago and Borkey (2021), there are three policy instruments used to tackle the use of single-use plastics including market-based instruments, regulatory instruments, and voluntary approaches.

The market-based instruments focus on the final plastic products, rather than the raw materials or the intermediate products. They usually apply taxes and mandatory charges on the price of single-use plastic products especially plastic bags to reduce their purchase. For instance, the Irish authorities have imposed taxes on single-use plastic bags which resulted in 90% reduction of their use. Yet, when the same principle was applied to other single-use plastic products, the impact on the reduction on their usage is less apparent. This is due to the fact that the market-based instrument depends largely on the consumer purchase frequency and willingness to buy. Also, the availability of other alternatives to plastic bags is an important factor to determine the success of this market approach. Last but not least, directing price signals at consumers' results in stronger behavioral change than if the price signals are directed toward the upper end of the plastic value chain where fewer stakeholders (e.g., manufacturers) are involved.

Whereas regulatory instruments related to single-use plastic sector involve two types of regulations: bans on certain products, intermediate goods or materials, and standards requiring the use of specified materials or manufacturing procedures. Plastic bag bans are the most prevalent regulation measure for preventing plastic waste. The evidence, however, of the effects of plastic bag bans is still limited due to insufficiency of data. On the contrary, other regulatory measures involving material requirements and technology standards are targeted toward producers. They are required to use certain materials (compostable/biodegradable materials or secondary materials from recycling) which have lower environmental impact. Voluntary approaches in regard to single-use prevention are commitments between industries and policymakers that sometimes are guided by the participation of third parties like environmental NGOs. The approach is usually more flexible and appealing than market-based approaches or regulatory instruments. Despite the lack of legal power, the credible prospect of more regulation, peer pressure, and civil society's examination of periodical progress reports may provide each signatory considerable incentives for compliance and openness. Voluntary initiatives in the past have been beneficial in reducing waste especially when such efforts included all packaging rather than just single-use plastics. Medium- and long-term quantitative goals for waste avoidance, product recyclability characteristics, and recycling rates were often included in these kinds of agreements. The voluntary alternatives are not likely to maximize the economic efficiency and are best complemented by extra governmental incentives.

4.2 Downstream Plastic Waste Flow in Line with Circular Economy Principles

4.2.1 Deposit-Return Schemes and Waste Banks

Currently, plastic packaging becomes the main contributor to the environmental degradation related to the unsustainable linear plastic production and consumption system because of the widespread use of single-use plastic (SUP) packaging which is used once before being thrown away. Bans and taxes on SUP have been enforced in more than 120 countries. However, the regulations cannot effectively reduce the overall pollution especially in developing countries. These countries are still struggling to find out the effective ways to reduce the use of SUP by introducing innovations. Both in developed and developing countries, the deposit-refund system (DRS) is commonly practiced nowadays. It is an option instrument for implementation of extended producer responsibility (EPR) in waste management. The trend shows that DRS is a potentially promising way to solve the problem of SUP (Walls, 2011).

Deposit-refund schemes (DRS) are commonly applied to plastic beverage bottles and cans. They aim to encourage the consumers that have to pay a deposit for the bottles/cans included into the final price, to collect the wasted bottles or cans, and bring them back to the shop or collection point for receiving the refund (usually for a small amount). This scheme can ensure one appropriate recycling method and increase the recycling rates. Through DRS, consumers will get the deposit back when they return the packaging back at the shop or collection point.

Yet, there are pros and cons about the benefits of DRS. The critics tend to mostly emphasize that the consumers might be more interested in separating SUP packaging than other plastic waste because they expect to have refund from bottle containers. Another issue is the argument about the high initiating costs for a DRS. However, this type of scheme may have both economic and environmental benefits such as cost reduction for collection system and landfill operation due to less waste and greenhouse gas emissions (GHGs) reduction or less waste transportation because of less trips of waste collection. Moreover, the studies of Reloopplatform (2022) and Hogg et al. (2011) showed that DRS contributes to more employments than other waste management options. While DRS is widespread implemented in both developed and developing countries, a community-based waste management with the similar system is uniquely applied in developing countries. It is called waste bank and has showed increasing trend indicating that the scheme is widely accepted in developing countries.

The first waste bank (WB) was founded in 2008 in Thailand, and the number of WBs is increasing since some developing countries consider the scheme to be effective in their countries. The adjustments made are adapted to the local conditions for more community acceptance of the scheme. Accordingly, these are important because WB is a bank-like institution founded by the community having various demographic, culture, and norm background from countries to countries. The social aspect becomes an important issue in WBs because it can affect the performance and community acceptance of the scheme (Meidiana et al., 2019). WB offers benefits especially to households (HHs), as the main waste producer, by buying their separated waste and giving the access for savings the money from the waste by issuing private account for each member (Pariatamby & Tanaka, 2014). WB consists of the same actors as the commercial banks, i.e., customers, bookkeeping, and its management. However, customer deposits money in commercial bank, while customers deposit waste in waste bank. In the same time, the waste bank is different by the commercial bank since it is operated in the public domain and the working system is based on the household level, by giving rewards to the ones that use it. By sorting, depositing, and selling the waste to the WB, a member of WB can get some extra money from the

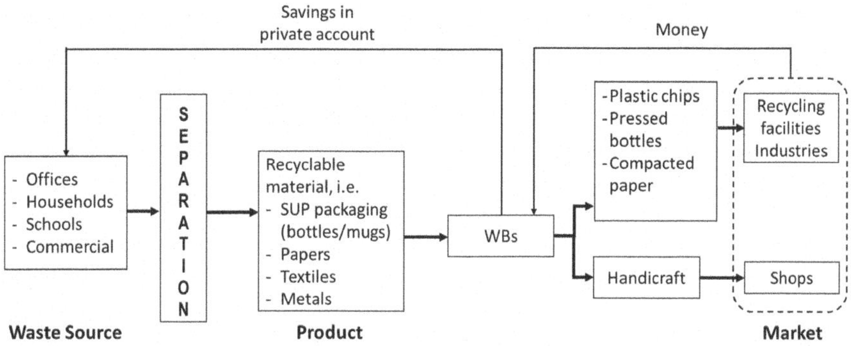

Fig. 3 Waste bank scheme

trash to get the value of certain facilities (Friedberg & Hilderbrand, 2017). WB determines the type of waste received, SUP packaging (bottles and mug), paper, plastic bags, electrical wires, and metals being the common recyclable waste allowed to be sold. Each recyclable waste has different price per unit mass, and it will be weighed, and the weight will be recorded. The amount of waste sold to the WB is recorded in a member's account book containing type of waste, weight, price per kilogram, and total amount. The member can claim their account for a certain period or exchange it with programs offered by WB such as health insurance, electricity/water supply coupons, staple food coupons, school tuition, etc. Some waste banks make partnerships with women, encouraging the initiatives to use the materials for creating handicrafts that end up being sold to the local retail shops (Asteria & Heruman, 2015) (Fig. 3).

4.2.2 Prepare for Reuse Practices

Initiating reuse practices in waste management asks for considering different factors. Reuse can be defined as part of the five Rs policy which seeks to minimize the quantity of waste generation through the sentience of the public and consists of: (i) reconsidering requirements for implementing the consumption, production, and discarding standards; (ii) rejecting the risks for redundant consumption and production that produce major effects; (iii) reducing to eluding waste, consuming fewer products, giving importance to those that offer less waste generation potential and have longer resilience; and (iv) reusing goods with good or acceptable conditions (Gomes et al., 2022). There are four hierarchies of desired waste management, namely prevention of waste generation, preparation for reuse, waste recovery, and lastly waste disposal in landfill (Singh & Sushil, 2017). In general, the materials can be reused through these two methods, namely: (i) direct reuse, meaning that materials can be reused as they are, and (ii) in an indirect way, which means the material needs to undertake major conversion to meet the standard and to be used again (Sañudo et al., 2022). It is also important to consider the methodology engaged to undertake the reuse concept, especially in regard to the processing costs and environmental factors such as the potential carbon footprint generation (Sañudo et al., 2022).

Sañudo et al. (2022) also proposed that the condition of the available reuse materials represents the possibility to commence the reuse practices, such as materials with no contamination, no change in size, and no visible wear. These are eligible for the direct use, whereas partial contamination or the alteration of size and signs of wear could be classified either acceptable, for indirect use, or having a bad condition for being reused. Moreover, Tsai et al. (2021) stated that the legal framework of sustainable waste management which influences the reuse practice should establish the clear definition of different types of items along with their specific reuse particularities, as well as the responsible authority. For instance, some parts of e-waste plastics could be reused for plastic products, plastic pyrolysis, or auxiliary fuel for cement or steel plants under the supervision of environmental authorities.

Therefore, regarding those mentioned specificities of reuse practice, it is clear that a clear legal framework to promote it, a strong commitment of related stakeholders in promoting the reuse action, a well-defined planning management system, and adequate resources to commence this circular practice are necessary for ensuring the success of the program.

4.2.3 Material Recycling Improvement

The plastic recycling process depends on the quality of plastic feedstock; therefore, clean and efficient source-separated collection schemes are required to fulfill this step. Mixed waste collection of municipal waste fraction is detrimental to recycling purposes favoring the landfilling process. There are different waste collection systems where post-consumer plastic waste is separately collected and further processed in material recovery facilities (MRFs) such as sorting stations. Here, sorting of plastics is done either manually or automated with associated refuses that will be sent to incineration or landfills sites (OECD, 2022). The plastic destination bins or containers are often contaminated with other items (e.g., food waste) that will increase the refusal rate at sorting plants, and ultimately plastics is disposed of in landfills. The technical capabilities of waste collection services and environmental awareness of community are crucial to provide qualitative plastic materials for MRFs. Sorting station prepares the plastic items in bales to be transported to recycling companies. At recycling centers, pet bottles are transformed into PET flakes to produce further new plastic materials such as r-PET, polyester synthetic fibers, and PET tapes.

PET bottles and HDPE materials have higher demand on recycling market, but plastic waste involves other packaging materials and polymers. Furthermore, plastic waste flows are included as part of other waste streams (e-waste, bulky waste, end-of life vehicles, construction and demolition waste) or industrial by-products.

Therefore, new cost-efficient recycling technologies are required to process other plastic waste flows instead of being sent to waste-to energy plants or landfills. Mechanical recycling has two dimensions: (i) upcycling where the plastic waste is recycled into same quality product (e.g., pet bottles to pet bottles) and replaced the virgin plastic materials or (ii) the plastic waste is recycled in items of poorer quality than the original product (downcycling), case in which the replacement of virgin materials is not assured. In line with circular economy principles, the upcycling mechanisms and reuse practices are main approaches to be scale-up compared to downcycling or energy recovery options. However, there is a debate if material recycling or energy recovery such as thermal energy with carbon capture is more feasible for mixed waste flows including plastics (Fellner & Brunner, 2022). Plastic pellets are processed through extrusion molding (pipes, tubes, or fibers), injection molding (various plastic items), or blow molding in the case of plastic bottles (Li et al., 2022). Other conventional technologies such as pyrolysis, liquefaction, and gasification raise environmental concerns beside significant economic costs and scalability, despite the role of using mixed plastics (Zhang et al., 2022). Table 1 reveals some emerging technologies that could improve mechanical and chemical recycling process of plastic materials. These technologies require more research for reducing their operational costs and for improving the scalability for industrial actors.

Table 1 Innovative and emerging technologies to develop plastic recycling industries

Emerging technologies	Benefits	Drawbacks	Source
Spectroscopy with chemometrics	Sorting technique improvement	Technological dependent	Adarsh et al. (2022)
Dissolution-based approaches	Producing virgin like-resins, multilayer items	Toxicity and energy consumption related to the use of solvents	Li et al. (2022)
Compatibilization	No pre-sorting requirement, mixed plastics	High costs	Zhao et al. (2022)
Plasma-assisted conversion	High conversion efficiency	Expensive to scale-up	Zhao et al. (2022)
Microwave assisted conversion	Catalyze chemical reactions, faster degradation	Affordability	Li et al. (2022)
Photoreforming	CO_2-to-fuel conversion coupled to plastic-to-chemical conversion using solar energy	Pilot research	Bhattacharjee et al. (2023)
Biocatalysis (enzymes)	Plastic-degrading enzymes for biocatalytic depolymerization	Short enzyme lifetimes and difficulty in enzyme recovery and reuse	Zhu et al. (2022)

4.2.4 Plastic Waste to Energy Recovery

For solving the problem associated with the end-of-life management of plastics, an appropriate method is necessary. Basically, two recycling methods can be applied, i.e., either mechanical or chemical recycling technologies (Ragaert et al., 2017). Geyer et al. (2017) estimated that the global plastic waste production is about 300 Mt yearly leading to significant effect on the environment. Many countries enforce the ban of certain plastic items or single-use plastic levies to reduce plastic waste. However, the enforcement may be a short-term solution for some countries with low enforcement law, while the plastic will remain a very significant material in the future. The dependence on plastics cannot be denied nowadays for its robustness, production efficiency, and duration. Therefore, other methods should be introduced, including the concept of circular economy, as it considers waste prevention, reuse, recycling, and recovery practices (Payne et al., 2019). However, alternatives for reusing the plastic products with the same type or function are often inadequate because of some factors, i.e., technical limitation or legal conditions (Coelho et al., 2020). Recycling is a reasonable option so far for end-of-life management of plastic waste (EC, 2018). Other alternatives are energy recovery through incinerator to produce heat and electricity as source of alternative energy. Through incineration, energy can be generated from plastic waste since it produces high pressure and temperature steam for power generation using steam turbine generator. Some disadvantages of incineration have been notified such as need of large amount of daily waste, high capital investment for construction, work, and maintenance, as well as capable human resources to operate with the high invested technology facility. These problems are met especially in developing countries where there is no funding, and public acceptance on incineration is relative low. However, environmental concerns related to air contamination risks and with the remaining ashes from incineration plants should be further examined.

Different studies of life cycle assessment (LCA) examined the environmental impacts of various end-of-life treatments of plastic waste. Yet, they presented different conclusions. Eriksson and Finnveden (2009) found that, in terms of the climate change impact, the disposal of non-recyclable plastic waste in landfill was a better option for the Swedish case than incinerating it. This result was confirmed by Demetrious and Crossin (2019) in an Australian case where landfilling of mixed plastic waste was also found to be more appropriate than the thermal treatments for CC, with three other impacts, i.e., eutrophication, acidification, and photochemical smog. Lazarevic et al. (2010) also concluded that landfilling of plastic has the least CC impact. On the other hands, the landfilling method was the least preferred waste management option for plastic waste for all other impact classifications. It is also found that mechanical recycling was generally a better option than incineration. Still, the recycling's environmental performance depends on the quality of recyclable materials affecting the substitution ratio for virgin material replacement. Rajendran et al. (2013) suggested that mechanical recycling had lower impacts than thermal treatment only if the substitution rate of virgin plastics was at least 70–80%. However, many studies observed that the LCA impacts of thermal treatments with energy recovery are sensitive to various assumptions, including the amount of electricity recovered, geographical location, and the electricity mix assumed for system credits (Antelava et al., 2019). Other study of Azapagic et al. (2021) showed that chemical recycling of mixed plastic waste via pyrolysis produces about 50% lower climate change (CC) impact and life cycle energy use than energy recovery from mixed plastic waste.

5 Best Practices in Plastic Waste Management in Selected Countries

5.1 Romania

Despite the recent improvements in the municipal and packaging waste management systems, the 2022 Country Report for Romania elaborated by the European Commission staff clearly points on the fact that this country has not made progress in circular secondary material usage over the past decade, decreasing, between 2015 and 2020, from 1.7 to 1.3%, while the EU average increased from 11.3 to 12.8% over the same period (EC, 2022). The same source specifically mentioned that Romania failed to decouple its economic growth from the generation of waste, with high landfilling rates and, consequently, with a municipal waste recycling rate equal to 13.7%, well below the EU average of around 48%, and far below the 2020 and 2025 EU targets of 50% and 55%, respectively (EC, 2022).

In terms of plastic packaging flow, the recycling rate declined from 51.7% in 2013 to 31.1% in 2019, but it is still above Austria, France, Ireland, or Malta, while EU-27 value is around 40.6% (Eurostat, 2022). However, the recycling of overall packaging waste in Romania is 44.9% in 2009 compared to France (65.5%) and Austria (65.4%). A new circular economy strategy is adopted, and ambitious targets were established: generally, to recycle 55% of municipal waste till 2025 and 60% till 2030; specifically, to recycle 50% of waste generated through plastic packaging till 2025 and 55% till 2030 in line with EU regulations (Romanian Ministry of Environment, Water and Forests, 2022). Furthermore, the EU's Single-Use Plastics Directive (SUP) is transposed by the national legislation, the single-use plastic products (e.g., plates, forks, sticks, or plastic cups for drinks) being banned to enter on the market. Moreover, the environmental crimes such as burying and burning waste are punishable by 3–5 years in prison; in the same way, the penalties for dumping waste were increased (EEA, 2022).

Plastic recycling market emerged in Romania involving sorting stations, recycling companies, and plastic manufactures. GreenTech is the largest plastic recycling company that operates here, and it is a leading producer of PET flakes, r-PET granules, and PET tape in the Eastern Europe. The operational recycling capacity is 100,000 tons of PET bottles per year closing the loop (bottle to bottle) with upgraded technologies to deliver sound upcycling services. The company received Gold Standard certified European company for "Reducing CO_2 emissions through PET recycling", reducing CO_2 emissions by over 45,380 t per year (GreenTech, n.d.). Future aims in terms of investments in plastic recycling development are (GreenTech, n.d.): (i) developing a new production line for materials used in the hygienic industry obtained from r-PET such as sanitary napkins, absorbents, diapers, and effective equipment for medical use; (ii) expanding r-PET capacity for the food industry; and (iii) expanding recycling capabilities in the Central and Eastern Europe to provide support for increasing plastic capture rate prospects in line with the EU targets.

However, the domestic plastic source-separation must improve in both urban and rural areas of Romania to reduce the importation of PET bottles bales from abroad in order to maintain the feedstock required for optimal use of the plastic recycling plant. Some municipalities succeeded to implement an efficient source-separated waste collection schemes including plastic fraction. For example, Targu Lapus has one of the highest recycling rates from Romanian cities (60%) due to a continuous effort of local administration since 2006, when a source-separated collection scheme and recycling systems were implemented through a Phare project in the city (Green News, 2020).

Salacea Commune (Bihor County) aims to be a Zero-Waste Certified Municipality by following the steps required by Zero-Waste Europe. First of all, the dual system (wet-dry fraction) of MSW stream was replaced with five fraction systems (plastic, paper/cardboard, glass, biowaste, and residual) combined with environmental awareness campaigns and multi-stakeholder cooperation (local authority-waste operator-sorting plant) (ZeroWaste Romania, 2022). After three months, the separate collection rate increased from 1 to 67%, recycling rate from 1 to 40%, while landfill rate decreased from 99 to 56% (ZeroWaste Romania, 2022). These results show that the small rural municipalities (< 5000 inhabitants) with community involvement can achieve progress toward a circular economy path by improving the source-separate performance levels in a quite short-term period. The five fractions systems proved to be more efficient in recovering plastic waste from households to be further transported to the sorting station.

In addition, innovative projects such as Plastic Circle involved European knowledge exchange between Alba Iulia County (Romania), Valencia (Spain), and Utrecht (Netherland) to support circularity of plastics at the municipal level (Plastic Circle, n.d.). This project aimed to demonstrate that innovations could be replicated and introduced into the middle of the citizens, as part of European transformation toward a more circular economy. This Smart CE solution for the management of plastic wastes in Alba Iulia was implemented in the period between June and September 2020. The project involved four key steps in regard to (i) innovative collection systems (smart containers identifying the quantity and the quality of packaging deposited into them, ensuring better separation of plastic waste, and reducing the amount of mixed waste generated); (ii) transport (using a special CAN-Buss device, optimized routes, skipping empty containers route, and eco-driving, all integrated into a mobile application); (iii) sorting (using innovative technologies—spectrometers, for increasing the quality of the recovered plastic); and (iv) reprocessing the materials into products (automotive parts, foam boards for wind turbines, roofing structures, rubbish bags, asphalt, fences, and benches). This initiative represents other concrete case that demonstrates the importance of reconnection in the circularity practice, as public authorities, innovators, and local business community were closely collaborating each other on collecting and recycling plastic waste. Moreover, being a pilot project, it has a high potential to be transferred in other communities/cities/regions.

5.2 Canada

Canadians should be highly concerned about the wastefulness of their industries and lifestyles, given that approximately three million tons of plastic waste are thrown away every year, of which approximately only 9% are recycled, while the rest is landfilled, incinerated, or littered in the environment (Molloy et al., 2022b). Canada produces more plastic waste per capita than any other country in the world, a fact that puts significant pressure on provinces and territories to curb plastic waste and plastic pollution (Molloy et al., 2022a).

In Canada, the federal government is responsible for product regulatory matters regarding plastics. However, it is the provinces, territories, and municipalities who regulate many of the environmental protection issues. While the municipal governments oversee the collection, recycling, composting, and disposal of household waste, the provinces have exclusive jurisdiction over municipal institutions. It is currently the responsibility of the province and territorial authorities to implement waste reduction policies and programs and to authorize and monitor waste management facilities and operations. These arrangements result in fragmented regulations and a lack of harmony among different administrative levels, with disparities between provinces and cities in terms of actions taken and timelines set to address plastic waste issues. While municipalities are often the drivers for innovative solutions, they experience power limitations, when it comes to regulate plastics use. In 2019, the city Victoria in British Columbia, for example, wanted to enact a bylaw banning plastic bags, but had not obtained the BC Minister's approval. Consequently, the ban was overturned by a B.C. court.

In 2018, building on the Ocean Plastics Charter (Government of Canada, 2021), the Canadian Council of Ministers of the Environment proposed a strategy to move forward with a Canada-wide strategy on zero plastic waste, by 2030. The Plan is developed in two phases, involving public hearing and expert analyses, defining specific actions and timelines to help prevent, reduce, reuse, recover, capture, and clean up plastic waste and pollution. The initiative follows a circular economy and life cycle approach to plastics and provides a framework for action in Canada.

At the federal level, the Canadian Environmental Protection Act, SC 1999 (CEPA) regulates plastics deemed to be toxic, by placing them on the List of Toxic Substances, which can then be properly managed and restricted. Under this act, the federal government prohibited in 2017, the manufacture and import of plastic microbeads frequently used in toiletries. Another concrete outcome is the Single-Use Plastic Prohibition Regulations ("SUPPR"), published in June 2022, which specifically prohibits the manufacture, import, and sale of the following six categories of single-use plastics: (i) checkout bags; (ii) cutlery; (iii) some of the foodservice ware; (iv) ring carriers; (v) stir sticks; and (vi) straws (Government of Canada, 2022).

Recent research into understanding public perception of plastic reduction legislation in the Atlantic provinces of Canada shows strong public support (77%, $n = 838$) for bans of single-use plastic bags at the consumer level and for additional legislation to curb plastic pollution (Molloy et al., 2022a).

A draft regulatory text for labeling rules of recyclability is to be expected by mid-2023, after the government had launched two consultations in support of its Zero Plastic Waste Agenda: (i) the development of labeling rules for recyclability and compostability and (ii) the development of a federal plastics registry for producers of plastic products. These measures will contribute to improve the collection and end-of-life management, for example, for reuse, repair, and recycling of plastics through better product design, helping keep these products out of landfills, incinerators, and the environment. The proposed regulations would also include rules requiring minimum levels of recycled plastic in certain products. In order to prevent contamination from unwanted products in recycling and composting facilities, the government plans to regulate the use of the terms "compostable", "degradable", and "biodegradable" in the labeling of plastic packaging and single-use items.

Canada is also developing a registry where producers will be required to report on the quantity of plastic products put on the market and will indicate the final destination of the plastics at the end of the life cycle. It is expected that these innovative regulations will expand and create new EPR programs for packaging, in addition to the already implemented five provincial EPR programs.

Advancing science and finding plastic substitutes and innovative forms to recover used plastics are necessary. Plastics derived from crude oil are widely used not only in food and product packaging, but also in agriculture and on livestock farms. Many of these plastics are single-use and difficult to recycle due to the contamination on their surface. Plant-based plastic replacements (for ex., using zein a corn protein, cutin from tomato peels, etc.) have good performance and could become environmentally friendly alternatives (Hood et al., 2021). However, we cannot put new materials on the market and design new products without having solved the cradle to cradle impact of each new material put into circulation.

The regulatory inconsistencies within different regions are still a major stumbling block in the implementation of a coherent EPR and recycling system for plastics in Canada. A multi-level governance approach seems not only appropriate to harmonize current and future legislation to curb plastic pollution but also necessary for the implementation of coherent EPR programs for packaging waste. Canada is taking an inclusive approach to implement plastic waste solutions, by working with all levels of government, industry, non-profit organizations, and communities. The federal government has invested over $5 million in projects focused on environmental education and awareness building, community science, and community projects. Canada is also collaborating on the development of an ambitious legally-binding, global instrument on plastic pollution.

Yet, the very grassroots level, for example composed of diverters, binners, Valoristes, or waste pickers (local denominations of the individuals recovering refundable beverage containers) remains at the margins of the circular economy, despite their central contribution of diverting recyclables toward reuse and recycling (Sholanke & Gutberlet, 2021). In order to improve overall plastic waste management and increase a circular plastics economy including all actors as well as maintaining a diverse agenda of actions and strategies (as described above) seems essential steps (Diggle & Walker, 2022).

5.3 *Egypt*

Egypt has been active in addressing waste management concerns, particularly those related to plastic waste. Egypt created its national waste management plan, which resulted in the approval of Egypt's waste management legislation, in 2020, to support scientific research, technical development, and innovation in waste management, including plastic pollution. This country has also proposed to sign the planned regional strategy for Mediterranean nations on marine litter control. Best practices in plastic waste management have been presented by certain government and private sector initiatives, as well as by several startups and non-governmental organizations (NGOs) (IAEA, 2021).

Firstly, regulation of single-use plastic bags has been a major focus of policy at the national and state levels. The government's Waste Management Law of 2020 prohibits the use of single-use plastic bags. The legislation attempts to minimize consumption by providing incentives and exemptions to the private sector, businesses, and manufacturers of single-use plastic bags. In 2019, a coalition of over 20 NGOs and social enterprises launched the movement Ban Plastic, whose primary goal was to advocate for an absolute ban on the use of single-use plastic bags and plastic products, firstly focused on the Red Sea Region, but then expanding to Cairo and other cities (Elkady et al., 2021).

Plastic bag bans were initially implemented in the Red Sea Governorate, focusing on plastic bags and commodities from hotels, restaurants, and cruise ships. Businesses that do not comply with the announced decree on plastic bags face penalties from the Governorate. Hurghada Environmental Protection and Conservation Association (HEPCA), a non-governmental organization (NGO) specialized in marine and land conservation, held public awareness campaigns and provided certification to companies that eliminated the single-use plastic goods (WES, n.d.).

Additionally, the EU's SwitchMed programme and the UN Environment Programme (UNEP) have sponsored activities to reduce Egypt's use of plastic bags. In 2017, Egypt's Ministry of Environment demonstrated a national initiative to decrease the use of plastic bags. Over 4500 alternatives to single-use plastic bags were provided in seven supermarket chains in Cairo and Alexandria. The initiative has drawn a lot of attention and has already been highlighted at various conferences across the world (SwitchMed in Egypt, n.d.).

Moreover, the Japan International Cooperation Agency (JICA) is contributing to a program to cut down on single-use plastics in Egypt that is being managed by the UN Industrial development Organization (UNIDO) in partnership with the Egyptian Ministry of Environment (Elkady et al., 2021). By demonstrating alternative, environmentally friendly methods through technical assistance to small and medium-sized businesses (SMEs), the project will assist government efforts to reduce the amount of plastic waste that leaks into the environment. It will also work to increase awareness of current sustainable plastic production and consumption patterns among stakeholders, producers, and consumers (United Nations Egypt, 2021).

Extended Producer Responsibility is another policy scheme that is adopted by the government and the private sector to minimize the plastic waste challenge in Egypt. The government gathered suggestions from both local and international partners as it gets ready to adopt EPR policy. GIZ, German government's agency for international cooperation has recently conducted an EPR study with two German consulting firms along with a local support, suggesting an EPR system for Egypt that combines the best practices from the EU (Black Forest Solutions GmbH, Landbell, A.G., 2021).

More importantly, EPR policy has been adopted on the ground through integrating the informal sector. Since the informal sector is responsible for recycling of 80% of what they collect, a large-scale plastic recovery scheme co-designed by a consulting company and Nestle (one of the largest food and beverages companies in the world) to target 100% recyclable and reusable packaging by the year 2025. A trial of this scheme started in 2019, with Nestle leading the scheme, followed by PepsiCo and Al-Ahram Beverages. Initially, the program focused on encouraging the informal sector to collect PET plastics; in the future, it will expand to include the collection of all post-consumer packaging. The initiative, which is called Dorna, employs a reverse credit approach in which companies involved for manufacturing plastic waste electronically pay collectors, sorters, traders, and processors for each batch collected and transported for recycling. The initiative has been further backed up by the government by establishing the Egyptian Plastics Pact that will be signed by companies as an affirmation of their commitment to national regulations and polices (including EPR) regarding plastic waste (Iskander, 2021).

Other initiatives are also aiming to integrate the informal sector in collecting, processing, and recycling plastic waste. One of these is maintained by Plastic Bank, which aims to establish collection centers of plastic waste in Egypt as well as creating a medical insurance program for collectors that has been created and proved in other countries where Plastic Bank operates. Individual components of the plan, such as the establishment of waste collection centers, have received assistance from global corporations such as Henkel (2021). In October 2022, Henkel and Plastic Bank celebrated the collection of 300 million plastic bottles, establishment of 33 centers in four governorates (Mahmoud, 2022).

Bassita, another social company, is taking a different approach to transforming the gathering and processing of post-consumer plastic, beginning on one Nile Island and encouraging fishermen and local communities to take responsibility for this plastic waste themselves (Elkady et al., 2021).

Equally important, Bariq is considered the first "bottle-to-bottle" business in the MENA region that turns post-consumer PET into recycled polyethylene terephthalate (RPET) pellets bottles; that would have otherwise gone to landfills. These pellets have proudly received approval from international authorities including the European Food Safety Authority (EFSA), REACH, and Health Canada, as well as the U.S. Food and Drug Administration (FDA). The leading worldwide makers of bottles and food containers as well as significant producers of recycled-material fibers in Europe and the USA are then supplied with these food-grade PET pellets (Hamdallah, 2021).

5.4 Indonesia

Policy and Regulations related to waste management in Indonesia have been transformed since the enactment of National Waste Law 18/2008. Beforehand, the paradigm of waste management (WM) was end-of-pipe solution focusing on waste treatment before discharging it to the environment. There was neither waste reduction nor waste-to-resource management for resource efficiency. The waste was just pollutants loads in line with linear economy principle. Due to the environmental loads brought by these unsustainable ways of WM, the government of Indonesia (GoI) enacted the Waste Law in 2018 whose articles include the practices of reuse, reduce, and recycling (3R), as well as producer responsibility through the scheme of Extended Producer Responsibility (EPR). The enactment was a milestone for waste management in Indonesia as it transformed the old paradigm in the new one with its principles of reduction at source and resource efficiency. The development of WM policy has been accelerating since then and evolving to be more concerned in regard to the waste utilization as an approach to circular economy. Therefore, the WM in Indonesia has changed from model of "take-make-dispose" toward a circular model of "make-use-recycle".

The GoI has introduced some innovations to support the implementation of circular economy in waste management. The involvement of informal sector in WM is an inimitable characteristic. According to Meidiana and Gamse (2010) and Fei et al. (2016), informal sector contributes to waste reduction, and the involvement of informal waste sector is common in many developing countries with different schemes. In Indonesia, this sector is played by some parties such as waste banks, community-based recycling centers, and scavengers. Fortunately, the GoI supports the involvement of informal waste sector by enactment of Ministerial Decree 13/2012 regulating the 3R activities through waste bank. WB is becoming popular in Indonesia, and the increasing number of WBs within the last decade indicates the important role of WB in waste reduction. According to the data from the Ministry of Environment and Forestry, there were 1640 WBs in 2011 and 11,566 WBs in 2021, indicating sevenfold increase within 10 years. Plastic is the most valuable waste for informal sector as well as for the formal one because it has the highest selling price.

Both sectors have interest for plastic since it can contribute to side income. Therefore, local government encourages the local community to become member of WB or to get involved in waste reduction through WB. This number is expected to grow for achieving the national waste reduction target in 2025. WB is fully supported by the local government as it represents community-based waste management model reasonable for current MSWM condition in many Indonesian cities, even in other developing countries with similar waste problems. To sum up, waste bank management model increases public awareness of waste problem and economic value of waste. It can motivate people to involve more in waste separation, reduction, reuse, and recycle (Meidiana et al., 2022).

Beside informal sector, innovation in waste management has been introduced in formal sector, where the local government is the main actor, supported by the community through mutual collaboration called 3R-transfer point (3R-TP). 3R-TP differs from WB although having the same goal, i.e., waste reduction. It requires community involvement without giving monetary rewards through refund or other similar buy-back system, but giving service of regular collection system. The scheme is commonly implemented in rural area where the daily collection system cannot be provided and the waste management service is lack due to geographical obstacles. These area will be provided with collection system when the community starts to separate the waste. In other words, waste separation is compulsory for the community since 3R-TP facilities usually cooperate with industries or recycling centers for buying the recyclable materials. The 3R-TP will receive payment for waste selling used for operating the waste collection and treatment.

This scheme is widely applied, especially in rural areas, and the results show tremendous trend since many regencies are introducing the concept of one village one 3R-TP (OVO-3R). However, there are not many studies presenting the evaluation of waste reduction target achievement in rural areas through this concept. Still, it is expected that, in the next years, more studies measuring the role of OVO-3R in waste reduction to be elaborated and to confirm its efficiency.

5.5 *Brazil*

Solid waste management has been a challenge for urban sustainability, especially regarding plastics, amplified by factors like the population growth and the increase in consumption, besides the main bottlenecks of waste governance, such as the lack of fiscal incentives and selective waste collection (Alfaia et al., 2017; Pacheco et al., 2012), and aggravated by the increased volume of waste during the pandemic (Forlani & Njie, 2022). Brazil is the fourth largest producer of plastic waste in the world, behind the USA, China, and India (Atlas do Plástico, 2020). In total, according to the data from a WWF report (WWF, 2019), 11.3 million tons of plastic were generated in 2019, but only 1.3% of this total, or 145,000 tons, were recycled. The absence of a specific national guideline for plastics, associated to lack of knowledge and accountability of the plastic chain production and stakeholders, contributes to this scenario (Pertussatti, 2020; WWF, 2019). Therefore, Brazil is below the global average in terms of plastic recycling (about 25%) (Gorziza et al., 2021).

The incorrect disposal of most of the plastic used in Brazil, coupled with the lack of education and awareness of the population and inequality, ends up generating serious social and environmental problems (Baia et al., 2020; Wiesenhütter et al. 2021). These vary from economic issues, health problems, to the clogging of drainages, which consequently causes or worsens flooding episodes, and to the pollution of rivers and marine ecosystems, affecting wildlife as well as local communities in many different ways, especially those living in remote regions and those who depends on water resources (Atlas do Plástico, 2020; WWF, 2019). For example, according to the Atlas do Plástico (2020), studies carried out on the amount of waste on 170 Brazilian beaches showed that most of these places (54%) were dirty. Also, according to this source, Brazil occupies the 16th position in the worldwide ranking of the biggest plastic polluters of the ocean. Finally, not only littering of plastic waste, but also the indiscriminate production and use of plastics generate health risks of varying severity, in addition to contributing to the release of greenhouse gas emissions, worsening the impacts of climate change (Atlas do Plástico, 2020).

The National Solid Waste Policy (PNRS), Law No. 12.305/2010, is the Brazilian legal framework governing waste in the country. Besides this, 135 Bills (Law Projects) were proposed between 1995 and 2019 by the National Congress. However, the legal framework is mainly focused on post-consumption waste management only, when waste has already been generated, and this fact aggravates the mentioned bottlenecks (Pertussatti, 2020). Targeting citizen education toward waste generation, some initiatives were created. One recent advance is the banning of plastic straws in some Brazilian states. Yet, these measures also face some resistance from the population. Rio de Janeiro and São Paulo Cities created a ban on the supply of plastic straws in commercial establishments (Law 17.123/2019) (Pertussatti, 2020; São Paulo, 2019, 2020). The archipelago of Fernando de Noronha, a UNESCO World Natural Heritage site in Brazil, is the first territory to ban disposable plastics. This is part of the Noronha Plástico Zero policy, aiming educational and communication actions that seek to engage and mobilize residents and also tourists to contribute to it (Atlas do Plástico, 2020; Pertussatti, 2020).

Plastic waste is a serious socio-environmental problem in Brazil, and more effective measures are needed to improve the current scenario (WWF, 2019). Thus, it is necessary to broaden the debate about waste, especially involving plastics, aiming to educate, mobilize, and engage stakeholders in this process (Elen et al., 2012; Wiesenhütter et al. 2021). As stated in the Atlas do Plástico (2020), approaching the plastic waste issue means to stimulate the debate on access to water, sanitation, selective waste collection in cities, and is an urgent need for countries with a high inequality index, such as Brazil. There is a number of social initiatives aiming to educate citizens regarding general and plastic waste, including the Zero-Waste concept, which is a driver to support better waste management by eliminating progressive waste increase and requires transforming infrastructures and policies, education, training, and research (Gutberlet et al., 2016). Several Brazilian cities and also schools are now aligned to these principles. Besides them, there is also the invaluable, but not recognized, role of waste pickers, active members of the country's economy. There are between 400,000 and 800,000 waste pickers in Brazil and around 1700 waste picker organizations, which daily contribute to the correct waste management, including managing plastics. The National Movement of Waste Pickers (MNCR) has also been co-creating proposals aimed toward the circular economy with the inclusion of waste pickers (Gutberlet et al., 2020). Key actors, such as waste pickers, have been demanding for many years their direct involvement in the local selective waste collection and recycling programs. Other civil society groups are also calling for the promotion of adequate policies and measures that reduce the production and use of plastics in general. Finally, there is an urgent need of better and effective policies that address the whole plastic life chain, not only the disposal phase, prioritizing the prevention of plastic waste, reuse, and repair.

6 Alternatives to the Use of Plastics

6.1 Plant-Based and Compostable Packaging Materials

Given the disadvantages of circular principles like reuse or recycling in regard to cycle loses in the case of plastics, although they are preferable to the linear approach, different possible alternatives for replacing the material used should be taken into consideration and attentively analyzed. Accordingly, renewable and bio-based sourcing of packaging have to be better explored, possible to positively contribute to the advancement of circular economy (Kaur et al., 2018). As it was pointed out in the section about circular business models, business initiatives should be the key in the process of change in regard to touching the zero-waste plastic objective. McNeish and Neufeldt (2023) also underlined this fact, indicating that companies must solve the global issue of plastic pollution rather than waiting for external directions to follow. In this regard, Retamal et al. (2021) stated that many businesses, especially from the packaging sector, started encouraging innovative ideas for substituting plastics, using (i) bioplastics made from natural substances instead of petroleum, but with clear other disadvantages such as not breaking down in home composts, landfills, or not losing in the environment in their majority or (ii) natural alternatives like plant-based bottles that require no fossil fuels, and that degrade much faster than plastic or paper bottles (but with the use of normal plastic in the interior) (McNeish & Neufeldt, 2023). These plant-by-products are obtained from fruits, vegetables, grains, legumes, and seeds such as sugarcane bagasse/waste, areca-nut leaves, or bamboo, all delivering distinct functions to the packaging matrix (Santhosh et al., 2021), playing an integral part in transitioning to a circular economy (Shogren et al., 2019).

Accordingly, this option is also met in the practical routine from the organizational level, not only within the academic literature, existing a real declared concern among big companies (such as Nestle, Coca-Cola, Danone, or PepsiCo) to reduce the amount of plastics used in their packaging, the achievement of net zero-waste packaging being established as an important objective (McNeish & Neufeldt, 2023; Shogren et al., 2019). The alternatives to plastic material in terms of packaging are called in different ways, i.e., green plastics (de Vargas Mores et al., 2018; Kuruppalil, 2011), eco-friendly/sustainable/biodegradable packaging (Koch et al., 2022; Nguyen et al., 2020), edible films and coatings (Galus et al., 2020), emphasizing "the new-age packaging trends" (Santhosh et al., 2021). The main benefits of this kind of packaging are in regard to their peculiarities as these are made by renewable, non-toxic, biodegradable, biocompatible, and cost-effective raw materials (Santhosh et al., 2021), having the potential of reducing global warming affected by plastic using (Jørgensen & Pedersen, 2018; McNeish & Neufeldt, 2023), or, in detail, with the capacity of waste reduction, lowering the greenhouse gas emissions, promoting rural investment, reducing the volume of harmful chemicals/pollutants, conserving ecosystems and biodiversity, and helping transition to the circular economy (Shogren et al., 2019). Still, other aspects to be taken into consideration should be the ones of the linked effects of initiatives dedicated to ameliorating plastic pollution problem registered on the social and economic dimensions, besides the environmental one. McNeish and Neufeldt (2023) observed that moving from plastic to aluminum, for example, would cause the increasing of organizational carbon footprint, while the using of plant-based packaging instead of plastic would increase the risk of food scarcity. However, plant-based materials have the potential to take over much of the plastics market, while support from government, industry, and consumers might constitute into significant drivers for the transition to a bio-based circular economy (Shogren et al., 2019).

One specific example of an innovative plastics manufacturer is given by Jørgensen and Pedersen (2018), i.e., a supplier of the technology company Dell, called Newlight that used to deliver a plastic certified carbon-negative. Besides this particularity, other advantages are in regard to its biodegradability, or competitive quality and price. These authors also offered the example of other company called Skipping Rocks Lab that, basing on biomimicry, developed an affordable alternative to traditional plastic bottles, i.e., a liquid packaging made from seagrass and other naturally occurring input factors completely edible (Jørgensen & Pedersen, 2018). In the same way, Körber (n.d.), a company that positions itself as a pioneer in sustainable production chains and supplier for international pharmaceutical customers with packaging that is totally made from renewable paper fibers, is other practical example showing that the change of the using plastic habit is possible to be applied in an industry such as the pharmaceutical one. Moreover, the trend of extending the number of business initiatives built on such ideas might be overseen, illustrating, in this way, the high potential of biomimicry in the problem of plastic pollution (Jørgensen & Pedersen, 2018). Consequently, if considering their multi-dimensional advantages, formulating novel food packaging materials using plant-by-products seems to be a feasible endeavor (Santhosh et al., 2021).

6.2 Glass Reuse and Recycling

Glass seems to represent other sustainable alternative to plastic. Still, if considering the effects of actions like transportation, retailing, and its return to manufacturer, concerns in regard to sustainability appear. Also including details in regard to peculiarities such as weight or fragility, it might be possible to conclude that glass is less economically and even less environmentally efficient (Kellogg, 2019; McNeish & Neufeldt, 2023). Moreover, the one-way glass bottles are considered in many studies as the most unfavorable packaging in term of associated environmental burdens, which either originates in higher specific weight or energy consumption during the entire life cycle (Cleary, 2013; Simon et al., 2016). Still, if comparing to many single-use plastic bottles that cannot be refilled, glass containers have the necessary characteristics for reuse as they are long-lasting, durable, and corrosion resistant, being, in the same time, the packaging material certified by the U.S. Food and Drug Administration (Glass Packaging Institute, n.d.) as "generally regarded as safe". If extended, this reuse action is able to reduce the contribution to landfills. Moreover, the glass recycling is other circular option contributing to less environmental pollution. In this regard, although improvements in container glass recycling across the globe were registered, energy savings is still an important objective intended to be achieved in the process of closing the glass production-consumption-reuse/recycling loop (Butler & Hooper, 2019), with certain advantages such as: the cost savings related to energy usage, reduction of related air pollution and related water pollution, and reduction of the space in landfills or bin (WWF, n.d.a, n.d.b). One example of scheme for reducing the process of reuse and recycle might be the one of "Deposit-and-Return" system. It implies a small deposit retained from customers which could be redeemed on return of a glass bottle after finishing the beverage. If looking into the past, when this system was frequently used, such a scheme is expected to be supported by local re-bottling operations. Accordingly, for encouraging refilling systems, initiatives of eco-friendly policy like the empty return policy need to be considered. For responding to the desideratum of sustainability, it is useful to take into consideration different "sensitive" parameters able to play a role in the case of the refillable bottles (Cleary, 2013, Detzel & Mönckert, 2009) such as: the number of refills, the transport distance from filler to point-of-sale, and the energy (and water) efficiency of bottle and crate cleaning (Simon et al., 2016). Particularly, as a result of high specific weight, the glass bottle shows high sensitivity with respect to the distribution distance, but this disadvantage might be improved by multiple refills of bottles and their distribution on a local scale (Detzel & Monckert, 2009; Simon et al., 2016), registering a positive effect on the overall environmental burden of bottles (Simon et al., 2016).

6.3 Cardboard Mono-Material

In terms of more eco-friendly packaging materials options, comparing to plastics, cardboard might be included on the small list of current candidates. For example, if comparing to biodegradable plastic that cannot be reused, cardboard is appropriate for recycling and composting. Accordingly, although deforestation is a major issue in the debate around environmental wellbeing, still, recycling might be one partial and available solution in this regard. In addition, given the fact that cardboard is completely compostable, not leaving behind harmful substances, it appears to be one more sustainable alternative to plastics in the packaging industry. In this way, in a paper of Albrecht et al. (2022), the claim of the European Federation of Corrugated Board Manufacturers (FEFCO), sustaining that corrugated cardboard boxes were advantageous over plastic-based reusable crates and that a modification of the waste hierarchy considering the life cycle analyses appears to be required, was specified. In the same way, Pajula and Sundqvist-Andberg (2022) support the idea of cardboard, with its high recycling rate, as a better environmental option for reducing plastic emissions. However, cardboard mono-material is the recommendation of different experts for a sustainable path because, as Albrecht et al. (2022) clarified, the goal cannot be to replace plastic littering with paper littering. In this respect, very frequently, containing printing inks or additives or increasingly coated with polymers, even cardboard material is not or only difficultly degradable, and, in the same time, more problematic to be recycled. In this context, the study of Foschi et al. (2020) is remembered. Using an integrated decision-making system at the design stage, the results of this paper emphasized different benefits at multiple levels when substituting the plastic "open and close" cap with a closing method entirely made of cardboard, such as better recyclability and lower energy consumption, nearby the contribution to the reduction in the consumption of plastics and, accordingly, the prevention of pollution. Although analyzing an isolated case study, the study supports the idea of using cardboard mono-material as a more circular and sustainable alternative to plastic packaging material.

7 Conclusions

This chapter highlighted the necessary of transforming the current linear economy and overconsumption into a more compatible to sustainability manner of living and developing in regard to one of major global concerns, i.e., plastic waste. Plastic waste invades all aquatic and terrestrial environments through mismanagement of municipal waste, packaging waste, or wastewaters in addition to plastics leaked into natural environment by industrial and agricultural practices around the world. Therefore, critical steps such as universal waste collection, rural-urban integration into sound waste management systems, and better traceability of plastic waste at national and international levels are required to reduce the post-consumer plastic waste contamination pressure on the natural environment. This chapter argues the shift of plastic waste management toward zero-waste approach through two main levels: (i) downstream level to increase the plastic capture rate from municipalities and reduce the plastic leakage into natural environment and disposed of in landfills (e.g., deposit return schemes, waste banks, reuse and recycling practices, recycling technologies, energy recovery) and (ii) upstream level to address the full life cycle of plastic reduction and increasing circularity levels of existing plastic waste supported by EPRs policies and regulatory frameworks.

The chapter reveals the current knowledge and solutions to improve plastic waste management in line with Rs policies such as refuse, reduce, reuse, recycle, recovery, or replace plastics. The mini case studies from five countries (Romania, Canada, Egypt, Indonesia, and Brazil) reveal the specific progress and barriers related to plastic waste management in various geographical regions. As general remarks from all the investigated national contexts, the following most important ones were noticed: (i) all the countries are making efforts for changing the model of "take-make-dispose" toward a more circular model; (ii) the progress in circular secondary material usage over the last years was not enough for appropriately responding to the plastic problem in each national context, failing in the decoupling of their economic growth from the generation of waste, with still high landfilling rates; (iii) the local community is more and more encouraged to become highly (and actively) involved in local waste management activities, of great importance in this regard being the necessity of more education and awareness of the population in terms of environmental problems. More specifically, the role of informal sector in capture of plastics should not be overlooked in countries like Indonesia or Brazil, while improvement of source-separated plastic waste is expected in Romania and Egypt. Better EPRs policies and traceability of plastic waste flows in Canada will reduce the risks of transboundary movement of plastic waste toward countries with deficient waste management systems, while improving own recycling capabilities combined with plastic waste reduction efforts.

In addition, new recycling technologies and alternatives to the use of plastics are proposed in the current research and development sectors to curb the plastic pollution levels in the future. However, increasing the level of awareness across societies in regard to the necessity of ameliorating the problem of plastic pollution might be able to play the same important role as the technical and technological progress, while available solutions for this problem exist, but with certain requirements and economic losses.

Acknowledgements The work of **FCM and SRU** is supported by the Ministry of Research, Innovation, Digitization (Romania) CNCS-UEFISCDI grant no PN-III-P1-1.1-TE-2021-0075 within PNCDI III.

References

Adarsh, U. K., Kartha, V. B., Santhosh, C., & Unnikrishnan, V.K. (2022). Spectroscopy: A promising tool for plastic waste management. *TrAC Trends in Analytical Chemistry,* 116534.

Albrecht, S., Bertling, J., Fischer, M., Gehring, F., Kabasci, S., Prescher, T., & Schulte, A. (2022). Reusable plastic crates vs. single-use cardboard boxes—Two packaging systems in competition. In *Oberhausen/Stuttgart 2022.*

Alfaia, R. G. D. S. M., Costa, A. M., & Campos, J. C. (2017). Municipal solid waste in Brazil: A review. *Waste Management & Research, 35*(12), 1195–1209.

Allen, S., Allen, D., Karbalaei, S., Maselli, V., & Walker, T. R. (2022). Micro(nano)plastics sources, fate, and effects: What we know after ten years of research. *Journal of Hazardous Materials Advances, 6,* 100057. https://doi.org/10.1016/j.hazadv.2022.100057

Antelava, A., Damilos, S., Hafeez, S., Manos, G., Al-Salem, S. M., Sharma, B. K., Kohli, K., & Constantinou, A. (2019). Plastic solid waste (PSW) in the context of life cycle assessment (LCA) and sustainable management. *Environmental Management, 64,* 230–244.

Asteria, D., & Heruman, H. (2015). Waste banks as an alternative of community-based waste management strategy in Tasikmalaya. *Jurnal Manusia dan Lingkungan, 23*(1), 136–141.

Atlas do Plástico. (2020). *Fatos e números sobre o mundo dos polímeros sintéticos/organização.* In M. Montenegro, M. Vianna, & D. B. Teles (Eds.) (1st ed.). Fundação Heirich Böll, 2020. ISBN 978-65-87665-02-3.

Azapagic, A., Jeswani, H., Krüger, C., Russ, M., Horlacher, M., Antony, F., & Hann, S. (2021). Life cycle environmental impacts of chemical recycling via pyrolysis of mixed plastic waste in comparison with mechanical recycling and energy recovery. *Science of the Total Environment, 769,* 144483.

Baia et al. (2020). Plásticos e seus impactos ambientais. International Studies Coepta 3–4. *International Studies on Law & Education,* 34–45. Cemoroc-Feusp/IJI-Univ. do Porto/Colégio Luterano São Paulo. http://www.hottopos.com/isle34_35/167-176JVernePlasticosF.pdf

Balaji, A. B., & Liu, X. (2020). Plastics in circular economy: A sustainable progression. In *An introduction to circular economy* (pp. 159–178). Springer Singapore.

Barford, A., & Ahmad, S. R. (2022). Levers for a corporate transition to a plastics circular economy. *Business Strategy and the Environment,* 1–15. https://doi.org/10.1002/bse.3182

BBC News. (2021). *Fast fashion: The dumping ground for unwanted clothes.* Retrieved January 12, 2023, from https://www.bbc.com/news/av/world-africa-58836618

Bhattacharjee, S., Rahaman, M., Andrei, V., Miller, M., Rodríguez-Jiménez, S., Lam, E., Pornrungroj, C., et al. (2023). Photoelectrochemical CO_2-to-fuel conversion with simultaneous plastic reforming. *Nature Synthesis.* https://www.nature.com/articles/s44160-022-00196-0

Black Forest Solutions GmbH, Landbell, A.G. (2021). *Extended producer responsibility scheme for packaging waste in Egypt.* Deutsche Gesellschaf für Internatonale Zusammenarbeit. https://www.giz.de/en/downloads/20210602_EPR%20Scheme%20for%20Packaging%20Waste%20in%20Egypt_barrierefrei.pdf

Butler, J. H., & Hooper, P. D. (2019). Glass waste. In *Waste* (pp. 307–322). Academic Press.

Casillas, G., Hubbard, B. C., Telfer, J., Zarate-Bermudez, M., Muianga, C., Zarus, G. M., Carroll, Y., Ellis, A., & Hunter, C. M. (2023). Microplastics scoping review of environmental and human exposure data. *Microplastics, 2*(1), 78–92.

Chen, J., Wu, J., Sherrell, P. C., Chen, J., Wang, H., Zhang, W. X., & Yang, J. (2022). How to build a microplastics-free environment: Strategies for microplastics degradation and plastics recycling. *Advanced Science, 9*(6), 2103764.

Circle Economy. (2023). *The circularity gap report 2023* (pp. 1–64). Circle Economy.

Cleary, J. (2013). Life cycle assessments of wine and spirit packaging at the product and the municipal scale: A Toronto, Canada case study. *Journal of Cleaner Production, 44,* 143–151. https://doi.org/10.1016/j.jclepro.2013.01.009

Coelho, P. M., Corona, B., ten Klooster, R., & Worrell, E. (2020). Sustainability of reusable packaging—Current situation and trends. *Resources, Conservation & Recycling: X, 6,* 100037.

Cook, E., & Velis, C. (2021). *Global review on safer end of engineered life*. Royal Academy of Engineering.

Cornago, E., & Borkey, P. (2021, October 13). *Preventing single-use plastic waste: Implications of different policy approaches* (Environment Working Paper, 182). OECD Environment Directorate.

de Jong, A. M., & Mellquist, A. C. (2021). The potential of plastic reuse for manufacturing: A case study into circular business models for an on-line marketplace. *Sustainability, 13*(4), 2007.

de Vargas Mores, G., Finocchio, C. P. S., Barichello, R., & Pedrozo, E. A. (2018). Sustainability and innovation in the Brazilian supply chain of green plastic. *Journal of Cleaner Production, 177*, 12–18.

Demetrious, A., & Crossin, E. (2019). Life cycle assessment of paper and plastic packaging waste in landfill, incineration, and gasification-pyrolysis. *Journal of Material Cycles and Waste Management, 21*, 850–860.

Detzel, A., & Monckert, J. (2009). Environmental evaluation of aluminium cans for beverages in the German context. *The International Journal of Life Cycle Assessment, 14*, 70–79. https://doi.org/10.1007/s11367-008-0057-1

Di Bartolo, A., Infurna, G., & Dintcheva, N. T. (2021). A review of bioplastics and their adoption in the circular economy. *Polymers, 13*(8), 1229.

Diggle, A., & Walker, T. R. (2022). Environmental and economic impacts of mismanaged plastics and measures for mitigation. *Environments, 9*, 15. https://doi.org/10.3390/environments9020015

Dijkstra, H., van Beukering, P., & Brouwer, R. (2020). Business models and sustainable plastic management: A systematic review of the literature. *Journal of Cleaner Production, 258*. https://doi.org/10.1016/j.jclepro.2020.120967

Dimitropoulos, A., Tijm, J., & Veld, D. (2021, July). *Extended producer responsibility: Design, functioning, and effects* [Background Document]. PBL Netherlands Environmental Assessment Agency. https://www.cpb.nl/sites/default/files/omnidownload/PBL-CPB-2021-Extended-Producer-Responsibility-Design-Functioning-Effects.pdf

EC. (2018). *A European strategy for plastics in a circular economy*. European Commission. https://ec.europa.eu/environment/circular-economy/pdf/plastics-strategy-brochure.pdf

EC. (2022). *2022 Country report—Romania*. European Commission. https://commission.europa.eu/system/files/2022-06/2022-european-semester-country-report-romania_en.pdf.

EEA. (2022). *Circular economy country profile—Romania. ETC CE Report 2022/5—Romania*. https://www.eionet.europa.eu/etcs/etc-ce/products/etc-ce-products/etc-ce-report-52022-country-profiles-on-circulareconomy/romania-ce-country-profile-2022_for-publication.pdf

Elen, B. A. V., Pacheco, L. M., & Ronchetti, E. M. (2012). An overview of plastic recycling in Rio de Janeiro. *Resources, Conservation and Recycling, 60*, 140–146. ISSN 0921-3449. https://doi.org/10.1016/j.resconrec.2011.12.010

Elkady, A., Habib, M., [Mostafa], Iskandar, L., Kamal, A., & Yassin, A. (2021, January 1). *Coming to grips with plastic: Egypt's waste management challenge*. Economist Impact. Retrieved January 13, 2023, from https://www.amcham.org.eg/bic/pdf/cop27/coming-to-grips-with-plastic.pdf

Ellen MacArthur Foundation. (n.d.). *Extended producer responsibility: A necessary part of the solution to packaging waste and pollution*. https://ellenmacarthurfoundation.org/extended-producer-responsibility/overview

EMF. (2015). *Towards a circular economy: Business rationale for an accelerated transition*. https://www.ellenmacarthurfoundation.org/assets/downloads/TCE_Ellen-MacArthur-Foundation_9-Dec-2015.pdf

Eriksson, O., & Finnveden, G. (2009). Plastic waste as a fuel—CO_2-neutral or not? *Energy & Environmental Science, 2*, 907–914.

Eurostat. (2022). *Recycling rates of packaging waste for monitoring compliance with policy targets, by type of packaging*. https://ec.europa.eu/eurostat/databrowser/view/ENV_WASPACR__custom_4246874/default/table

Fei, F., et al. (2016). How to integrate the informal recycling system into municipal solid waste management in developing countries: Based on a China's case in Suzhou urban area. *Resources, Conservation & Recycling, 110*, 74–86.

Fellner, J., & Brunner, P. H. (2022). Plastic waste management: Is circular economy really the best solution? *Journal of Material Cycles and Waste Management, 24*, 1–3. https://doi.org/10.1007/s10163-021-01340-2

Forlani, G., & Njie, L. (2022). *In COVID-era trash surge, waste management ingenuity, circularity, and investments are key*. World Bank Blogs. https://blogs.worldbank.org/sustainablecities/covid-era-trash-surge-waste-management-ingenuity-circularity-and-investments-are

Foschi, E., Zanni, S., & Bonoli, A. (2020). Combining eco-design and LCA as decision-making process to prevent plastics in packaging application. *Sustainability, 12*(22), 9738.

Friedberg, E., & Hilderbrand, M. E. (2017). *Observing policy-making in Indonesia*. Springer Nature Singapore Pte Ltd.

Galus, S., Arik Kibar, E. A., Gniewosz, M., & Kraśniewska, K. (2020). Novel materials in the preparation of edible films and coatings—A review. *Coatings, 10*(7), 674.

Geissdoerfer, M., et al. (2020). Circular business models: A review. *Journal of Cleaner Production, 277*, 123741. https://doi.org/10.1016/j.jclepro.2020.123741

Geyer, R., Jambeck, J. R., & Law, K. L. (2017). Production, use, and fate of all plastics ever made. *Science Advances, 3*, e1700782.

Glass Packaging Institute. (n.d.). *What is glass?* Retrieved January 25, 2023, from https://www.gpi.org/what-is-glass

Gomes, A. R., Lazaro, J. C., & Leocadio, A. (2022). Should I reuse it or throw it out? Analysis of the management of household plastic waste by Brazilian consumers during the COVID-19 pandemic through practice lens. *Sustainability, 14*(14). https://doi.org/10.3390/su14148512

Gorziza et al. (2021). *Apenas 1,3% do lixo plástico Gerado no Brasil é reciclado*. Retrieved February 3, 2023, from https://piaui.folha.uol.com.br/apenas-13-do-lixo-plastico-gerado-no-brasil-e-reciclado

Government of Canada. (2021). *Ocean plastics charter* (June, 9, 2018). https://www.canada.ca/en/environment-climate-change/services/managing-reducing-waste/international-commitments/ocean-plastics-charter.html

Government of Canada. (2022, October 19). *Single-use plastics prohibition regulations—Overview* [Overview of SUPPR]. https://www.canada.ca/en/environment-climate-change/services/managing-reducing-waste/reduce-plastic-waste/single-use-plastic-overview.html#toc0

Green News. (2020). Targu Lapus is recycling 10 times more than the rest of the country. Retrieved February 9, 2023, from https://www.greennews.ro/article/la-targu-lapus-se-recicleaza-de-10-ori-mai-mult-decat-in-restul-tarii

GreenTech. (n.d.). *Recycled plastics. Cleaner environment*. Retrieved February 6, 2023, from https://www.greentechglobal.com/assets/app/green_group_tech_brosura_sustenabilitate_v7_RO.pdf

Grodzińska-Jurczak, M., Krawczyk, A., Akhshik, A., Dedyk, Z., & Strzelecka, M. (2022). Contradictory or complementary? Stakeholders' perceptions of a circular economy for single-use plastics. *Waste Management, 142*, 1–8.

Gutberlet, J., Baeder, A. M., Pontuschka, N. N., Felipone, S. M. N., dos Santos, T. L. F., & de Souza, A. M. (2016). Pesquisa-ação em educação ambiental e saúde dos catadores: estudo de caso realizado com integrantes de cooperativas de coleta seletiva e reciclagem na região metropolitana de São Paulo. In *Catadores de materiais recicláveis: um encontro nacional*. IPEA.

Gutberlet, J., Besen, G. R., & Morais, L. (2020). Participatory solid waste governance and the role of social and solidarity economy: Experiences from São Paulo, Brazil. *Detritus, 13*, 167–180. https://doi.org/10.31025/2611-4135/2020.14024

Hamdallah, D. (2021, July 21). *EBRD and EU support Egyptian plastic recycling industry*. https://www.ebrd.com/news/2021/ebrd-and-eu-support-egyptian-plastic-recycling-industry.html

Henkel. (2021, March 31). *Henkel and Plastic Bank open first collection centers in Egypt* [Press release]. https://www.henkel.com/press-and-media/press-releases-and-kits/2021-03-31-henkel-and-plastic-bank-open-first-collection-centers-in-egypt-1168118

Hogg, D., Fletcher, D., Eye, M., Mulcahy, K., & Elliott, T. (2011). *From waste to work: The potential for a deposit refund system to create jobs in the UK*. Eunomia Research & Consulting Ltd.

Hood, C., Laredo, T., Marangoni, A. G., & Pensini, E. (2021). Water-repellent films from corn protein and tomato cutin. *Journal of Applied Polymer Science, 138*, 50831.

IAEA. (2021). *IAEA annual report 2021*. Accessed February 9, 2023, from https://www.iaea.org/sites/default/files/publications/reports/2021/gc66-4.pdf

Iskander, L. (2021, June 14). *How tech is helping Egypt's informal recyclers build a circular economy*. World Economic Forum. https://www.weforum.org/agenda/2021/06/technology-egypt-recycling-circular-economy/

Jørgensen, S., & Pedersen, L. J. T. (2018). *RESTART sustainable business model innovation* (p. 253). Springer Nature.

Kaur, G., Uisan, K., Ong, K. L., & Lin, C. S. K. (2018). Recent trends in green and sustainable chemistry & waste valorisation: Rethinking plastics in a circular economy. *Current Opinion in Green and Sustainable Chemistry, 9*, 30–39.

Kellogg, K. (2019, September, 12). Which is better for the environment? Glass or plastic? *Going Zero Waste*.

Kibria, M. G., Masuk, N. I., Safayet, R., Nguyen, H. Q., & Mourshed, M. (2023). Plastic waste: Challenges and opportunities to mitigate pollution and effective management. *International Journal of Environmental Research, 17*(1), 20.

Klotz, M., Haupt, M., & Hellweg, S. (2022). Limited utilization options for secondary plastics may restrict their circularity. *Waste Management, 141*, 251–270.

Koch, J., Frommeyer, B., & Schewe, G. (2022). Managing the transition to eco-friendly packaging—An investigation of consumers' motives in online retail. *Journal of Cleaner Production, 351*, 131504.

Körber. (n.d.). *Sustainability: Packaging without plastic*. Retrieved January 28, 2023, from https://www.koerber-pharma.com/en/solutions/packaging-materials/sustainable-packaging

Kruse, C., Boyda, E., Chen, S., Karra, K., Bou-Nahra, T., Hammer, D., Mathis, J., Maddalene, T., Jambeck, J., & Laurier, F. (2023). Satellite monitoring of terrestrial plastic waste. *PLoS ONE, 18*(1), e0278997.

Kuruppalil, Z. (2011). Green plastics: An emerging alternative for petroleum-based plastics. *International Journal of Engineering Research & Innovation, 3*(1), 59–64.

Lazarevic, D., Aoustin, E., Buclet, N., & Brandt, N. (2010). Plastic waste management in the context of a European recycling society: Comparing results and uncertainties in a life cycle perspective. *Resources, Conservation and Recycling, 55*, 246–259.

Li, H., Aguirre-Villegas, H. A., Allen, R. D., Bai, X., Benson, C. H., Beckham, G. T., ... Huber, G. W. (2022). Expanding plastics recycling technologies: Chemical aspects, technology status and challenges. *Green Chemistry*.

Liro, M., van Emmerik, T. H., Zielonka, A., Gallitelli, L., & Mihai, F. C. (2023). The unknown fate of macroplastic in mountain rivers. *Science of the Total Environment, 865*, 161224.

Mahmoud, H. (2022, October 30). Plastic Bank, Henkel celebrate collecting 300 Mn plastic bottles in less than 2 years. *The Egyptian Gazette*. https://egyptian-gazette.com/cop-27/plastic-bank-henkel-celebrate-collecting-300-mn-plastic-bottles-in-less-than-2-years/

Mahmoudnia, A., Mehrdadi, N., Kootenaei, F. G., Deiranloei, M. R., & Al-e-Ahmad, E. (2022). Increased personal protective equipment consumption during the COVID-19 pandemic: An emerging concern on the urban waste management and strategies to reduce the environmental impact. *Journal of Hazardous Materials Advances, 7*, 100109.

Material Economics. (2022). *Europe's missing plastics—Taking stock of EU plastics circularity*. Commissioned by Agora Industry.

Mattos, C. A., Scur, G., & Albuquerque, T. L. (2022). Evaluation of circular business model: Theory of change approach. *Evaluation and Program Planning, 92*, 102069.

McNeish, J. E., & Neufeldt, J. (2023). Can Nestlé transform from a corporation that 'talks about doing good' to 'doing good' for the environment? In *Dealing with socially responsible consumers: Studies in marketing* (pp. 429–456). Springer Nature Singapore.

Meidiana, C., & Gamse, T. (2010). Development of waste management practices in Indonesia. *European Journal of Scientific Research, 40*(2), 199–209.

Meidiana, C., Kurniawan, T. A., Yudono, A., & Surjono, S. (2022). *Modern challenges and approaches to humanitarian engineering* (pp. 98–113). IGI Global.

Meidiana, C., Pamungkas, Y. A., & Sari, K. E. (2019). Effectiveness and efficiency as a tool of measurement of waste bank's performance. *IOP Conference Series: Earth and Environmental Science, 328*(1), 012014.

Meijer, L. J. J., van Emmerik, T., van der Ent, R., Schmidt, C., & Lebreton, L. (2021). More than 1000 rivers account for 80% of global riverine plastic emissions into the ocean. *Science Advances, 7*(18). https://doi.org/10.1126/sciadv.aaz5803

Mihai, F. C. (2017). One global map but different worlds: Worldwide survey of human access to basic utilities. *Human Ecology, 45*, 425–429. https://doi.org/10.1007/s10745-017-9904-7

Mihai, F. C. (2018). Rural plastic emissions into the largest mountain lake of the Eastern Carpathians. *Royal Society Open Science, 5*(5), 172396.

Mihai, F.-C. (2023). Circular economy and sustainable rural development. *Sustainability, 15*, 2139. https://doi.org/10.3390/su15032139

Mihai, F.-C., Gnoni, M. G., Meidiana, C., Schneider, P., Ezeah, C., & Elia, V. (2022b). A global outlook on implementation of Basel convention and transboundary movement of E-waste. In *Paradigm shift in E-waste management: Vision for future* (pp. 49–75). CRC Press. https://doi.org/10.1201/9781003095972-4

Mihai, F.-C., Gündoğdu, S., Markley, L. A., Olivelli, A., Khan, F. R., Gwinnett, C., Gutberlet, J., Reyna-Bensusan, N., Llanquileo-Melgarejo, P., Meidiana, C., Elagroudy, S., Ishchenko, V., Penney, S., Lenkiewicz, Z., & Molinos-Senante, M. (2022a). Plastic pollution, waste management issues, and circular economy opportunities in rural communities. *Sustainability, 14*, 20. https://doi.org/10.3390/su14010020

Mihai, F. -C., Markley, L. A., Khan, F. R., Suaria, G., & Gündoğdu, S. (2023). Microplastics in the ecosphere—Air, water, soil and food. In M. Vithanage & M. N. V. Prasad (Eds.), *Microplastics in freshwater environments* (pp. 163–175). Wiley. https://doi.org/10.1002/9781119879534.ch10.

Mihai, F. C., & Ulman, S. -R. (2022). Post-consumer plastic waste pollution under linear economy mechanisms in Romania. In *SWS International Scientific Conference on Social Sciences (ISCSS) Conference Proceedings. Section Planning and Development, 9*(1). https://doi.org/10.35603/sws.iscss.2022/s15.129

Molloy, S., Medeiros, A. S., Walker, T. R., & Saunders, S. J. (2022a). Public perceptions of legislative action to reduce plastic pollution: A case study of Atlantic Canada. *Sustainability, 14*, 1852. https://doi.org/10.3390/su14031852

Molloy, S., Varkey, P., & Walker, T. R. (2022b). Opportunities for single-use plastic reduction in the food service sector during COVID-19. *Sustainable Production and Consumption, 30*, 1082–1094. https://doi.org/10.1016/j.spc.2022.01.023

Nguyen, A. T., Parker, L., Brennan, L., & Lockrey, S. (2020). A consumer definition of eco-friendly packaging. *Journal of Cleaner Production, 252*, 119792.

OECD. (2014, June 17–19). *The state of play on extended producer responsibility (EPR): Opportunities and challenges* [Issues Paper]. Global Forum on Environment: Promoting Sustainable Materials Management Through Extended Producer Responsibility (EPR).

OECD. (2022). *Global plastics outlook: Economic drivers.* OECD Publishing. https://doi.org/10.1787/de747aef-en

Ormazabal, M., Prieto-Sandoval, V., Puga-Leal, R., & Jaca, C. (2018). Circular economy in Spanish SMEs: Challenges and opportunities. *Journal of Cleaner Production, 185*, 157–167. https://doi.org/10.1016/j.jclepro.2018.03.031

Pacheco, E. B. A. V., Ronchetti, L. M., & Masanet, E. (2012). An overview of plastic recycling in Rio de Janeiro. *Resources, Conservation and Recycling, 60*, 140–146. https://doi.org/10.1016/j.resconrec.2011.12.010

Pajula, T., & Sundqvist-Andberg, H. (2022). *A critical view on packaging recycling and reuse in the European circular economy*. VTT Technical Research Centre of Finland Ltd.

Paletta, A., Leal Filho, W., Balogun, A. L., Foschi, E., & Bonoli, A. (2019). Barriers and challenges to plastics valorisation in the context of a circular economy: Case studies from Italy. *Journal of Cleaner Production, 241*, 118149.

Pariatamby, A., & Tanaka, M. (2014). *Municipal solid waste management in Asia and the Pacific Island: Challenges and strategic solutions*. Springer-Verlag.

Payne, J., McKeown, P., & Jones, M. D. (2019). A circular economy approach to plastic waste. *Polymer Degradation and Stability, 165*, 170–181.

Pertussatti, C. A. (2020). *Gestão Ambiental De Resíduos Plásticos No Brasil: Subsídios para uma Diretriz Nacional*. ENAP (Escola Nacional de Administração Pública). https://repositorio.enap.gov.br/bitstream/1/5134/1/Caroline%20Alvarenga%20Pertussatti.pdf

Pieroni, M. P., McAloone, T. C., & Pigosso, D. C. (2019). Business model innovation for circular economy and sustainability: A review of approaches. *Journal of Cleaner Production, 215*, 198–216.

Plastic Circle. (n.d.). Material recycling. Retrieved January 30, 2023, from https://plasticircle-albaiulia.ro/

Ragaert, K., Delva, L., & Van Geem, K. (2017). Mechanical and chemical recycling of solid plastic waste. *Waste Management, 69*, 24–58.

Rajendran, S., Hodzic, A., Scelsi, L., Hayes, S., Soutis, C., AlMa'adeed, M., & Kahraman, R. (2013). Plastics recycling: Insights into life cycle impact assessment methods. *Plastics, Rubber and Composites, 42*(1), 1–10.

Reloopplatform. (2022). *Global deposit book 2022. An overview of deposit return systems for single-use beverage containers*.

Retamal, M., Pai Panandiker, A., Talwar, S., Sah, S., & King, S. (2021). *Circular business models for plastics in India: Literature and practice review* (Report Number 2021-3). UTS, TERI, Development Alternatives and CSIRO.

Roebroek, C. T., Harrigan, S., Van Emmerik, T. H., Baugh, C., Eilander, D., Prudhomme, C., & Pappenberger, F. (2021). Plastic in global rivers: Are floods making it worse? *Environmental Research Letters, 16*(2), 025003.

Romanian Ministry of Environment, Water and Forests (2022). *Strategia națională privind economia circulară/National strategy regarding circular economy*. Monitorul Oficial al României, Partea I, Nr. 943 bis/27.IX.2022.

Santhosh, R., Nath, D., & Sarkar, P. (2021). Novel food packaging materials including plant-based byproducts: A review. *Trends in Food Science & Technology, 118*, 471–489.

Sañudo, R., Goswami, R. R., Ricci, S., & Miranda, M. (2022). Efficient reuse of railway track waste materials. *Sustainability, 14*(18). https://doi.org/10.3390/su141811721

São Paulo (Município). (2019). *Lei nº 17.123, de 25 de junho de 2019. Dispõe sobre a proibição de fornecimento de canudos confeccionados em material plástico, nos locais que especifica, e dá outras providências*. https://www.legisweb.com.br/legislacao/?id=378875

São Paulo (Município). (2020). *Lei nº 17.261, de 13 de janeiro de 2020. Dispõe sobre a proibição de fornecimento de produtos de plástico de uso único nos locais que especifica*. http://legislacao.prefeitura.sp.gov.br/leis/lei-17261-de-13-de-janeiro-de-2020

Shogren, R., Wood, D., Orts, W., & Glenn, G. (2019). Plant-based materials and transitioning to a circular economy. *Sustainable Production and Consumption, 19*, 194–215.

Sholanke, D., & Gutberlet, J. (2021). Call for participatory waste governance: Waste management with binners in Vancouver. *Journal of Environmental Policy & Planning, 24*(2), 94–108. https://doi.org/10.1080/1523908X.2021.1956308

Simon, B., Amor, M. B., & Földényi, R. (2016). Life cycle impact assessment of beverage packaging systems: Focus on the collection of post-consumer bottles. *Journal of Cleaner Production, 112*, 238–248.

Singh, A., & Sushil. (2017). Developing a conceptual framework of waste management in the organizational context. *Management of Environmental Quality: An International Journal, 28*(6), 786–806. https://doi.org/10.1108/MEQ-07-2016-0045

Sridharan, S., Kumar, M., Singh, L., Bolan, N. S., & Saha, M. (2021). Microplastics as an emerging source of particulate air pollution: A critical review. *Journal of Hazardous Materials, 418*, 126245.

Suzuki, G., Uchida, N., Tanaka, K., Matsukami, H., Kunisue, T., Takahashi, S., Viet, P. H., Kuramochi, H., & Osako, M. (2022). Mechanical recycling of plastic waste as a point source of microplastic pollution. *Environmental Pollution, 303*, 119114.

SwitchMed in Egypt. (n.d.). *Reducing plastic bag consumption in Egypt* [Fact sheet]. United Nations Environment Programme. https://switchmed.eu/wp-content/uploads/2020/11/Factsheet-SwitchMed-demo-Egypt-Plastic-Bag-initative-07.04.2020.pdf

Tsai, C. H., Shen, Y. H., & Tsai, W. T. (2021). Reuse of the materials recycled from renewable resources in the civil engineering: Status, achievements and government's initiatives in Taiwan. *Materials, 14*(13). https://doi.org/10.3390/ma1413

UNEA. (2022). *End plastic pollution: Towards an international legally binding instrument*. Retrieved February 7, 2023, from https://wedocs.unep.org/bitstream/handle/20.500.11822/38522/k2200647_-_unep-ea-5-l-23-rev-1_-_advance.pdf?sequence=1&isAllowed=y

United Nations Egypt. (2021, March 9). *UNIDO and Japan to support Egypt with alternative approach to single-use plastics production and consumption* [Press release]. https://egypt.un.org/en/115456-unido-and-japan-support-egypt-alternative-approach-single-use-plastics-production-and

United Nations Environmental Programme. (2018). *Single-use plastics—A roadmap for sustainability: Fact-sheet for policymakers*. https://wedocs.unep.org/bitstream/handle/20.500.11822/25523/singleUsePlastic_sustainability_factsheet_EN.pdf?sequence=1&isAllowed=y

Urbinati, A., Chiaroni, D., & Chiesa, V. (2017). Towards a new taxonomy of circular economy business models. *Journal of Cleaner Production, 168*, 487–498.

Walls, M. (2011). Deposit-refund systems in practice and theory. Retrieved January 20, 2023, from https://media.rff.org/documents/RFF-DP-11-47.pdf

Water and Environment Support (WES). (n.d.). *Single-use plastic ban by the Red Sea Governorate. Best practices in achieving waste reduction in the Mediterranean*. Retrieved February 1, 2023, from https://bestpractices-waste-med.net/single-use-plastic-ban-by-the-red-sea-governorate/

Watkins, E., & Gionfra, S. (2020). *How to implement extended producer responsibility (EPR): A briefing for governments and businesses*. World Wide Fund (WWF). https://wwfint.awsassets.panda.org/downloads/how_to_implement_epr___briefing_for_government_and_business.pdf

Watkins, E., Gionfra, S., Schweitzer, J., Pantzar, M., Janssens, C., & Brink, P. (2017, November 9). *EPR in the EU plastics strategy and the circular economy: A focus on plastic packaging*. Institute for European Environmental Policy. https://zerowasteeurope.eu/wp-content/uploads/2019/11/zero_waste_europe_IEEP_EEB_report_epr_and_plastics.pdf

Wiesenhütter, L. L., Meinerz, N. Z., & Griebler, T. F. (2021). O consumo de plástico no Brasil: impactos e alternativas ao seu uso. *Salão do Conhecimento, 7*(7).

WWF. (2019). *Extended producer responsibility (EPR) for plastic packaging*. Circular Economy. https://wwfeu.awsassets.panda.org/downloads/wwf_epr_position_paper.pdf

WWF. (n.d.a). *Recycling glass—How it helps environment*. Retrieved January 30, 2023, from https://wwf.panda.org/discover/knowledge_hub/teacher_resources/project_ideas/recycling_glass/

WWF. (n.d.b). *Fundo Mundial para a Natureza*. ISBN 978-2-940529-93-3.

Yadav, V., Sherly, M. A., Ranjan, P., Prasad, V., Tinoco, R. O., & Laurent, A. (2022). Risk of plastics losses to the environment from Indian landfills. *Resources, Conservation and Recycling, 187*, 106610.

Yoshida, A. (2022). China's ban of imported recyclable waste and its impact on the waste plastic recycling industry in China and Taiwan. *Journal of Material Cycles and Waste Management, 24*(1), 73–82.

Zero Waste Europe. (2017). *Seizing the opportunity: Using plastic only where it makes sense.* Retrieved January 30, 2023, from https://www.zerowasteeurope.eu/wp-content/uploads/2017/09/ZWE-Plastics-Reduction-Paper-FINAL.pdf

ZeroWaste Romania. (2022). *Salacea towards zero-waste.* Retrieved January 30, 2023, from https://www.zerowasteromania.org/salacea-spre-zero-waste/

Zhang, F., Wang, F., Wei, X., Yang, Y., Xu, S., Deng, D., & Wang, Y. Z. (2022). From trash to treasure: Chemical recycling and upcycling of commodity plastic waste to fuels, high-valued chemicals and advanced materials. *Journal of Energy Chemistry, 69*, 369–388.

Zhao, X., Korey, M., Li, K., Copenhaver, K., Tekinalp, H., Celik, S., Kalaitzidou, K., Ruan, R., Ragauskas, A. J., & Ozcan, S. (2022). Plastic waste upcycling toward a circular economy. *Chemical Engineering Journal, 428*, 131928.

Zhu, B., Wang, D., & Wei, N. (2022). Enzyme discovery and engineering for sustainable plastic recycling. *Trends in Biotechnology, 40*(1), 22–37.

Chapter 7
Value Addition to Waste for Circular Economy and Sustainable Development

Younis Ahmad Hajam, Rajesh Kumar, and Neelam

Abstract The growing global concern over environmental degradation and resource depletion has intensified the search for innovative solutions that promote sustainable development. The concept of a circular economy has emerged as a viable approach to counteract the traditional linear economic model that relies on "take-make-dispose" principles. In the circular economy, waste is seen as a valuable resource rather than a burden, and its transformation into useful products and materials is central to achieving sustainability. It highlights the adverse impacts of waste generation, such as pollution, depletion of natural resources, and landfill overflow, which necessitate a paradigm shift towards circularity. Sustainable development goals, including reduced consumption of finite resources, increased resilience against supply chain disruptions, and improved waste management infrastructure in developing regions, are discussed. Therefore, it can be concluded with a call to action for policymakers, businesses, and individuals to embrace value addition to waste as a vital step towards achieving a more sustainable and prosperous future. Collaborative efforts in research, investment, and policy implementation are essential to accelerate the transition to a circular economy that maximises resource efficiency, minimises environmental impacts, and fosters long-term sustainable development.

Keywords Circular economy · Waste valorisation · Sustainable development · Recycling · Waste-to-energy · Environmental impact

---

Y. A. Hajam (✉) · R. Kumar · Neelam
Department of Life Sciences and Allied Health Sciences, Sant Baba Bhag Singh University, Jalandhar, Punjab, India
e-mail: younismajeed64@mail.com

R. Kumar
Department of Biosciences, Himachal Pradesh University, Shimla, Himachal Pradesh, India

R. A. Bhat et al. (eds.), *Zero Waste Management Technologies*,
https://doi.org/10.1007/978-3-031-57275-3_7

1 Introduction

Clean environment is the basic requirement for the survival of the all living organisms (Bhat et al. 2021). With the advancement and development, new inventions and discoveries were being made during the last century to make human life more comfortable, relaxed and vigorous. In the era of green revolution, chemical fertilisers, preservatives, pesticides were introduced and used to support and increase crop production (Hajam et al., 2023; Unsworth et al., 2019). So, that sufficient food could be made available to increasing human population in the world. Further, automatic methods and use of machines replaced manual labour to minimise the labour costs. Tractors, harvesters, sprayers, and other tools replaced the involvement of human beings and animals. With time, the significance of organic products (manure, slurry and green compost) for the maintenance of soil fertility has become more or less negligible (Krauss et al., 2020).

2 Modernisation: Serious Cause of Waste Generation

Increase in global population, modernisation and technological advancement developed growing need of food as well as need of daily items (Bhat et al., 2012, 2014). Due to this generation of waste has also increased in our homes (Chalak et al., 2019). Various studies have evaluated that there is direct relation between amount of waste production and the economic population explosion, growth of the nation including urbanisation and industrialisation (Sarkodie et al., 2020). In developing countries, population in urban areas is increasing very intensely, it has been predicted that up to 2025 that two-third population may prefer to urban life (Bhat et al., 2018b; Sharma & Jain, 2020). In 2016, 2010 million tonnes of municipal solid waste has been generated in the world and it may increase up to 3400 MT up to 2050 if the problem remained unresolved (Kanwal et al., 2022). It indicates nearly 70% increase in production of MSW within 34 years (Khandelwal et al., 2019). According to the reports, around the world 242 MT of plastic waste has generated in 2016, which constituted 12% of MSW (Kibria et al., 2023). Plastics consist of organic and inorganic raw materials such as hydrogen, silicon, nitrogen, carbon and oxygen, which are generally take out from natural gas, oil, and coal (Daligaux et al., 2021). In the past, natural polymers are replaced by synthetic polymers and consequently it has now become an important part of our day to day life (Lett et al., 2021). Progressively, with the rise in the robustness, durability, stability, and strength, plastics have become unaffected to various environmental variables. It has been studied that 10% of waste generated in our homes constitutes plastic waste, which is mainly dumped in landfills (Ibrahim et al., 2023). Nonetheless, it has been studied that 60–80% of the waste were found on beaches, floats in ocean or sealed in plastic (Barnes & Milner, 2005; Derraik, 2002). Mostly plastics are non-biodegradable and as a result have becomes accumulated into the environment in extensive quantities due to the inappropriate management of waste

(Sharma & Dhingra, 2016). The increasing consumption of plastic waste is polluting oceans due to this many developing countries does not have proper plastic waste management systems (Spranghers et al., 2017). The wildlife species were adversely affected by the huge quantities of synthetic plastics as young animals occasionally get intertwined in the plastic which results in major injury to them. Additionally, in birds and animal ingestion of plastic debris which persists in their digestive system causes several problems including reduced "feeding stimulus, reduced level of steroid hormones, gastric enzymes", and also leads to GIT obstruction (Ahmed et al., 2018). Moreover, the non-biodegradable nature of the synthetic plastics and their long-term accumulation in the soil leads to soil fertility loss which results in decreased growth and development of plant. These synthetic plastics also cause harmful effect in the aquatic flora and fauna as their amount has raised six times higher than the planktons present in the water bodies (Comăniță et al., 2016). When aquatic animals consume manmade plastics and its becomes stored in their body (bio-magnification) which later on produce substantial harm effects to the human beings (Orose et al., 2023).

3 Impact of Waste Management

Improper disposal and management of municipal solid waste (MSW) causes pollution in air, water, and soil. Non-selective dumping of wastes pollutes surface as well as ground water (Bhat et al., 2021). MSW blocks the drains, produces standing water, which favours the breeding of insect during rainy season. Improper burning and incineration of municipal solid waste (MSW) leads to the air pollution (Sreenivasan, 2020). Decomposition of organic wastes in landfills produces greenhouse gases while untreated leach ate pollutes surrounding area mainly soil and water bodies (Ghosh et al., 2023). Improper municipal solid waste management (MSWM) creates serious health (cholera, diarrhoea, typhoid fever) and safety issues. According to Gajalakshmi and Abbasi (2008), the waste materials can be categorised in various ways such as on the basis of their source of origin (domestic, kitchen, market, agricultural, industrial, municipal, temple waste); on the basis of their consistency (gaseous, liquid or solid) or on the basis of their chemical nature (organic or inorganic). By using polluted water for drinking, bathing, food irrigation, etc., can also introduce individuals to various disease organisms and other contaminants. In India about 50 million tonnes of municipal waste produces yearly (Hazra & Goel, 2009; Saha et al., 2010; Sharholy et al., 2008). This huge amount of waste being has produced challenging issues for its proper management (Idris et al., 2004). In India about 960 million tonnes of solid waste is produced per year such as "industrial waste, municipal, mining, agricultural sector" (Pappu et al., 2007). In addition to this about 350 MT was produced because of the improper management that also generates different environmental and health issues (Sharholy et al., 2008).

Manufacturing and practice of chemical products, fertilisers, pesticides, preservatives, colours and dyes, threads and textiles, additives, carry bags, plastic utensils, packing stuff, processed foods, disposals, thermacol, pharmaceuticals, vitamins, transportation and communication technologies, health care and medical services made life relaxed and smooth. The major drawback of these synthetic chemicals includes environmental pollution (air, soil and water), emission of greenhouse gases, greenhouse effect, ozone layer depletion, global warming and global climatic changes (Yoro & Daramola, 2020). In addition to this, synthetic chemicals also cause disruption in natural equilibrium, depletion of ground water level, crisis of potable water, decreasing soil fertility, biodiversity loss, human health hazards, etc. (Mishra, 2023). In recent years due to increasing awareness about environmental problems, emphasis is given to take instant action and means for declining in environmental pollution, promotion of organic farming, adoption of green technologies, and sustainable agriculture.

4 Waste Management

Scientific management of wastes is major issue that society is facing today. Large amount of organic wastes is produced every day, mainly waste is thrown in open spaces (water bodies (rivers, streams or drainage system) or on the roads which contaminates the environment (Bhat et al., 2012). Most of these organic deposits are dumped on barren land or used as land fillings or burned (Yaashikaa et al., 2022). These organic wastes create major environmental and disposal problems. Therefore, maintaining healthy environment on biosphere and safe disposal of different types of wastes has become very important. However, unscientific disposal of waste in open places can cause offensive odour, ground water contamination and soil pollution in environment (De & Debnath, 2016). This can also adversely impact the human health (Pappu et al., 2007; Rathi, 2006; Sharholy et al., 2005, 2008). One of the best methods for safe and eco-friendly disposal and management of biodegradable waste is the "11 R formula Rethink, Refuse, Reduce, Reuse, Repurpose, Recycle, Regift, Repair, Refill, Rot and Repeat" (Fig. 1).

5 Production of Plastic Waste

Production of plastic-based items has changed the complete situation. Various synthetic or semi-synthetic organic materials are used to manufacture plastic that are soft and can be shaped into solid objects of different forms (Kowsari et al., 2023). Plastics consists of organic polymers having high molecular mass. Plastic constitute an estimated 10% of household waste, maximum of which is inclined in landfill. Despite that, "60–80% of the waste found on beaches and floating on ocean". Plastic materials are mostly non-biodegradable with a low mass density which makes them unsuitable for discarding in landfills (Mohanaprasadh et al., 2022). In developing countries use of plastic grows continuously. There is no report regarding the specific time period for degradation of plastic but it could be take hundreds or thousands

Fig. 1 11R strategy for waste management

of years to degrade completely. We are using plastic carry bags, utensils, clothing, packing material, pipes along with artificial food colours, additives, preservatives and medicines. According to European Union reports, that maximum utilisation and generation of plastic in developing countries has become the main cause for environmental degradation (Ajibade et al., 2021).

6 Green Revolution and Chemical Farming

The era of "Green Revolution and Chemical Farming" was necessity of the day. Man-made chemical products changed the natural environment. The practice of chemical farming is still continued. During the era of "Green Revolution" and "Chemical Farming" industrial development increased production of new chemicals used in every sphere of human life, both of these has changed the life style by improving living standard and makes life more comfortable. Currently farmers are using chemical fertilisers and pesticides in agricultural fields to increase the crop productivity. These chemicals induce hazardous effects on environment including depletion of biodiversity, ozone layer depletion, global climatic changes earthquakes, volcanoes, enhancement of incidences of land-slides, irregular pattern of rains, scarcity of drinking water, contamination of resources (water, air, food), deterioration of quantity and quality agricultural production, declining tendency of soil fertility. Moreover, these chemical fertilisers effects living organisms such as increase in diseases including diabetes, cardiovascular diseases, blood pressure, cardiac and neural infection, asthma, cancer, paralysis, etc. (Mukah et al., 2023). Toxicological assessment revealed that various synthetic chemicals induce harmful effects in laboratory animals (rats, rabbits, guinea

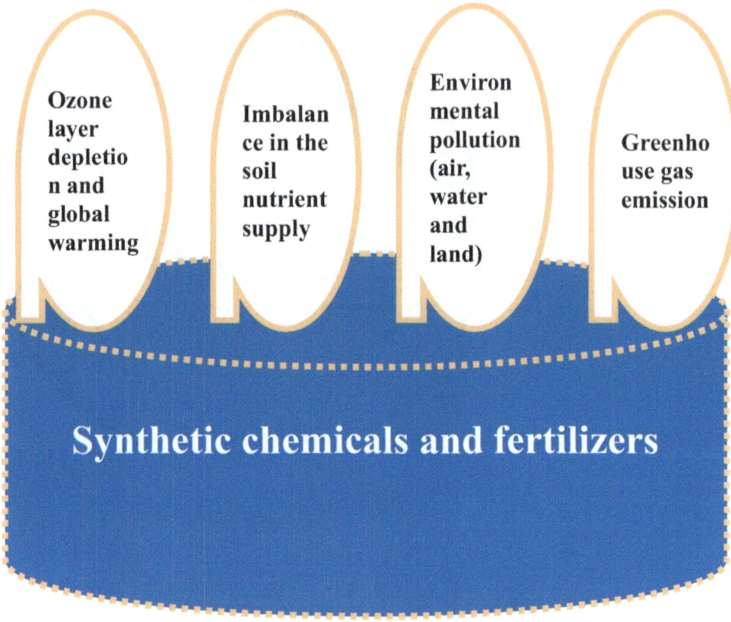

Fig. 2 Effect of synthetic chemicals and fertilisers on biosphere

pig) under laboratory conditions. The toxic effects of these chemicals have been evaluated through experimental studies. One experimental study demonstrated that synthetic chemicals and their breakdown products may become either more or less toxic as a result of their reactions or interactions with a range of other "chemicals in natural environments and consequence of transformations by organisms or exposure to natural light" (Fatta-Kassinos et al., 2011). Synthetic chemicals and their by-products develops various environmental issues (contamination of air, earth, water, and sea) (Fig. 2).

7 Toxicology of Synthetic Products

Various pesticides and pharmaceuticals persist in the environment, which indicates that these chemicals show slow degradation rate and are continuously released into the environment at excess quantities then their rate of degradation (Bose et al., 2021). When a molecule is resistant to effective biological breakdown, its effects on the environment linger long after its toxicity is known and, consequently, well after effective steps can be done to stop its creation and proliferation. The ecosystem's balance is harmed by overuse of chemicals and pesticides, which ultimately causes these chemicals to be bio-magnified in the food chain (Mondal, & Kole, 2020). Such long-term effects have been well documented for polychlorinated biphenyls (PCBs), which will continue to build up in fish, fish-eating birds, and humans for many decades (Beyer & Biziuk, 2009), and chlorofluorocarbons (CFCs), which will continue to deplete the stratospheric ozone layer for the foreseeable future (Adams & Halden, 2010). The most commonly used synthetic compounds, however, are not well sufficiently studied to determine their environmental persistence, pseudo-persistence, or long-term effects (Li et al., 2017; Muir & Howard, 2006). Chemical technology-based development has been shown to be hazardous for non-sustainable development (Lozano & García-Verdugo, 2023). Solid Waste Management (SWM) is currently one of the main concerns for contemporary society as a result of abandoned urbanisation, industrialisation, and changes in consumption patterns (Fadugba et al., 2022). Household, institutional, commercial, building and demolition, street sweeping, and sanitary garbage are all included in MSW (Municipal Solid garbage) (Ding et al., 2021). In addition to recyclables like paper, plastic, glass, and metals, municipal solid waste (MSW) also contains toxic materials like pesticides, paints, and old batteries, as well as compostable organic matter like fruit and vegetable peels and food waste (Sharholy et al., 2008) and soiled waste like sanitary napkins, cotton that has been stained with blood, and disposable syringes (Rajalakshmi et al., 2023).

8 Status of Waste Generation

8.1 International Status of MSW Generation

Global production of municipal solid waste causes several environmental impacts, comprising ocean plastic accumulation, greenhouse gas emissions, and nitrogen pollution. The production of waste is an obvious consequence of most of the anthropogenic activities or processes (Khanet al., 2022). According to Wilson and Velis (2015), 7–9 billion tonnes of garbage are produced annually over the world. According to Wilson and Velis (2015), municipal solid waste (MSW) is a specific type of garbage generated by households that may also include commercial and industrial waste. In 2016, MSW made up 2 billion tonnes of the total amount of waste produced. However, considering its environmental effects at the local, national, and international levels, it merits special consideration (Kaza et al., 2018; Wilson & Velis, 2015). However, according to Kaza et al. (2018), 70% of the world's trash now is disposed of in landfills and dumps, which are collectively referred to as Solid trash Disposal Systems (SWDS). It is challenging to quantify global estimates of waste-related pollution, in part because of the dearth of reliable data.

9 National Status of MSW Waste Generation

According to UNDESA (2016) and Jolliet et al. (2004), India is experiencing the emergence of densely populated megacities with rapid environmental degradation and health risks for city residents (Galimova et al., 2022). According to Sharholy et al. (2006), the amount of solid trash produced in Indian cities increased from 6 million tonnes in 1947 to 48 million tonnes in 1997 at an annual growth rate of 4.25%. By 2047, this amount is projected to reach 300 million tonnes. The amount of waste produced per person has increased annually at a rate of 1% to 1.33% over the past few decades (Shekdar, 1999). If this pace of growth persists, India's trash production would likely increase from less than 40,000 tonnes annually to over 125,000 tonnes by 2030 (Bhat et al., 2018a, 2018b). India's condition is concerning because only 12.45% of the total garbage collected is handled properly, and the remainder is dumped in open pits (Dutta & Jinsart, 2020). Municipal solid waste (MSW) created in India is notably different from MSW generated in Western countries in terms of both composition and quantity (Jalan & Srivastava, 1995), especially with regard to hazardous features. According to Sharholy et al. (2008), a significant portion of MSW in urban areas is made up of biodegradable materials. It has been noted that while the total generation of MSW is dependent on population size, the physical and chemical makeup of MSW is affected by a number of variables, including dietary preferences, style of living, intensity of commercial activity, and seasons. The use of paper, plastic, and metal is at its pinnacle, which makes disposal more challenging. Municipalities have not been able to gather and get rid of the massive amount of rubbish being produced. Since approximately 90% of waste is currently disposed of by open dumping, waste management and disposal is a critical problem that India is currently experiencing. Composting, landfilling, and incineration are some often employed techniques for managing trash (Malav et al., 2020).

10 Available Methods for Waste Management

10.1 Incineration

Today, there are many methods for getting rid of waste, including incineration, which involves burning waste materials and produces ash residue and air emissions. However, these garbage incinerators actually create more waste than they remove. According to Narayana (2009), an incinerator really transforms the original waste materials into a variety of new forms, including ash, air emissions, and liquid discharge. The treatment of these new waste forms is far more challenging than the initial raw garbage. The world is experiencing an increase in dispute over incineration because to the restriction on certain technologies, particularly in the USA, Europe, and Japan. Due to dwindling sales and popularity, the incineration businesses in these nations are searching for new markets. In poorer nations, the trash is primarily organic and has high moisture content, making incineration a subpar solution (Naworio, 2022). A cancer-causing agent has been identified in the air emissions from garbage incinerators due to the presence of heavy metals, dioxins, and other volatile organic compounds. Ash is a byproduct of incineration; landfill ash deposits may eventually allow its pollutants to seep into soil and water. Using incinerators has lower economic benefits but necessitates a significant financial expenditure. Medical and quarantine waste should not be disposed of safely by incinerating it. According to Sangkham (2020), just 15% of medical waste is infectious. Although incineration definitely eliminates infections, it also eliminates the materials on which the germs are present, such as metal, glass, paper, and plastic, which transforms a potentially dangerous biological risk into a confusing array of biochemical effects (Vesley, 1999). In India, it is illegal to dispose of waste through incineration, although garbage must be disposed of in a way that has a low calorific value. The technology for incineration, however, is unavailable domestically and import alternatives are exceedingly expensive.

11 Landfilling

A landfill is a location where rubbish is placed after being carefully chosen. Landfilling is primarily used to prevent contact between garbage and the nearby environment, particularly groundwater. Select locations where the disposed-off waste is crushed and a topsoil cover is daily applied to prevent annoyances are known as partially controlled or functional landfills. Wastes from all sectors—municipal, industrial, clinical, and hospital—are disposed of together. Municipal dump sites produce leachate, a contaminated liquid that includes concentrated harmful substances (Denison & Ruston, 1990). The main causes of leachate formation include rainfall, surface water, and groundwater entering the landfill site. In addition, it contaminates the ecosystems of both surface and groundwater in cities all over the world (Danthurebandara et al., 2012). Leachate from landfills has been shown to include a variety of contaminants, including PCBs, cadmium, chromium, toluene, mercury, lead, organophosphates, and dioxins. People who live close to landfills where landfill gas transfers with soil have reported developing several cancer types. Numerous studies have documented the occurrence of cancer, low birth weights, and congenital abnormalities in residents who live close to garbage sites. The majority of rubbish in India and other developing nations is dumped in yards or landfilled. All of these techniques—burning, dumping, land filling, etc.—are inadequate, inappropriate, and unsafe for both the environment and animals (Siddiqua et al., 2022). Additionally, the ecosystem is degrading due to the burning of fossil fuels, production of greenhouse gases like carbon dioxide (CO_2) and others, contaminating water supplies, and widespread use of chemical pesticides and fertilisers (Ahmad et al., 2022). A sustainable method of managing all of these wastes, treating them on-site with some biological agents and creating usable goods (Priya et al., 2021). These researches have shown that there are numerous ways to decompose different waste types. However, there hasn't been much research done on waste management strategies for temples.

12 Temple Waste: A Big Jam for Aquatic Pollution and Vector-Borne Diseases

The improper management of organic waste, especially temple garbage, is one of the main causes of environmental degradation. Due to higher trash production, a lack of landfills, dumping grounds, and water resources, as well as less vegetation, a greater usage of poly carry bags and disposals, and so-called contemporary lifestyles, waste problems in cities are much worse than in villages (Mihai et al., 2021). There are numerous temples in India that receive a lot of pilgrimage. The temples receive a large number of everyday visitors who come to pray and worship.

These visitors come and offer several items to the God including flowers, floral garlands, fruits, agarbati, dhoopbati, dhatura, prasaad, milk, coconuts, sweets, other edibles, cash money, etc. Stems, leaves and flowers are also used for decoration purposes in the temples. Ordinarily, edible items are taken out to be given to the worshippers as "Prasad" and eaten by the priests, temple personnel, and other employees. Waste includes non-consumable resources like plants, leaves, and floral garlands. Large quantities of waste generated from the temples. The environment around the temple area becomes polluted and it creates pollution, foul and unbearable gases which influence human health. On special occasions "Havan", "Yagns", "Pooja", "Bhagwat", "AkhandRamayan", "Raas Lila", "KrishnLeela", "Jagran", etc., and festivals like "Holi", "Dusehara", "Deepawali", "Maha Bhagwat Pooja", "Ganesh Chaturthi", "Ram Navmi", "Navratri" etc., the number strength of visitors increases enormously and variety of offerings including flowers also increases resulting in huge amount of waste. At several times temples are decorated with lots of flowers, leaves and cut banana trees (PhoolBunglow) and ChhappanBhog (~56 types of food recipes) is offered to the God. Flowers are also generously used at several other occasions and rituals by people of all religion. Customs of welcoming and felicitating the guests with garlanding and showering of loose flower petal on the path or on special guests are common practices. Wedding gardens, hotels, homes, and other locations where flowers are utilised for a variety of purposes also produce a substantial amount of floral waste (Pirani & Arafat, 2014). One of the largest sources of solid waste and a significant contributor to the production of municipal solid waste (MSW) is flower trash (Erdogan et al., 2020). At several places such waste is considered to be sacred by religious people and members of temple administration. Special arrangements are made for its collection, short term (3–7 days) transport and release in nearby rivers, ponds, wells, and water bodies. But usually temple waste is disposed-off like other wastes and no special ways and means are used for their disposal and dumping. These waste materials are thrown at the road side or dumped or washed off in drainage or nalla or they are simply burned to get rid of.

The existing methods for disposing of waste—dumping, releasing into the water, and burning—are unsafe and contribute to pollution, bad odours, an unsanitary environment, and other issues. Therefore, there is a growing need for attention to be paid to the clean, odourless, and disease-free disposal and management of temple trash, and practical, easy solutions must be created. Therefore, it was determined to investigate the potential for managing temple flower waste.

13 Traditional Methods for Recycling of Biodegradable Waste

13.1 Composting

Organic wastes are transformed into a uniform, useful, nutrient-rich, and environmentally beneficial material through the biological process of composting. According to Atalia et al. (2015), the conversion process takes place in aerobic (oxygen-containing) conditions with the appropriate moisture and temperature. In the course of composting, numerous intricate metabolic processes carried out by various microorganisms (bacteria, fungus, and Actinomycetes) consume available nitrogen (N) and carbon (C) to create their own biomass. Additionally, these bacteria create compost, a stable solid substrate with low levels of nitrogen and carbon (Azim et al., 2018). According to research by Antizar Ladislao et al. (2006), composting speeds up the biological, aerobic transformation of organic materials into compost. With adjustable strength dependent on temperature, moisture content, C/N ratio, and composition of fresh organic materials, several communities of microorganisms, mostly bacteria, fungus, and protozoa, are actively involved in the composting process (Hameed et al., 2021; Hassen et al., 2001; Mustin, 1987; Tuomela et al., 2000). The two main stages of the composting process are as follows: the first stage is characterised by microbial activity during the breakdown of the biodegradable material and the stability of the final organic product; the second stage is characterised by the conversion of a portion of the leftover organic material into humic substances (Adani et al., 1999). Given the significance of humic compounds in soil ecology, fertility, and structure, humification is thought to be a key component in increasing compost quality (Chen & Aviad, 1990). Composting is the greatest option for proper disposal of floral waste due to the features of floral waste, which are rich in organic compounds, macronutrients, and micronutrients. Thus, floral waste can be transformed into organic fertilisers that are valuable for agriculture, potentially reducing the need for non-renewable chemical fertilisers and pesticides. In order to assist the free passage of air during the composting process, bulking agents such wood shaving, dry leaves, newspaper, wheat bran, and pumice are typically added (Sharma et al., 2017). Composting floral waste is a much slower process than composting kitchen trash, according to Mulay et al. (2020). Therefore, earthworms, a biological agent, could be deployed to speed up the decomposition of the temple's floral waste. Because it is simple to use and can be done in a space that is specifically designated for it, composting is the most affordable and environmentally friendly method for managing organic waste. It is a natural method of treating organic waste that is now used with some changes. Humus that is rich in nutrients is created by biological processes that break down organic matter. The Latin verb "componere" (which means to put together) is the root of the English word "compost." Composting is the ideal method for handling municipal solid waste, as evidenced (Waqas et al., 2023). With regard to composting's advantages, it can be utilised to convert organic material into valuable products by utilising the degradative properties of both macro- and microorganisms present in the contaminated matrix. It is a common technique for turning organic waste into organic manure (compost), which recycles minerals including nitrogen (N), phosphorus (P), and potassium (K) that may be used for farming (Irshad et al., 2013). As a result, compost is a successful rival in the fertiliser market. By eradicating harmful organisms found in garbage, the thermophilic conditions associated with composting allow for complete hygienisation of the compost (Kulikowska, 2016). Humification (the creation of humic acid and other humic compounds) and the decomposition of organic waste are the outcomes of the composting process. Therefore, it plays a key function as a stabiliser (immobilises metals in the soil) if it is utilised as a bioremediation alternative. Due to the presence of humic compounds that may be employed in the treatment of soil contaminated with heavy metals, it is also utilised as a washing agent (Kulikowska et al., 2015).Composting is becoming more and more popular as an ethical approach to get rid of and use organic waste. Recent research has shown that commercial inorganic fertilisers cannot compete with the quality of compost made from organic wastes (Mupambwa & Mnkeni, 2018). Composts can take the role of soil conditioners in order to encourage humus production, which is a benefit that cannot be artificially produced, according to Hermann et al. (2011). The biotic community, notably biodegrading microbes and associated soil-dwelling animals like earthworms, have undergone radical alterations as a result of the accumulation of significant amounts of toxicants, pollutants, plastic polymers, and residues (Fan et al., 2022).

14 Role of Earthworms in the Recycling of Biodegradable Waste

14.1 Earthworms

Because of their capacity to cycle different nutrients in soil and modify its structure, earthworms are regarded as ecosystem engineers (Paramdeep et al., 2022). Earthworms actively encourage the decomposition of organic matter, the mineralisation of nitrogen (N), and water infiltration, all of which have a significant impact on the soil's physical characteristics (Hättenschwiler & Gasser, 2005). These worms are crucial to the development of the services provided by the soil ecosystem. The term "drilosphere" refers to the area of soil that is directly impacted by earthworms (Bouché, 1977; Lavelle, 2002). The earthworm community, which is an important functional component of the soil, develops various structures such middens, burrows, tunnels, and casts. Additionally, through creating organo-minerals, these building processes create a conduit for the stabilisation of soil organic carbon as well as an input of organic matter to the soil (Deeb et al., 2017). Thus, earthworms are known as biochemical engineers (Lavelle et al., 2016) in addition to modifying soil structure because they have a substantial impact on soil organic matter and microorganisms in their gut, casts, and drilosphere. Since ancient times, earthworms have been used in agriculture and to benefit human health, attracting the interest of philosophers and naturalists all around the world. Earthworms are known to exist in about 3200 different species worldwide; on the Indian subcontinent, they are represented by 500 different species that fall under 67 different genera. From various habitats, 374 of these species have been documented in India (Senapati & Sahu, 1993). Sir Charles Darwin's vast work on earthworms greatly influenced Dr. Thomas J. Barrett in (1959), and he tried to employ them to transform his arid, dry fields into lush, fertile farms. He referred to earthworms as "Earth master Systems" and suggested introducing particular earthworm species to underdeveloped agricultural land. He regarded earthworms as "Humus Factories of Manure" as well. Barrett (1959) called earthworms the "Greatest Servant of Nature" in his book "Harnessing the Earthworms" and emphasised their ability to create compost, which is dark-coloured fertile soil. The idea to raise some earthworm species on a large scale was initially put forth by him. Later, other researchers (Edwards & Heath, 1963; Heungens, 1969) showed the importance of earthworms in the recycling of organic materials (leaf litter). According to Macfadyen (1963), the idea of using soil organisms, such as earthworms, to process vast amounts of waste biomass as well as the potential and range of vermin-stabilisation of organic waste have been explored and analysed.

According to Stockli (1949), earthworm castings have a significant impact on boosting microorganism development, soil fertility, and plant productivity. Earthworms and microbes should have a symbiotic connection for the breakdown of organic matter, which aids in the realisation of mineral nutrients in the soil. Cast has a greater microbial flora as a result than the nearby soil (Parle, 1963). The intestines and castings of earthworms contain nitrogen-fixing bacteria (French et al., 1976; Satchell, 1983). Yeates (1981) claimed that while certain microbe species are eliminated when they transit through the gut, others multiply. In order to turn garbage into compost, earthworms and some kinds of bacteria "Hand-in-Hand" work together. According to Pandit et al., the biology of earthworm species that are good for composting has been reported (Pandit et al., 2012). According to research (Appelhof, 1980; Hartenstein et al., 1979; Senapati & Dash, 1982), earthworms may break down industrial wastes (paper, wool, and food) as well as home waste, city, refuge, and sewage sludge. Earthworms' potential as a treatment or management tool for various organic waste streams has been studied by a number of researchers (Chauhan et al., 2010; Dandotiya & Agrawal, 2014; Degefe et al., 2016; Hand et al., 1988; Kalamdhad et al., 2012; Mehta & Karnwal, 2013). Earthworms consume sewage sludge, solids from water wastes, processed paper industry waste, brewery waste, and coffee pulp

(Sinha et al., 2010a, 2010b), among other materials (Butt, 1993). Epigeic earthworms efficiently degrades the organic wastes produced by various industries has been including the dairy industry (Gratelly et al., 1996), the wood and chip industry (Maboeta & Rensburg, 2003), the winery and distillery industry (Nogales et al., 2005), the paper and pulp industry (Kumar & Jagadeesan, 2010). However, because to their great tolerance for ecological variables including temperature, pH, and moisture content, *Eisenia fetida* is the most popular and beneficial species of earthworms utilised for vermicomposting of vegetable waste (Suthar, 2009). Using earthworms to break down various kinds of organic wastes has been tried and studied by a number of researchers (Aalok et al., 2009; Allevi et al., 1987; Shanthi et al., 1993). In a medium including horse dung and activated sludge, Kumar et al. found that Eisenia fetida reached its maximum weight and thrived best at temperatures between 20 °C and 29 °C and moisture levels between 70 and 85%. According to research by Ranganathan and Vinotha (1998), the digestive epithelium of earthworms' simple, straight tubular gut secretes a number of enzymes, including cellulase, amylase, invertase, protease, and phosphatase. A study found that some species, including *Eisenia foetida, Dendrobaenavenata,* and *Lumbricusrubellus* from temperate regions and *Eudrillus eugeniae* and *Perionyx excavates* from the tropics, are highly useful in the breakdown of organic wastes. The ideal temperature and moisture levels for *E. fetida* growth in various animal and vegetable wastes are 25–30 °C and 75–90%, respectively (Mashologu, 2016). *Eisenia fetida* can metabolise 75% of their body weight per day when fed bio-solids as a feed substrate. Earthworms have been used as food and therapy for a number of human diseases since ancient times (about 4000 years, 2600 B.C.), including Japan, Vietnam, India, Korea, China, Cambodia, Myanmar (Burma), and Iran. They can be utilised in vermiculture, vermicomposting, vermin wash, and organic waste management. It is now appropriate for India to consider vermin technology commercially in order to support organic farming.

15 Earthworms: Friends of the Farmers

Since ancient times, earthworms have been regarded as "friends of the farmer." They are phylum Annelida members of the invertebrate animal kingdom. There are roughly 1800 species of earthworms, which are divided into five groups and have a global distribution (Talavera et al., 2020). The family Lumbricidae, which has roughly 220 species, is home to the majority of worms in North America, Europe, and Asia. According to Tripathi and Prasad (2005), the Indian subcontinent is home to 509 different species of earthworms, and 374 of those have been identified in India alone from a variety of habitats. One of the world's most useful organisms for humans is the earthworm. They live in soil and make tunnels and burrows in soil, consume soil mixed with organic matter and release their excreta or casts. They are capable to influence physical, chemical, and biological properties of soil to make it more productive. They turn soil up and down and enhance porosity in the soil which helps in passage for water and air, easy and prolific growth of roots of the plants and trees. This action also supports activities of aerobic microbial community of the soil which is necessary to convert organic inputs into compost. Earthworms cooperate with microorganisms mainly bacteria, fungi, and ciliates and play a vital role in soil biology, destroy pathogens, biological waste inputs into useful bio-fertilisers, enzymes, and growth factors to maintain soil fertility (Paramdeep et al., 2022). Earthworms and these microbes work in harmony and help one another. In traditional agriculture, when fertilisers are not utilised, such a condition has been successful. Earthworms are crucial to the complicated terrestrial ecosystem's nutrient cycling. According to how they affect the soil compartment and the microbial communities that live there, earthworms can be divided into three primary functional or ecological groupings (Thakuria et al., 2010). These include (i) epigeic earthworms, which live on the soil's surface and consume litter; (ii) endogeic earthworms, which live underground and create horizontal tunnels while consuming mineral soil and partially decomposed material; and (iii) anecic earthworms, which create permanent vertical burrows and consume litter that they drag into the burrows to be pre-decomposed by microorganisms while leaving casts at the burrow entrance (Bouché, 1977; Lavelle, 1983; Lee, 1985).

Worms, according to Aristotle, are the earth's intestines. There are several different species of earthworms, including *Eisenia fetida, Eudrilus eugeniae, Perionyxex cavatus*, and *Lumbricus rubellus* (Mainoo et al., 2009; Suthar, 2009). The majority of these species are employed in vermicomposting. Although due to their high tolerance to environmental variables like temperature, pH, and moisture content, *Eisenia fetida* is the most suitable and frequently employed species of earthworms for the purpose of vermicomposting of temple floral waste (Suthar, 2009). It is easy, cheap, and energy-efficient to use them to transform organic waste that is decaying into fertiliser that is rich in nutrients. It has been well known that earthworms can eat a variety of organic wastes, including sewage sludge, animal waste, crop residues, and industrial waste (Chan & Griffiths, 1988; Hartenstein & Bisesi, 1989). According to reports, high-quality vermicompost can be produced using a unique breed of epigeic earthworms. Microorganisms breakdown organic substances, which include carbohydrates, lipids, cellulose, and lignin, in both composting and vermicomposting (Zhou et al., 2023). While lignin is more resistant to breakdown than carbohydrates, they are both easily broken down. Microorganisms that are aerobic require oxygen, water, and nutrients for their internal metabolism. Temperature rises as a result of heat being released from the composting mass due to microbial activity. The mesophilic phase gives way to the thermophilic phase, which is followed by the return to the mesophilic phase. The microbial population changes during these transitions, which has an impact on how quickly organic waste decomposes. In the traditional way of farming, composting organic wastes on-site was given enough consideration to maintain soil fertility (Onwosi et al., 2017). However, as artificial fertilisers and pesticides developed, earthworms lost some of their significance in the effort to increase food production and meet the daily demands of the expanding human population (Chávez-Dulanto et al., 2021).

16 Vermitechnology: An Eco-Friendly Way to Recycle Temple Waste

The management of garbage and vermicomposting have been the subject of extensive investigation (Ducasse et al., 2022). Vermitechnology has been seen to have an impact on the overall, sustainable growth of human society. It involves employing earthworms and bacteria to biodegrade or stabilise organic waste (either naturally occurring or anthropogenically produced) (Garg & Gupta, 2011; Hand et al., 1988; Mainoo et al., 2009; Suthar, 2007).The effects of population density on the growth and production of E. fetida during the vermicomposting of activated sludge have been investigated (Hait & Tare, 2011). Vermicomposting facilities are currently available for industrial and home use in numerous nations, including the USA, Canada, Italy, and Japan. Initially developed in Ontario (Canada) in 1970, vermicomposting presently processes roughly 75 tonnes of garbage every week.

In Japan, there are also close to 3000 vermicomposting facilities with a monthly waste handling capacity of 5 to 50 tonnes. According to some researchers, it may be possible to compost a variety of animal and plant wastes utilising *Eisenia foetida, Eudrilus eugeniae*, and *Perionyx excavatus* under semi-natural circumstances (Kale et al., 1982). Methane gas concentration ranged from 60 to 72%. People's interest in using organic manures has been piqued by the harmful consequences of chemical fertilisers. In trays, wooden boxes, plastic crates, cement tanks, etc., Kale et al. (1982) and Parlekar et al. (1992) researched the laboratory and semi-natural conditions of waste decomposition employing several species of earthworms.

Vermicomposting is regarded as a sustainable supply of plant growth hormones, macro and micronutrients, suppressive microorganisms and enzymes, etc. (Kale and Karmegam, 2010), which aids in stimulating microbial population and holding onto nutrients for longer periods of time (Ndegwa and Thompson, 2001). Auxins, gibberellins, and cytokinins are plant growth regulators found in vermin compost, together with humic acids, which are responsible for improved plant development and crop output (Atiyeh et al., 2002). However, these microscopic microbes—such as fungi, bacteria, Actinomycetes, and others—form these plant growth-regulating substances (Pathma & Sakthivel, 2012). Earthworm population development benefits from physical characteristics such aeration, drainage, high porosity, water holding capacity, and microbial activity (Atiyeh et al., 2002; Suthar, 2009). Additionally, earthworms reduce the quantity of human pathogens, which is accomplished in conventional composting by increasing temperature. Animal manure, sewage sludge, and agro-industrial wastes have all been employed as sources for vermicomposting (Bernal et al., 1998; Gomez-Brandon et al., 2011; Paredes et al., 2002). However, the goal of the current study was to use the vermicomposting process to monitor the deterioration of temple floral waste.

Vermicomposting is a straightforward biotechnological method of composting in which several kinds of earthworms are employed to speed up the conversion of waste and create a higher-quality final product. Earthworms and bacteria are used in the stabilisation and biodegradation of organic waste (Hand et al., 1988; Mainoo et al., 2009; Suthar, 2007). The microorganisms that break down and mineralise organic wastes, primarily bacteria, fungus, and ciliates, are the main ingredients in vermicomposting. The vermicomposting food web's most numerous and diversified component, microorganisms include thousands of distinct species (Ducasse et al., 2022). These related bacteria also create extracellular enzymes that aid in the breakdown of a variety of substances, including cellulose and phenolic compounds. It is a mesophilic technique that makes use of microorganisms and earthworms that are active between 10 and 32 degrees Celsius (not the surrounding air temperature, but the temperature inside the pile of moist organic material). Because the material passes through the earthworm's gut, which serves as a bioreactor and undergoes a considerable transformation, the process is quicker than composting. The resulting earthworm castings (worm manure) are fortified with pest-repelling properties as well as being high in microbial activity and plant growth regulators. Through the use of earthworms, a bio-transformation process that can turn trash into "gold" is carried out (Sinha et al., 2010a, 2010b). A relatively recent technique for composting called vermicomposting includes stabilising organic solid waste into earthworm castings (Ducasse et al., 2022).

The first stage in vermicomposting is selecting the proper kind of earthworm since it influences how quickly trash stabilises. Thus, it may be claimed that composting is a necessary stage in the process of vermicomposting, which turns trash into palatable food for earthworms. The medium hosting earthworms should have a balanced nutrition (C/N ratio), enough moisture (40%), and reasonable bulk density (porosity) to allow for simple air and water flow as well as the movement of earthworms. Earthworms efficiently complete the waste breakdown or composting process in the vermicomposting process, which is started by microorganisms (Ducasse et al., 2022). Vermicomposting is the method by which "Nature's black magic" is created, in which particular species of earthworms act as the genuine magicians (bio-managers) and the accompanying microorganisms as their employees. The "black gold" (vermicompost), a special and amazing organic soil amendment produced from a variety of bio-wastes, is the real product here. Earthworms are used in vermiculture biotechnology, also known as vermin technology, to produce useful products like vermicompost, vermin wash as bio-fertiliser, and worm biomass for use as nutrient-rich animal feed, as well as to monitor soil fertility and maintain environmental quality by keeping an eye on pollution. It falls within the area of agricultural, environmental, and rural biotechnology because it is largely focused on agriculture, the reduction of environmental pollution, and the growth of the rural sector. Many researchers have noted that vermicompost is enriched with nitrogen, phosphorus, and potassium (NPK) (Manyuchi et al., 2012; Quaik et al., 2012). Vermicompost is one of the richest soil conditioners useful for farming and gardening practises and is more than simply simple fertilisers. It restores the structure of the soil and improves its capacity to hold onto water. It supplies crucial nutrients that are available to plants for a long time as well as helpful microbial activities (Ducasse et al., 2022). Vermicomposting increases the productivity and disease and parasite resistance of plants. Vermicompost is proving to be a highly nutritious organic fertiliser that protects farm inputs and is a more potent growth promoter than traditional composts. It enhances the soil's physical, chemical, and biological characteristics and brings back the natural fertility that has been lost due to the long-term usage of chemical fertilisers (Ducasse et al., 2022).

17 Role of Trichoderma in the Recycling of Biodegradable Waste

A genus of filamentous fungi that lives in the soil and bears teleomorphs is called Trichoderma. It belongs to the Ascomycota division's Hypocreales order. According to several studies (Ahamed & Vermette, 2009; Azevedoet al., 2000; Chulalaksananukul, 2008; Contreras-Cornejo et al., 2009; Harman et al., 2004; Lorito et al., 2010), the genus Trichoderma is essentially highly different and has a variety of skills. Various plant-parasitic nematodes can be effectively controlled biologically by Trichoderma spp., a free-living soil-dwelling fungus (Jatala, 1986; Spiegel & Chet, 1998).

The compatibility of the fungal isolate, host cultivar, and soil substrate directly correlated with the growth and persistence of Trichoderma spp. in the soil. Significant biocontrol isolates can successfully multiply on newly created roots and swiftly survive in the environment (Sariah et al., 2005). They can also supply ongoing earnings while annual crops are being grown (Harman, 2000). Due to its ability to suppress disease and compost, trichostome is regarded as an efficient bio-control agent in agricultural settings. This agent promotes plant growth and boosts soil fertility (Contreras-Cornejo et al., 2009; Harman et al., 2004; Lorito et al., 2010). In order to prepare the phospho compost, Trichoderma viridae has been successfully using rice straw (Sreenivas & Narayanasamy, 2003). *Trichoderma viridae* cellulose has been found to biodegrade waste paper, according to a number of researchers (Van-Wyk & Mohulatsi, 2003). Trichoderma fungus has degrading properties, but it also functions as an efficient bio-pesticide by exerting an opposite influence on other fungi and microorganisms that cause numerous crop diseases (Freeman et al., 2004; Elad, 2000; Naseby et al., 2000; Srivastva, 2004). According to reports (Ezzi & Lynch, 2005; Harman et al., 2004; Kredics et al., 2001), the fungus Trichoderma is capable of cleaning up contaminated soil and water. Additionally, its application in the creation of biofilms in the field of nanotechnology has been proven (Vahabi et al., 2011). It is widely known that Trichoderma engages in mycoparasitism and antibiosis to target soil-borne plant pathogens (Harman et al., 2004; Howell, 2003; Lorito et al., 2010). It is an effective plant growth agent because of its rhizosphere competence (Rigot & Matsumura, 2002), capability to promote production of hormones in plants (Contreras-Cornejo et al., 2009; Harman et al., 2004; Lorito et al., 2010), liberation of nutrients from soil (Altomare et al., 1999; Harman et al., 2004), and boost development of root system (Contreras-Cornejo et al., 2009; Harman et al., 2004).

Numerous studies have shown how well microbes can tolerate and detoxify hazardous heavy metals, and among them, fungi are the most crucial and prevalent elements that have the ability to break down harmful compounds because they produce a significant amount of microbial biomass (Chaudhry et al., 2005). According to Lorito et al. (2010) and Harman et al. (2004), the genus Trichoderma contains fungi with high biodegradation capacities that are actively colonising the soil. Trichoderma is among the genetically varied strains with significant industrial and agricultural relevance. Trichoderma spp. is typically found in soils and on plant roots, harm fungi, and has high agrochemical resistance.

Additionally, according to Ezzi and Lynch and Harman et al. (2004), they exhibit remarkable resistance to a variety of toxicants, particularly heavy metals, tannery effluents, organometallic compounds, and harmful chemicals like cyanide (CN). As a result, this fungus genus may provide access to fresh genetic resources for bioremediation of various harmful contaminants. Genus Trichoderma is a good producer of numerous hydrolytic and industrially important enzymes, including production of cellulases from T. reesei (Ahamed & Vermette, 2009; Gupta et al., 2014), 1, 3 bglucanases from T. harzianum, T. koningii (Monteiro & Ulhoa, 2006; Noronha & Ulhoa, 2000), chitinases from T. aureoviridae and T. harzianum (Sandhya et al., 2004) and amylases from T. harzianum (Harman et al., 2004). Numerous filamentous organisms, including bacteria (Jin et al., 2011) and fungi (Neosartorya, Cunninghamellaelegans, Amorphoteca, Phanerochaetechrysosporium, and Talaromyces), are recognised for degrading polycyclic aromatic hydrocarbons (PAHs). On the other hand, nothing is known about Trichoderma spp.'s resistance to and ability to degrade polycyclic aromatic hydrocarbons (PAH). Through the use of in-vitro systems, several investigations have documented the tolerance of eleven Trichoderma isolates to naphthalene (NAPH), crude oil (COIL), and phenanthrene (PHE) (Argumedo-Delira et al., 2012). A high genetic diversity of Trichoderma with stress tolerance to diverse stimuli may be employed to support sustainable agriculture in severely polluted soil and environmental circumstances with the aid of molecular and proteomic approaches. Vermicomposting and Trichodermaviride inoculation, however, may be highly beneficial in managing temple trash in an environmentally acceptable manner by obtaining nutrient-rich manure (Singh, 2023).

18 Temple Waste: A Magical Source for Development of Value-Added Products

In India, people worship their gods in gurudwaras, mosques, and temples. The offerings that people make to God in temples—floral garlands, agarbatti, fruits, milk, prasad, dhatura, dhoopbati, coconut, sweets, etc.—are turned into garbage known as temple waste. About 70% of all solid waste is created from temples and is primarily organic in nature (Singh et al., 2018). According to Kumar et al. (2020), 300 million tonnes of floral waste are produced daily in India. Each temple's makeup differs, but temple garbage is a heterogeneous collection of various elements donated to God. The majority of the garbage is organic (both biodegradable and non-biodegradable). The contributions of visitors, such as floral garlands, milk, agarbatti, dhatura, belpatra, and coconut, among others, determine the makeup of temple garbage. The following collections of temple waste elements can be identified: (i) decomposing waste refuge (floral, fruits, agarbatti, and coconut); (ii) non-biodegradable materials, including textiles, Chola, plastic bags, glass frames, diya, and plastic strips; (iii) organic pollutants, such as those that contain cadmium (Cd), chromium (Cr), arsenic (As), lead (Pb), mercury (Hg), and others, which may have a negative effect on plant growth. People primarily present flowers and agarbatti to our revered God during the festive season in exchange for greater life survival. Chrysanthemum, China Rose, Jasmine, Marigold, and Rose are a few of the most common flowers that people typically gift to God. Flower garbage, which is different from other waste and dumped in water bodies, is thought to be sacred residue. In general, floral trash is disregarded and continuously harms the ecosystem and various waterways. These waste products emit an intolerably terrible odour and release a lot of gases, which lead to a variety of infectious diseases in people (Gupta et al., 2023).

But it was discovered that temple wastes are the primary cause of environmental contamination and a number of dangerous ailments. Incense burning is seen by all religions as a common spiritual practise for adoring the gods and a component of daily life (Jhala, 2000). There are almost two million temples in India. According to Dewangan (2016), burning incense is a significant source of both cancer-causing Polycyclic Aromatic Hydrocarbons (PAHs) and particulate matter. Due to the prolonged, sluggish, and incomplete combustion involved in this practise, smoke is continuously produced. Numerous investigations showed that those who were exposed to this smoke had a significant chance of developing acute throat and upper respiratory tract irritation. A WHO (World Health Organisation) assessment from 2005 states that 36% of the pesticides used by Floraverde crops are thought to be highly harmful. Moving flowers from one location to another creates a significant amount of pollution, and driving causes the release of numerous hazardous chemicals into the atmosphere. According to Joshi et al. (2009), 16% of all river pollution is attributed to floral waste. When floral waste begins to decompose, the excess organic content in the flowers increases the creation of algal blooms and causes eutrophication in water bodies. Additionally, it could affect the oxygen levels in the water bodies and result in the death of aquatic organisms. But the floral waste that temples produce is becoming a serious menace to both the environment and people. The organic makeup of temple flower waste makes it suitable for a variety of biological management techniques, such as vermicomposting, which can be used to manage waste instead of dumping it in landfills, dumping it in the open, or using other pricy and environmentally risky methods (Adhikary, 2012).

19 Value-Added Products Derived from Floral Wastes

Floral waste can be transformed into a variety of products with added value using the solid state fermentation method, including biofuels, bioethanol, organic acids, pigments, dyes, food items, bio-surfactant manufacture, incense sticks, compost, and more (Gupta et al., 2023).

20 Production of Biofuels, Bioethanol and Biogas

Fermentation biotechnologists employ a variety of bacteria as a tool to transform sugar into ethyl alcohol. The ethanol produced by submerged fermentation from Mahua flowers has enormous commercial advantages (Johar & Kumar, 2020). (Ranjitha et al., 2014) Temple flower waste may be a significant source for the creation of biogas. An approach to encourage the creation of electricity from the waste of temple flowers has been developed. In addition to conducting an experimental trial (under anaerobic conditions) to manufacture fuel from the mixture of floral and vegetable waste, Ranjitha and Vijyalakshmi did a study on the generation of fuel from floral and vegetable waste mixtures.

21 Organic Acids, Dyes, and Pigments

The main source of raw materials for the manufacturing of oxalic acid in India is sugarcane. The Mahua (*Madhu calongifolia*) flower's higher sugar content created an alternate potential for oxalic acid generation (Das et al., 2010). Oxalic acid is widely employed as a chelating agent and preservative. According to a study, Aurangabad temples' flower debris, particularly marigold and rose petals, has been utilised to make dyes (Karolia & Dilliwar, 2004). Because they are more environmentally friendly than synthetic dyes, biodegradable dyes are the best option. The Pashmina shawl is dyed using saffron flower petal extract (Raja et al., 2012). Hibiscus (Hibiscus rosa-sinensis) has reportedly been used as a natural dye for colouring textiles (Vankar & Shukla, 2011). The extraction of patuletin dye from floral wastes like marigold (Tageteserecta L.) and French marigold (Tagetespatula L.) and its usage in the textile industry as well as in antioxidant treatment have been described by a number of studies (Waghmode et al., 2018). The pigments that give flowers their colour are carotenoids, anthocyanins, and betalains (Grotewold, 2006). The antioxidant and radical-scavenging properties of betalains point to their potential utility in medicine (Gandía-Herrero et al., 2012; Escribano et al., 1998). Red (carthamidin) and yellow (carthamin), two safflower pigments, are utilised as dyes (textile colouring). According to Machewad et al. (2012), safflower petals contain 0.83% red and 30% yellow pigments. These pigments are widely used in printing, staining, cosmetics, natural food colouring, and beverages as additives. Additionally, ice cream and many other culinary products are prepared with carthamidin pigment. A significant source of carotenoids and lutein is the African marigold flower (Tageteserecta L.), which is also grown as a garden flower and cut flower. Marigold flowers are a strong source of lutein, a carotenoid pigment, and range in colour from yellow to orange-red. Lutein is currently gaining popularity as an active component utilised in food production and textile colouring (Bureau & Bushway, 1986). Due to this pigment's great colour value, its significance has increased. The use of Tagetuserecta, a waste flower, for dyeing cotton, wool, and silk on an industrial scale was reported by Vankar and Shanker (2011).

22 Food Products

Mahua (*Madhu calongifolia*) flower extract is frequently utilised in the culinary industry. According to Patel and Naik (2010), it is used to produce jams, jellies, biscuits, as well as various nutritious ingredients like vitamins, sugars, amino acids, organic acids, enzymes (protease), and other substances like tannins, betaine, crude pigments, etc. A natural source of xanthophylls, marigold flower extract is employed in a variety of culinary sectors.

23 Bio-Surfactant Production

The surface-active biomolecules known as bio-surfactants exhibit a range of beneficial properties, including potent antibacterial, antifungal, and antiviral activity. They function as anti-adhesive agents against a variety of infections and are also helpful as therapeutic and probiotic agents in the treatment of a variety of ailments (Gharaei-Fathabad, 2011). In 2018, Waghmode et al. investigated the generation of surface-active compounds by Microbispora sp. utilising *Madhucalatifolia* L. flower extract. The surface-active molecule produced has bio-surfactant properties. The glycoprotein bio-surfactant exhibits high antifungal, antibacterial, and antiviral action in addition to having good wetting properties.

24 Medicinal Value of Flowers

Flowers from the Mahua (*Madhu calatifolia* L.) plant are used to treat bronchitis and to increase nursing mothers' milk supply (Sunita & Sarojini, 2013). Calendula oil is a good source of massage oil and a value-added product made from the fusion of olive oil with dried calendula flowers. Lily (Lilium) flower is used to treat jaundice, pulmonary and gastrointestinal conditions, while Rhododendron arboretum flower juice is used to control high blood pressure. Passionflower (Passiflora) decreases anxiety, sleeplessness, and stress. The essential oil of chamomile is effective in treating a variety of skin conditions, including dermatitis, as well as cramping in the muscles. The presence of rose blossoms throughout the year in the majority of India creates a significant opportunity for commercial rose cultivation for the manufacturing of floral extracts. About 80% of rose blooms are used to make rose water, 10% to make ittar, 1% to 2% to make pankhuri, and the remainder to make gulkand. In the creation of lotions, ointments, creams, soaps, and perfumes, among other products, rose oil serves as the primary scent element. Gulkand, a concoction made from crumpled rose petals, is used as a tonic and medicine to treat a variety of ailments. Mountain arnica, or Arnica montana L., is a plant that has been used in traditional and homoeopathic medicine (Puhlmann et al., 1991). According to Lanigan (2001), it has antibacterial, antiphlogistic, anti-inflammatory, and analgesic effects. Liniments are medicinal products intended to treat pain and stress that contain plant extract (Lanigan, 2001). The anthocyanin pigments found in saffron petals belong to the parent class of chemicals known as flavonoids. Flavonoids and anthocyanins are examples of phenolic chemicals that are mostly associated with advantages to both human and animal health. Flower pigments with antioxidant qualities, such as carotenoid, anthocyanin, and betala, are helpful in preventing human cancer (Escribano et al., 1998; Johnson, 2002).

25 Incense Sticks, Dhoop Sticks, and Rose Water Using Floral Wastes

The most popular products made from temple garbage are incense and dhoop sticks. Incense sticks are usually made from flowers like roses, genda, and lilies. Rose water is another product made from roses. Daily, a large number of flowers are offered to the deity at the Chandrika Devi Temple in Lucknow's Kathwara Village. The women of this region used the flower debris to create a variety of value-added products, such as incense sticks, dhoop sticks, and other items, which they then sold to local traders in the village markets to earn money (Waghmode et al., 2018). This method of manufacturing incense sticks is both humane and environmentally friendly (Fig. 3).

Fig. 3 Conversion of floral waste into value-added products

References

Aalok, A., Soni, P., & Tripathi, A. K. (2009). Role of Earthworms in breakdown of different organic wastes into manure: A review. *Dynamic Soil, Dynamic Plant, 3*(2), 13–20.

Adams, D. E. C., & Halden, R. U. (2010). Fluorinated chemicals and the impacts of anthropogenic use. *Contaminants of Emerging Concern in the Environment: Ecological and Human Health Considerations* (pp. 539–560).

Adani, F., Genevini, P. L., Gasperi, F., & Tambone, F. (1999). Composting and humification. *Compost Science and Utilization, 7*(1), 24–33.

Adhikary, S. (2012). Vermicompost, the story of organic gold: A review. *Journal of Agricultural Sciences, 3*(7), 13.

Ahamed, A., & Vermette, P. (2009). Effect of culture medium composition on Trichoderma reesei's morphology and cellulase production. *Bioresource Technology, 100*(23), 5979–5987.

Ahmad, F., Saeed, Q., Shah, S. M. U., Gondal, M. A., & Mumtaz, S. (2022). Environmental sustainability: Challenges and approaches. *Natural Resources Conservation and Advances for Sustainability* (pp. 243–270).

Ahmed, W., Zhang, Q., Lobos, A., Senkbeil, J., Sadowsky, M. J., Harwood, V. J., Saeidi, N., Marinoni, O., & Ishii, S. (2018). Precipitation influences pathogenic bacteria and antibiotic resistance gene abundance in storm drain outfalls in coastal sub-tropical waters. *Environment International, 116*, 308–318.

Ajibade, F.O., Adelodun, B., Lasisi, K.H., Fadare, O.O., Ajibade, T.F., Nwogwu, N.A., Sulaymon, I.D., Ugya, A.Y., Wang, H.C., & Wang, A. (2021). Environmental pollution and their socioeconomic impacts. In *Microbe Mediated Remediation of Environmental Contaminants* (pp. 321–354). Woodhead Publishing.

Allevi, P., Anastasia, M., Ciuffreda, P., Fiecchi, A., & Scala, A. (1987). The first direct method for C-glucopyranosyl derivatization of 2, 3, 4, 6-tetra-O-benzyl-D-glucopyranose. *Journal of the Chemical Society, Chemical Communications, 16*, 1245–1246.

Altomare, C., Norvell, W. A., Björkman, T. H. O. M. A. S., & Harman, G. (1999). Solubilization of phosphates and micronutrients by the plant-growth-promoting and biocontrol fungus Trichoderma harzianum Rifai 1295–22. *Applied and Environmental Microbiology, 65*(7), 2926–2933.

Antizar-Ladislao, B., Lopez-Real, J., & Beck, A. J. (2006). Degradation of polycyclic aromatic hydrocarbons (PAHs) in an aged coal tar contaminated soil under in-vessel composting conditions. *Environmental Pollution, 141*(3), 459–468.

Appelhof, M. (1980). Vermicomposting on a household scale. In *Soil Biology as Related to Land use Practices: Proceedings Vlllntn. Colloque Soil. Zoologica* (pp. 157–160).

Argumedo-Delira, R., Alarcón, A., Ferrera-Cerrato, R., Almaraz, J. J., & Peña-Cabriales, J. J. (2012). Tolerance and growth of 11 Trichoderma strains to crude oil, naphthalene, phenanthrene and benzo [a] pyrene. *Journal of Environmental Management, 95*, S291–S299.

Atalia, K. R., Buha, D. M., Bhavsar, K. A., & Shah, N. K. (2015). A review on composting of municipal solid waste. *Journal of Environmental Science, Toxicology and Food Technology, 9*(5), 20–29.

Atiyeh, R. M., Arancon, N. Q., Edwards, C. A., & Metzger, J. D. (2002). The influence of earthworm-processed pig manure on the growth and productivity of marigolds. *Bioresource Technology, 81*(2), 103–108.

Azevedo, J. L., Maccheroni, W., Jr., Pereira, J. O., & De Araújo, W. L. (2000). Endophytic microorganisms: A review on insect control and recent advances on tropical plants. *Electronic Journal of Biotechnology, 3*(1), 15–16.

Azim, K., Soudi, B., Boukhari, S., Perissol, C., Roussos, S., & Thami Alami, I. (2018). Composting parameters and compost quality: A literature review. *Organic Agriculture, 8*, 141–158.

Barnes, D. K., & Milner, P. (2005). Drifting plastic and its consequences for sessile organism dispersal in the Atlantic ocean. *Marine Biology, 146*, 815–825.

Barrett, D. (1959). *Thomas J. Harnessing the Earthworm*. Boston.

Bernal, M. P., Sanchez-Monedero, M. A., Paredes, C., & Roig, A. (1998). Carbon mineralization from organic wastes at different composting stages during their incubation with soil. *Agriculture, Ecosystems and Environment, 69*(3), 175–189.

Beyer, A., & Biziuk, M. (2009). Environmental fate and global distribution of polychlorinated biphenyls. *Reviews of Environmental Contamination and Toxicology, 201*, 137–158.

Bhat, R. A., Dar, G. A., Jehangir, A., Bhat, B. A., & Yousuf, A. R. (2012). Municipal solid waste generation and present scenario of waste management during Yatra season in Pahalgam: A tourist health resort of Kashmir valley. *International Journal of Current Research, 4*, 004–009.

Bhat, R. A., Dervash, M. A., Mehmood, M. A., & Hakeem, K. R. (2018a). Municipal solid waste generation and its management, a growing threat to fragile ecosystem in Kashmir Himalaya. *American Journal of Environmental Sciences*. https://doi.org/10.3844/ajessp.2018

Bhat, R. A., Dar, S. A., Dar, D. A., & Dar, G. H. (2018b). Municipal Solid Waste generation and current scenario of its management in India. *International Journal of Advance Research in Science and Engineering, 7*(2), 419–431.

Bhat, R. A., Nazir, R., Ashraf, S., Ali, M., Bandh, S. A., & Kamili, A. N. (2014). Municipal solid waste generation rates and its management at Yusmarg forest ecosystem, a tourist resort in Kashmir. *Waste Management and Research, 32*, 165–169.

Bhat, R. A., Singh, D. V., Qadri, H., Dar, G. H., Dervash, M. A., Bhat, S. A., Unal, B. T., Ozturk, M., Hakeem, K. R., & Yousaf, B. (2021). Vulnerability of municipal solid waste: An emerging threat to aquatic ecosystems. *Chemosphere, 287*, 132223. https://doi.org/10.1016/j.chemosphere.2021.132223

Bose, S., Kumar, P. S., Vo, D. V. N., Rajamohan, N., & Saravanan, R. (2021). Microbial degradation of recalcitrant pesticides: A review. *Environmental Chemistry Letters, 19*, 3209–3228.

Bouché, M. B. (1977). Strategies lombriciennes. *Ecological Bulletins* (pp. 122–132).

Bureau, J. L., & Bushway, R. J. (1986). HPLC determination of carotenoids in fruits and vegetables in the United States. *Journal of Food Science, 51*(1), 128–130.

Butt, K. R. (1993). Utilisation of solid paper-mill sludge and spent brewery yeast as a feed for soil-dwelling earthworms. *Bioresource Technology, 44*(2), 105–107.

Chalak, A., Abiad, M. G., Diab, M., & Nasreddine, L. (2019). The determinants of household food waste generation and its associated caloric and nutrient losses: The case of Lebanon. *PLoS ONE, 14*(12), e0225789.

Chan, P. L., & Griffiths, D. A. (1988). The vermicomposting of pre-treated pig manure. *Biological Wastes, 24*(1), 57–69.

Chaudhry, Q., Blom-Zandstra, M., Gupta, S. K., & Joner, E. (2005). Utilising the synergy between plants and rhizosphere microorganisms to enhance breakdown of organic pollutants in the environment (15 pp). *Environmental Science and Pollution Research, 12*, 34–48.

Chauhan, A., Kumar, S., Singh, A. P., & Gupta, M. (2010). Vermicomposting of vegetable wastes with cowdung using three earthworm species Eisenia foetida, Eudrilus eugeniae and Perionyx excavatus. *Nature and Science, 8*(1), 34–42.

Chávez-Dulanto, P. N., Thiry, A. A., Glorio-Paulet, P., Vögler, O., & Carvalho, F. P. (2021). Increasing the impact of science and technology to provide more people with healthier and safer food. *Food and Energy Security, 10*(1), e259.

Chen, Y., & Aviad, T. (1990). Effects of humic substances on plant growth. *Humic Substances in Soil and Crop Sciences: Selected Readings* (pp. 161–186).

Chulalaksananukul, W. (2008). Construction of cellulase hyperproducers of Trichoderma reesei by genetic techniques. In *Sci Forum* (vol. 2008, p. 100).

Comăniţă, E. D., Hlihor, R. M., Ghinea, C., & Gavrilescu, M. (2016). Occurrence of plastic waste in the environment: ecological and health risks. *Environmental Engineering & Management Journal (EEMJ), 15*(3).

Contreras-Cornejo, H. A., Macías-Rodríguez, L., Cortés-Penagos, C., & López-Bucio, J. (2009). Trichoderma virens, a plant beneficial fungus, enhances biomass production and promotes lateral root growth through an auxin-dependent mechanism in Arabidopsis. *Plant Physiology, 149*(3), 1579–1592.

Daligaux, V., Richard, R., & Manero, M. H. (2021). Deactivation and regeneration of zeolite catalysts used in pyrolysis of plastic wastes—a process and analytical review. *Catalysts, 11*(7), 770.

Dandotiya, P., & Agrawal, O. P. (2014). Stabilization of vegetable market waste through vermicomposting. *International Journal of Science and Research, 3*(6), 50–55.

Danthurebandara, M., Van Passel, S., Nelen, D., Tielemans, Y., & Van Acker, K. (2012). Environmental and socio-economic impacts of landfills. *Linnaeus Eco-Tech, 2012*, 40–52.

Das, K., Tiwari, R. K. S., & Shrivastava, D. K. (2010). Techniques for evaluation of medicinal plant products as antimicrobial agent: Current methods and future trends. *Journal of Medicinal Plants Research, 4*(2), 104–111.

De, S., & Debnath, B. (2016). Prevalence of health hazards associated with solid waste disposal-A case study of Kolkata, India. *Procedia Environmental Sciences, 35*, 201–208.

Deeb, M., Desjardins, T., Podwojewski, P., Pando, A., Blouin, M., & Lerch, T. Z. (2017). Interactive effects of compost, plants and earthworms on the aggregations of constructed Technosols. *Geoderma, 305*, 305–313.

Degefe, G., Mengistou, S., & Mohammed, S. (2016). Physico chemical evaluation of coffee husk, wastes of enset (Enset ventricosum), vegetable and khat (Catha edulis) through vermicomposting employing an epigeic earthworm Dendrobaena veneta (Rosa, 1886). *African Journal of Biotechnology, 15*(20), 884–890.

Denison, R., & Ruston, J. (Eds.) (1990). *Recycling and Incineration: Evaluating the Choices*. Island press.

Derraik, J. G. (2002). The pollution of the marine environment by plastic debris: A review. *Marine Pollution Bulletin, 44*(9), 842–852.

Dewangan, S. (2016). *Study on Emission Estimation of Selected Carbonaceous Pollutants During Ritual Based Stuff Burning Activities*.

Ding, Y., Zhao, J., Liu, J. W., Zhou, J., Cheng, L., Zhao, J., Shao, Z., Iris, Ç., Pan, B., Li, X., & Hu, Z. T. (2021). A review of China's municipal solid waste (MSW) and comparison with international regions: Management and technologies in treatment and resource utilization. *Journal of Cleaner Production, 293*, 126144.

Ducasse, V., Capowiez, Y., & Peigné, J. (2022). Vermicomposting of municipal solid waste as a possible lever for the development of sustainable agriculture. A review. *Agronomy for Sustainable Development, 42*(5), 89.

Dutta, A., & Jinsart, W. (2020). Waste generation and management status in the fast-expanding Indian cities: A review. *Journal of the Air and Waste Management Association, 70*(5), 491–503.

Edwards, C. A., & Heath, G. W. (1963). The role of soil animals in breakdown of leaf material. *Soil Organisms* (pp.76–84).

Elad, Y. (2000). Biological control of foliar pathogens by means of Trichoderma harzianum and potential modes of action. *Crop Protection, 19*(8–10), 709–714.

Elvira, C., Sampedro, L., Dominguez, J., & Mato, S. (1997). Vermicomposting of wastewater sludge from paper-pulp industry with nitrogen rich materials. *Soil Biology and Biochemistry, 29*(3–4), 759–762.

Erdogan, I. G., Fosso-Kankeu, E., Ntwampe, S. K., Waanders, F. B., & Hoth, N. (2020). Management of metalliferous solid waste and its potential to contaminate groundwater: A case study of O'Kiep, Namaqualand South Africa. *Recovery of Byproducts from Acid Mine Drainage Treatment* (pp. 1–21).

Escribano, J., Pedreño, M. A., García-Carmona, F., & Muñoz, R. (1998). Characterization of the antiradical activity of betalains from Beta vulgaris L. roots. *Phytochemical Analysis: An International Journal of Plant Chemical and Biochemical Techniques, 9*(3), 124–127.

Ezzi, M. I., & Lynch, J. M. (2005). Biodegradation of cyanide by Trichoderma spp. and Fusarium spp. *Enzyme and Microbial Technology, 36*(7), 849–854.

Fadugba, G. O., Yusoff, M. S., Arogundade, S., Adam, N. H., Wang, L. K., & Wang, M. H. S. (2022). Sustainable solid waste management. *Solid Waste Engineering and Management* (Vol. 2, pp. 1–70). Springer International Publishing.

Fan, P., Yu, H., Xi, B., & Tan, W. (2022). A review on the occurrence and influence of biodegradable microplastics in soil ecosystems: Are biodegradable plastics substitute or threat? *Environment International, 163*, 107244.

Fatta-Kassinos, D., Hapeshi, E., Achilleos, A., Meric, S., Gros, M., Petrovic, M., & Barcelo, D. (2011). Existence of pharmaceutical compounds in tertiary treated urban wastewater that is utilized for reuse applications. *Water Resources Management, 25*, 1183–1193.

Freeman, S., Minz, D., Kolesnik, I., Barbul, O., Zveibil, A., Maymon, M., Nitzani, Y., Kirshner, B., Rav-David, D., Bilu, A., & Dag, A. (2004). Trichoderma biocontrol of Colletotrichum acutatum and Botrytis cinerea and survival in strawberry. *European Journal of Plant Pathology, 110*, 361–370.

French, V., Bryant, P. J., & Bryant, S. V. (1976). Pattern regulation in Epimorphic fields: Cells may make use of a polar coordinate system for assessing their positions in developing organs. *Science, 193*(4257), 969–981.

Gajalakshmi, S., & Abbasi, S. A. (2008). Solid waste management by composting: State of the art. *Critical Reviews in Environmental Science and Technology, 38*(5), 311–400.

Galimova, T., Ram, M., & Breyer, C. (2022). Mitigation of air pollution and corresponding impacts during a global energy transition towards 100% renewable energy system by 2050. *Energy Reports, 8*, 14124–14143.

Gandía-Herrero, F., Escribano, J., & García-Carmona, F. (2012). Purification and antiradical properties of the structural unit of betalains. *Journal of Natural Products, 75*(6), 1030–1036.

Garg, V. K., & Gupta, R. (2011). Optimization of cow dung spiked pre-consumer processing vegetable waste for vermicomposting using Eisenia fetida. *Ecotoxicology and Environmental Safety, 74*(1), 19–24.

Gharaei-Fathabad, E. (2011). Biosurfactants in pharmaceutical industry: A mini-review. *American Journal of Drug Discovery and Development, 1*(1), 58–69.

Ghosh, A., Kumar, S., & Das, J. (2023). Impact of leachate and landfill gas on the ecosystem and health: Research trends and the way forward towards sustainability. *Journal of Environmental Management, 336*, 117708.

Gomez-Brandon, M., Aira, M., Lores, M., & Domínguez, J. (2011). Epigeic earthworms exert a bottleneck effect on microbial communities through gut associated processes. *PLoS ONE, 6*(9), e24786.

Gratelly, P., Benitez, E. E. L. V. I. R. A., Elvira, C., Polo, A. & Nogales, R. (1996). Stabilization of sludges from a dairy processing plant using vermicomposting. In *Fertilizers and Environment:*

Proceedings of the International Symposium Fertilizers and Environment, Held in Salamanca, Spain, 26–29, September, 1994 (pp. 341–343). Springer Netherlands.

Grotewold, E. (2006). The genetics and biochemistry of floral pigments. *Annual Review of Plant Biology, 57*, 761–780.

Gupta, V. G., Schmoll, M., Herrera-Estrella, A., Upadhyay, R. S., Druzhinina, I. & Tuohy, M. (Eds.). (2014). *Biotechnology and Biology of Trichoderma*. Newnes.

Gupta, S., Tewari, S. K. & Pathak, S. (2023). Temple floral waste for various bio-products in India. In *Recent Trends in Solid Waste Management* (pp. 293–307). Elsevier.

Hait, S., & Tare, V. (2011). Optimizing vermistabilization of waste activated sludge using vermicompost as bulking material. *Waste Management, 31*(3), 502–511.

Hajam, Y. A., Kumar, R. & Kumar, A. (2023). Environmental waste management strategies and Vermi transformation for sustainable development. *Environmental Challenges*, 100747.

Hameed, M., Bhat, R. A., Pandit, B. A., Ramzan, S., Dijoo, Z. K., & Wani, M. A. (2021). Qualitative assessment of compost engendered from municipal solid waste and green waste by indexing method. *Journal of the Air and Waste Management Association*. https://doi.org/10.1080/10962247.2021.1959466

Hand, P., Hayes, W. A., Satchell, J. E. & Frankland, J. C. (1988). The vermicomposting of cow slurry, Earthworms in waste and environmental management. In C. A. Edwards, & E. F. Neuhauser (Eds.), SPB Academic Publishing, Netherlands. ISBN, 90, (pp. 5103–0177).

Harman, G. E. (2000). Myths and dogmas of biocontrol changes in perceptions derived from research on Trichoderma harzinum T-22. *Plant Disease, 84*(4), 377–393.

Harman, G. E., Howell, C. R., Viterbo, A., Chet, I., & Lorito, M. (2004). Trichoderma species—Opportunistic, avirulent plant symbionts. *Nature Reviews Microbiology, 2*(1), 43–56.

Hartenstein, R., & Bisesi, M. S. (1989). Use of earthworm biotechnology for the management of effluents from intensively housed livestock. *Outlook on Agriculture, 18*(2), 72–76.

Hartenstein, R., Neuhauser, E. F., & Kaplan, D. L. (1979). Reproductive potential of the earthworm Eisenia foetida. *Journal of Oecologia, 43*(3), 329–340.

Hassen, A., Belguith, K., Jedidi, N., Cherif, A., Cherif, M., & Boudabous, A. (2001). Microbial characterization during composting of municipal solid waste. *Bioresource Technology, 80*(3), 217–225.

Hättenschwiler, S., & Gasser, P. (2005). Soil animals alter plant litter diversity effects on decomposition. *Proceedings of the National Academy of Sciences, 102*(5), 1519–1524.

Hazra, T., & Goel, S. (2009). Solid waste management in Kolkata, India: Practices and challenges. *Waste Management, 29*(1), 470–478.

Hermann, B. G., Debeer, L., De Wilde, B., Blok, K., & Patel, M. K. (2011). To compost or not to compost: Carbon and energy footprints of biodegradable materials waste treatment. *Polymer Degradation and Stability, 96*(6), 1159–1171.

Heungens, A. (1969). The physical decomposition of pine litter by earthworms. *Plant and Soil*, 22–30.

Howell, C. R. (2003). Mechanisms employed by Trichoderma species in the biological control of plant diseases: The history and evolution of current concepts. *Plant Disease, 87*(1), 4–10.

Ibrahim, I. A., Rawindran, H., Alam, M. M., Leong, W. H., Sahrin, N. T., Ng, H. S., Chan, Y. J., Abdelfattah, E. A., Lim, J. W., Aliyu, U. S. A. & Khoo, K. S. (2023). Mitigating persistent organic pollutants from marine plastics through enhanced recycling: A review. *Environmental Research*, 117533.

Idris, Z., Orgéas, L., Geindreau, C., Bloch, J. F., & Auriault, J. L. (2004). Microstructural effects on the flow law of power-law fluids through fibrous media. *Modelling and Simulation in Materials Science and Engineering, 12*(5), 995.

Irshad, M., Eneji, A. E., Hussain, Z., & Ashraf, M. (2013). Chemical characterization of fresh and composted livestock manures. *Journal of Soil Science and Plant Nutrition, 13*(1), 115–121.

Jalan, R. K., & Srivastava, V. K. (1995). Priorities in municipal waste water management in India. *Wastes Management*, 26–7.

Jatala, P. (1986). Biological control of plant-parasitic nematodes. *Annual Review of Phytopathology, 24*(1), 453–489.
Jhala, J. (2000). Pūjā, pūjāri and Prabhu: Religious worship in the Hindu home. *Visual Anthropology, 13*(2), 103–128.
Jin, L., Hill, K. K., Filak, H., Mogan, J., Knowles, H., Zhang, B., Perraud, A. L., Cambier, J. C., & Lenz, L. L. (2011). MPYS is required for IFN response factor 3 activation and type I IFN production in the response of cultured phagocytes to bacterial second messengers cyclic-di-AMP and cyclic-di-GMP. *The Journal of Immunology, 187*(5), 2595–2601.
Johar, V., & Kumar, R. (2020). Mahua: A versatile Indian tree species. *Journal of Pharmacognosy and Phytochemistry, 9*(6), 1926–1931.
Johnson, M. J. (2002). The medication adherence model: A guide for assessing medication taking. *Research and Theory for Nursing Practice, 16*(3), 179–192.
Jolliet, O., Müller-Wenk, R., Bare, J., Brent, A., Goedkoop, M., Heijungs, R., Itsubo, N., Peña, C., Pennington, D., Potting, J., & Rebitzer, G. (2004). The LCIA midpoint-damage framework of the UNEP/SETAC life cycle initiative. *The International Journal of Life Cycle Assessment, 9*, 394–404.
Joshi, D. M., Kumar, A., & Agrawal, N. (2009). Assessment of the irrigation water quality of river Ganga in Haridwar district. *Rasayan Journal of Chemistry, 2*(2), 285–292.
Kalamdhad, A. S., Khwairakpam, M., & Kazmi, A. A. (2012). Drum composting of municipal solid waste. *Environmental Technology, 33*(3), 299–306.
Kale, R. D., & Karmegam, N. (2010). The role of earthworms in tropics with emphasis on Indian ecosystems. *Applied and Environmental Soil Science, 2010.*
Kale, R. D., Bano, K., & Krishnamoorthy, R. V. (1982). Potential of Perionyxexcavatus for utilizing organic wastes. *Pascal Francis, 23*(6), 419–425.
Kanwal, F., & Torriero, A. A. (2022). Biohydrogen—A green fuel for sustainable energy solutions. *Energies, 15*(20), 7783.
Karolia, A., & Dilliwar, S. (2004). Natural yellow dyes from marigold flowers for leather. *Colourage, 51*, 31–38.
Kaza, S., Yao, L., Bhada-Tata, P., & Van Woerden, F. (2018). *What a Waste 2.0: A Global Snapshot of Solid Waste Management to 2050*. World Bank Publications.
Khan, S., Anjum, R., Raza, S. T., Bazai, N. A., & Ihtisham, M. (2022). Technologies for municipal solid waste management: Current status, challenges, and future perspectives. *Chemosphere, 288*, 132403.
Khandelwal, H., Dhar, H., Thalla, A. K., & Kumar, S. (2019). Application of life cycle assessment in municipal solid waste management: A worldwide critical review. *Journal of Cleaner Production, 209*, 630–654.
Kibria, M. G., Masuk, N. I., Safayet, R., Nguyen, H. Q., & Mourshed, M. (2023). Plastic waste: Challenges and opportunities to mitigate pollution and effective management. *International Journal of Environmental Research, 17*(1), 20.
Kowsari, E., Ramakrishna, S., Gheibi, M., & Chinnappan, A. (2023). Marine plastics, circular economy, and artificial intelligence: A comprehensive review of challenges, solutions, and policies. *Journal of Environmental Management, 345*, 118591.
Krauss, M., Berner, A., Perrochet, F., Frei, R., Niggli, U., & Mäder, P. (2020). Enhanced soil quality with reduced tillage and solid manures in organic farming—A synthesis of 15 years. *Scientific Reports, 10*(1), 4403.
Kredics, L., Antal, Z., Manczinger, L., & Nagy, E. (2001). Breeding of mycoparasitic Trichoderma strains for heavy metal resistance. *Letters in Applied Microbiology, 33*(2), 112–116.
Kulikowska, D. (2016). Kinetics of organic matter removal and humification progress during sewage sludge composting. *Waste Management, 49*, 196–203.
Kulikowska, D., Gusiatin, Z. M., Bułkowska, K., & Klik, B. (2015). Feasibility of using humic substances from compost to remove heavy metals (Cd, Cu, Ni, Pb, Zn) from contaminated soil aged for different periods of time. *Journal of Hazardous Materials, 300*, 882–891.

Kumar, D. S., & Jagadeesan, L. (2010). Comparative decomposting efficacy of three earthworm species on papermill solid waste and macronutrient analysis. *Environment and Ecology, 28*(1), 212–215.

Kumar, V., Kumari, S., & Kumar, P. (2020). Management and sustainable energy production using flower waste generated from temples. *Environmental Degradation: Causes and Remediation Strategies, 1*, 154.

Lanigan, R. S. (2001). Final report on the safety assessment of Cocoyl Sarcosine, Lauroyl Sarcosine, Myristoyl Sarcosine, Oleoyl Sarcosine, Stearoyl Sarcosine, Sodium Cocoyl Sarcosinate, Sodium Lauroyl Sarcosinate, Sodium Myristoyl Sarcosinate, Ammonium Cocoyl Sarcosinate, and Ammonium Lauroyl Sarcosinate. *International Journal of Toxicology, 20*, 1–14.

Lavelle, P. (1983). The structure of earthworm communities. In *Earthworm Ecology*, pp. 449–466.

Lavelle, P. (2002). Functional domains in soils. *Ecological Research, 17*(4), 441–450.

Lee, K. E. (1985). *Earthworms: Their Ecology and Relationships with Soils and Land Use*. Academic Press Inc.

Lett, J. A., Sagadevan, S., Fatimah, I., Hoque, M. E., Lokanathan, Y., Léonard, E., Alshahateet, S. F., Schirhagl, R., & Oh, W. C. (2021). Recent advances in natural polymer-based hydroxyapatite scaffolds: Properties and applications. *European Polymer Journal, 148*, 110360.

Li, H., Cheng, F., Wei, Y., Lydy, M. J., & You, J. (2017). Global occurrence of pyrethroid insecticides in sediment and the associated toxicological effects on benthic invertebrates: An overview. *Journal of Hazardous Materials, 324*, 258–271.

Lorito, M., Woo, S. L., Harman, G. E., & Monte, E. (2010). Translational research on Trichoderma: From'omics to the field. *Annual Review of Phytopathology, 48*, 395–417.

Lozano, P., & García-Verdugo, E. (2023). From green to circular chemistry paved by biocatalysis. *Green Chemistry, 25*(18), 7041–7057.

Maboeta, M. S., & Van Rensburg, L. (2003). Vermicomposting of industrially produced woodchips and sewage sludge utilizing Eisenia fetida. *Ecotoxicology and Environmental Safety, 56*(2), 265–270.

Macfadyen, A. (1963). Heterotrophic productivity in the detritus food chain in soil. *International Congress of Zoology* (pp. 318–329).

Machewad, G. M., Ghatge, P., Chappalwar, V., Jadhav, B. & Chappalwar, A. (2012). Studies on extraction of safflower pigments and its utilization in ice cream. *Journal of Food Processing and Technology, 3*(8).

Mainoo, N. O., Barrington, S., Whalen, J. K., & Sampedro, L. (2009). Pilot-scale vermicomposting of pineapple wastes with earthworms native to Accra, Ghana. *Bioresource Technology, 100*(23), 5872–5875.

Malav, L. C., Yadav, K. K., Gupta, N., Kumar, S., Sharma, G. K., Krishnan, S., Rezania, S., Kamyab, H., Pham, Q. B., Yadav, S., & Bhattacharyya, S. (2020). A review on municipal solid waste as a renewable source for waste-to-energy project in India: Current practices, challenges, and future opportunities. *Journal of Cleaner Production, 277*, 123227.

Manyuchi, M. M., Phiri, A., Chirinda, N., Muredzi, P., Govhaand, J., & Sengudzwa, T. (2012). Vermicompostingof waste corn pulp blended with cow dung manure using Eisenia fetida. *International Journal of Chemical and Molecular Engineering, 6*, 753–756.

Mashologu, B. (2016). *Chemical Characteriistics, Fertilizer Value and Escherichia coli and Salmonella Composition of Vermicomposted Sewage Sludge* (Doctoral dissertation).

Mehta, N., & Karnwal, A. (2013). Solid waste management with the help of vermicomposting and its applications in crop improvement. *Journal of Biology and Earth Sciences, 3*(1), 8–16.

Mihai, F. C., Gündoğdu, S., Markley, L. A., Olivelli, A., Khan, F. R., Gwinnett, C., Gutberlet, J., Reyna-Bensusan, N., Llanquileo-Melgarejo, P., Meidiana, C., & Elagroudy, S. (2021). Plastic pollution, waste management issues, and circular economy opportunities in rural communities. *Sustainability, 14*(1), 20.

Mishra, R. K. (2023). Fresh water availability and its global challenge. *British Journal of Multidisciplinary and Advanced Studies, 4*(3), 1–78.

Mohanaprasadh, S. V., Singh, P., & Bahukhandi, K. D. (2022). Disposal of non-biodegradable waste using eco-friendly methods. *Environmental Pollution and Natural Resource Management* (pp. 345–363). Springer International Publishing.

Mondal, R., & Kole, R. K. (2020). Monitoring and evaluation of pesticide residues in aquatic systems. In *Sustainable Agriculture Reviews 47: Pesticide Occurrence, Analysis and Remediation Vol. 1 Biological Systems* (pp. 91–143). Springer International Publishing.

Monteiro, V. N., & Ulhoa, C. J. (2006). Biochemical characterization of a β-1, 3-glucanase from Trichoderma koningii induced by cell wall of Rhizoctonia solani. *Current Microbiology, 52*, 92–96.

Muir, D. C., & Howard, P. H. (2006). Are there other persistent organic pollutants? A challenge for environmental chemists. *Environmental Science and Technology, 40*(23), 7157–7166.

Mukah, F. E., Chinedu-Ndukwe, P. A., Imarhiagbe, O. & Nwaubani, D. A. (2023). Agrochemical use and emerging human and animal diseases. In *One Health Implications of Agrochemicals and their Sustainable Alternatives* (pp. 53–76). Springer Nature Singapore.

Mulay, Y., Owal, S., Chougule, P., & Pandit, A. (2020). Composting of floral waste by using indigenously isolated microbial consortium: An approach towards the Environment sustainability and waste management. *International Journal Environment Agricultural Research, 6*(4), 20–26.

Mupambwa, H. A., & Mnkeni, P. N. S. (2018). Optimizing the vermicomposting of organic wastes amended with inorganic materials for production of nutrient-rich organic fertilizers: A review. *Environmental Science and Pollution Research, 25*, 10577–10595.

Mustin, M. (1987). Le compost: Gestion de la matière organique, Ed. *François Dubusc, Paris.*

Narayana, T. (2009). Municipal solid waste management in India: From waste disposal to recovery of resources? *Waste Management, 29*(3), 1163–1166.

Naseby, D. C., Pascual, J. A., & Lynch, J. M. (2000). Effect of biocontrol strains of Trichoderma on plant growth, Pythium ultimum populations, soil microbial communities and soil enzyme activities. *Journal of Applied Microbiology, 88*(1), 161–169.

Naworio, N. B. (2022). WUYI21KY_thesis_waste-to-energy project_an integrated sustainable waste management in Yenagoa Metropolis_Numondogiyo Benjamin Naworio.

Ndegwa, P. M., Thompson, S. A., & Das, K. C. (2000). Effects of stocking density and feeding rate on vermicomposting of biosolids. *Bioresource Technology, 71*(1), 5–12.

Nogales, R., Cifuentes, C., & Benitez, E. (2005). Vermicomposting of winery wastes: A laboratory study. *Journal of Environmental Science and Health Part B, 40*(4), 659–673.

Noronha, E. F., & Ulhoa, C. J. (2000). Characterization of a 29-kDa β-1, 3-glucanase from Trichoderma harzianum. *FEMS Microbiology Letters, 183*(1), 119–123.

Onwosi, C. O., Igbokwe, V. C., Odimba, J. N., Eke, I. E., Nwankwoala, M. O., Iroh, I. N., & Ezeogu, L. I. (2017). Composting technology in waste stabilization: On the methods, challenges and future prospects. *Journal of Environmental Management, 190*, 140–157.

Orose, E., Wokeh, O. K., & Okey-Wokeh, C. G. (2023). Some behavioural and physiological effects of plastics (Polyethylene) on fish. *Tropical Aquatic and Soil Pollution, 3*(1), 46–57.

Pandit, N. P., Ahmad, N., & Maheshwari, S. K. (2012). Vermicomposting biotechnology an eco-loving approach for recycling of solid organic wastes into valuable biofertilizers. *Journal of Biofertilizers and Biopesticides, 3*, 1–8.

Pappu, A., Saxena, M., & Asolekar, S. R. (2007). Solid wastes generation in India and their recycling potential in building materials. *Building and Environment, 42*(6), 2311–2320.

Paramdeep, K. A. U. R., Singh, K., & Singh, B. (2022). Microplastics in soil: Impacts and microbial diversity and degradation. *Pedosphere, 32*(1), 49–60.

Paredes, C., Bernal, M. P., Cegarra, J., & Roig, A. (2002). Bio-degradation of olive mill wastewater sludge by its co-composting with agricultural wastes. *Bioresource Technology, 85*(1), 1–8.

Parle, J. N. (1963). Micro-organisms in the intestines of earthworms. *Microbiology, 31*(1), 1–11.

Parlekar, G. Y., Kale, S. P., & Makar, P. V. (1992). Simple techniques of rearing, breeding and production of earthworms for vermicomposting. In *Processing Natural Seminar on Organic Farming, Mahatma Phule Vidyapeeth, College of Agriculture, Pune* (pp. 47–48).

Patel, M., & Naik, S. N. (2010). Flowers of Madhuca Indica JF Gmel.: Present status and future perspectives.

Pathma, J., & Sakthivel, N. (2012). Microbial diversity of vermicompost bacteria that exhibit useful agricultural traits and waste management potential. *Springerplus, 1*(1), 1–19.

Pirani, S. I., & Arafat, H. A. (2014). Solid waste management in the hospitality industry: A review. *Journal of Environmental Management, 146*, 320–336.

Priya, A. K., Pachaiappan, R., Kumar, P. S., Jalil, A. A., Vo, D. V. N., & Rajendran, S. (2021). The war using microbes: A sustainable approach for wastewater management. *Environmental Pollution, 275*, 116598.

Puhlmann, J., Zenk, M. H., & Wagnert, H. (1991). Immunologically active polysaccharides of Arnica montana cell cultures. *Phytochemistry, 30*(4), 1141–1145.

Quaik, S., Embrandiri, A., Rupani, P. F., Singh, R. P., & Ibrahim, M. H. (July, 2012). Effect of vermiwash and vermicomposting leachate in hydroponics culture of Indian Borage (Plectranthus ambionicus) plantlets. In *UMT 11th International Annual Symposium on Sustainability Science and Management* (pp. 210–214).

Raja, A. S. M., Pareek, P. K., Shakyawar, D. B., Wani, S. A., Nehvi, F. A., & Sofi, A. H. (2012). *Extraction of natural dye from saffron flower waste and its application on pashmina fabric.*

Rajalakshmi, S., Amzad Basha, K., & Asif Jamal, G. A. (2023). *A manual on waste management audit.*

Ranganathan, L. S., & Vinotha, S. P. (1998). Influence of pressmud on the enzymatic variations in the different reproductive stages of Eudrilus eugeniae (Kinberg). *Current Science, 74*(7), 634–635.

Ranjitha, J., Vijayalakshmi, S., Kumar, V., & Ralph, N. (2014). Production of bio-gas from flowers and vegetable wastes using anaerobic digestion. *International Journal Research Engineering Technology, 3*(8), 279–283.

Rathi, S. (2006). Alternative approaches for better municipal solid waste management in Mumbai, India. *Waste Management, 26*(10), 1192–1200.

Rigot, J., & Matsumura, F. (2002). Assessment of the rhizosphere competency and pentachlorophenol—metabolizing activity of a pesticide-degrading strain of Trichoderma harzianum introduced into the root zone of corn seedlings. *Journal of Environmental Science and Health, Part B, 37*(3), 201–210.

Saha, S., Moorthi, S., Pan, H. L., Wu, X., Wang, J., Nadiga, S., Tripp, P., Kistler, R., Woollen, J., Behringer, D., & Liu, H. (2010). The NCEP climate forecast system reanalysis. *Bulletin of the American Meteorological Society, 91*(8), 1015–1058.

Sandhya, C., Adapa, L. K., Nampoothiri, K. M., Binod, P., Szakacs, G., & Pandey, A. (2004). Extracellular chitinase production by Trichoderma harzianum in submerged fermentation. *Journal of Basic Microbiology: An International Journal on Biochemistry, Physiology, Genetics, Morphology, and Ecology of Microorganisms, 44*(1), 49–58.

Sangkham, S. (2020). Face mask and medical waste disposal during the novel COVID-19 pandemic in Asia. *Case Studies in Chemical and Environmental Engineering, 2*, 100052.

Sariah, M., Choo, C. W., Zakaria, H., & Norihan, M. S. (2005). Quantification and characterisation of Trichoderma spp. from different ecosystems. *Mycopathologia, 159*, 113–117.

Sarkodie, S. A., Owusu, P. A., & Leirvik, T. (2020). Global effect of urban sprawl, industrialization, trade and economic development on carbon dioxide emissions. *Environmental Research Letters, 15*(3), 034049.

Satchell, J. E. (1983). Earthworm ecology in forest soils. *Earthworm Ecology: From Darwin to Vermiculture* (pp. 161–170). Springer, Netherlands.

Senapati, B. K., & Dash, M. C. (1982). Earthworms as waste conditioner. *Journal of Industrial Engineering, 11*(2), 53.

Senapati, B. K., & Sahu, S. K. (1993). *Reproductive Biology (Cocoon Morphology, Life Cycle Pattern and Life Table Analysis) in Earthworms* (pp. 79–96). Earthworm Resources and Vermiculture. Zoological Survey of India.

Shanthi, N. R., Bhoyar, R. V., & Bhide, A. D. (1993). Vermicomposting of vegetable waste. *Compost Science and Utilization, 1*(4), 27–30.

Sharholy, M., Ahmad, K., Mahmood, G., & Trivedi, R. C. (2005). December. Analysis of municipal solid waste management systems in Delhi–a review. In *Book of Proceedings for the Second International Congress of Chemistry and Environment, Indore, India* (pp. 773–777).

Sharholy, M., Ahmad, K., Mahmood, G., & Trivedi, R. C. (2006). January. Development of prediction models for municipal solid waste generation for Delhi city. In *Proceedings of National Conference of Advanced in Mechanical Engineering* (pp. 1176–1186).

Sharholy, M., Ahmad, K., Mahmood, G., & Trivedi, R. C. (2008). Municipal solid waste management in Indian cities–A review. *Waste Management, 28*(2), 459–467.

Sharma, M., & Dhingra, H. K. (2016). Poly-β-hydroxybutyrate: A biodegradable polyester, biosynthesis and biodegradation. *Microbiology Research Journal International*, (pp. 1–11).

Sharma, K. D., & Jain, S. (2020). Municipal solid waste generation, composition, and management: the global scenario. *Social Responsibility Journal, 16*(6), 917–948.

Sharma, P., Laor, Y., Raviv, M., Medina, S., Saadi, I., Krasnovsky, A., Vager, M., Levy, G. J., Bar-Tal, A., & Borisover, M. (2017). Green manure as part of organic management cycle: Effects on changes in organic matter characteristics across the soil profile. *Geoderma, 305*, 197–207.

Shekdar, A. V. (1999). Municipal solid waste management–the Indian perspective. *Journal of Indian Association for Environmental Management, 26*(2), 100–108.

Siddiqua, A., Hahladakis, J. N., & Al-Attiya, W. A. K. (2022). An overview of the environmental pollution and health effects associated with waste landfilling and open dumping. *Environmental Science and Pollution Research, 29*(39), 58514–58536.

Singh, B. (2023). Utilization of lignocellulosic plant residues for compost formation and its role in improving soil fertility. *Pedosphere, 33*(5), 700–716.

Singh, P., Singh, R., Borthakur, A., Madhav, S., Singh, V. K., Tiwary, D., Srivastava, V. C., & Mishra, P. K. (2018). Exploring temple floral refuse for biochar production as a closed loop perspective for environmental management. *Waste Management, 77*, 78–86.

Sinha, J., Biswas, C. K., Ghosh, A., & Saha, A. (2010a). Efficacy of vermicompost against fertilizers on cicer and pisum and on population diversity of N. *Journal of Environmental Biology, 31*, 287–292.

Sinha, R. K., Herat, S., Bharambe, G., & Brahambhatt, A. (2010b). Vermistabilization of sewage sludge (biosolids) by earthworms: Converting a potential biohazard destined for landfill disposal into a pathogen-free, nutritive and safe bio-fertilizer for farms. *Waste Management and Research, 28*(10), 872–881.

Spiegel, Y., & Chet, I. (1998). Evaluation of Trichoderma spp. as a biocontrol agent against soilborne fungi and plant-parasitic nematodes in Israel. *Integrated Pest Management Reviews, 3*, 169–175.

Spranghers, T., Ottoboni, M., Klootwijk, C., Ovyn, A., Deboosere, S., De Meulenaer, B., Michiels, J., Eeckhout, M., De Clercq, P., & De Smet, S. (2017). Nutritional composition of black soldier fly (Hermetia illucens) prepupae reared on different organic waste substrates. *Journal of the Science of Food and Agriculture, 97*(8), 2594–2600.

Sreenivas, C., & Narayanasamy, G. (2003). Preparation and characterization of phospho-composts: Effect of rock phosphate source, P enrichment and pyrites level. *Journal of the Indian Society of Soil Science, 51*(3), 262–267.

Sreenivasan, E. (2020). Solid waste management and its role in mitigating environmental degradation with special reference to recycling of wood wastes. *Environmental Degradation: Causes and Remediation Strategies*, (pp. 166–182).

Srivastava, V. K. (2004). Trichoderma spp-a boon for better crop health. *Pestology, 28*(8), 41–45.

Stockli, A. (1949). *Verwertungsmöglichkeiten des Kehrichtes in der Landwirtschaft.*

Sunita, M., & Sarojini, P. (2013). Madhuca lonigfolia (Sapotaceae): A review of its traditional uses and nutritional properties. *International Journal of Humanities and Social Science Invention, 2*(5), 30–36.

Suthar, S. (2007). Vermicomposting potential of Perionyx sansibaricus (Perrier) in different waste materials. *Bioresource Technology, 98*(6), 1231–1237.

Suthar, S. (2009). Vermicomposting of vegetable-market solid waste using Eisenia fetida: Impact of bulking material on earthworm growth and decomposition rate. *Ecological Engineering, 35*(5), 914–920.

Talavera, J. A., Cunha, L., Arévalo, J. R., Talavera, I. P., Kille, P., & Novo, M. (2020). Anthropogenic disturbance and environmental factors drive the diversity and distribution of earthworms in São Miguel Island (Azores, Portugal). *Applied Soil Ecology, 145*, 103301.

Thakuria, D., Schmidt, O., Finan, D., Egan, D., & Doohan, F. M. (2010). Gut wall bacteria of earthworms: A natural selection process. *The ISME Journal, 4*(3), 357–366.

Tripathi, G., & Prasad, R. (2005). Biodynamics of below-ground animal resources and interaction. *Basic Research and Applications of Mycorrhizae*, (pp. 297–325).

Tuomela, M., Vikman, M., Hatakka, A., & Itävaara, M. (2000). Biodegradation of lignin in a compost environment: A review. *Bioresource Technology, 72*(2), 169–183.

Unsworth, J., Nakagawa, Y., Harris, C., & Kleter, G. (2019). The battle for a sustainable food supply. *Sustainable Agrochemistry: A Compendium of Technologies*, (pp. 13–84).

Vahabi, K., Mansoori, G. A., & Karimi, S. (2011). Biosynthesis of silver nanoparticles by fungus Trichoderma reesei (a route for large-scale production of AgNPs). *Insciences Journal, 1*(1), 65–79.

Van Wyk, J. P. H., & Mohulatsi, M. (2003). Biodegradation of wastepaper by cellulase from Trichoderma viride. *Bioresource Technology, 86*(1), 21–23.

Vankar, P. S., & Shukla, D. (2011). Natural dyeing with anthocyanins from Hibiscus rosa sinensis flowers. *Journal of Applied Polymer Science, 122*(5), 3361–3368.

Vesley, D. (1999). *Human Health and the Environment: A Turn of the Century Perspective*. Springer Science & Business Media.

Waghmode, M. S., Gunjal, A. B., Nawani, N. N., & Patil, N. N. (2018). Management of floral waste by conversion to value-added products and their other applications. *Waste and Biomass Valorization, 9*, 33–43.

Waqas, M., Hashim, S., Humphries, U. W., Ahmad, S., Noor, R., Shoaib, M., Naseem, A., Hlaing, P. T., & Lin, H. A. (2023). Composting processes for agricultural waste management: A comprehensive review. *Processes, 11*(3), 731.

Wilson, D. C., & Velis, C. A. (2015). Waste management–still a global challenge in the 21st century: An evidence-based call for action. *Waste Management and Research, 33*(12), 1049–1051.

Yaashikaa, P. R., Kumar, P. S., Nhung, T. C., Hemavathy, R. V., Jawahar, M. J., Neshaanthini, J. P., & Rangasamy, G. (2022). A review on landfill system for municipal solid wastes: Insight into leachate, gas emissions, environmental and economic analysis. *Chemosphere*, 136627.

Yeates, G. W. (1981). Soil nematode populations depressed in the presence of earthworms. *Pedobiologia, 22*(1), 191–195.

Yoro, K. O., & Daramola, M. O. (2020). CO_2 emission sources, greenhouse gases, and the global warming effect. In *Advances in Carbon Capture* (pp. 3–28). Woodhead Publishing.

Zhou, Y., Liu, H., Chen, H., Awasthi, S. K., Sindhu, R., Binod, P., Pandey, A., & Awasthi, M. K. (2023). Introduction: Trends in composting and vermicomposting technologies. In *Current Developments in Biotechnology and Bioengineering* (pp. 1–28). Elsevier.

Chapter 8
Waste-To-Energy Technology for the Recycling of Municipal Solid Waste into Energy Producing Sources

Sabeeqa Samad, Gowhar Hamid Dar, and Zahid Bashir Naik

Abstract Fast expansion of urbanization and development is rapidly increasing the waste everywhere and land area left for dumping such huge waste is hardly available. Therefore, handling of such waste has now become a big challenge. This issue demands a kind of solution where the waste can be minimized without compromising environmental sustainability. Several methods are already available for the management of waste with each of them having some kind of drawback. The only method with complete solution to the waste handling in modern times is the process of recycling of waste and simultaneously derive good proportion of energy and value additive products. Waste to energy (WTE) technology involves eco-friendly methods including physical, thermal and biological method for the treatment of the waste and does not only minimizes the waste up to significant extent but offers a great benefit in terms of energy generation as well. Besides solving the challenging issue of waste generation, WTE technology reduces the energy demand as well.

Keywords Waste · Landfill · Energy · Technology · Urbanization · Thermal degradation · Incineration

1 Introduction

Fast growth of urbanization, population explosion, modern living trends and rapid economic development are responsible for significant waste generation that in turn causes environmental as well as socioeconomic issues (Malinauskaite et al., 2017). Dumping sites are not the only spots that are harmfully effected by municipal solid waste (MSW), but it does also affect the surrounding locations up to a wide range (Siddiqua et al., 2022). Dumping of waste is now a challenging factor for waste handling authority and if the situation of growing waste will remain the same, the

S. Samad · G. H. Dar (✉) · Z. B. Naik
Department of Environmental Science, Sri Pratap College, Srinagar, Jammu and Kashmir 190001, India
e-mail: dargowharhamid@gmail.com

R. A. Bhat et al. (eds.), *Zero Waste Management Technologies*,
https://doi.org/10.1007/978-3-031-57275-3_8

problem will be left with no solution. Hence, the situation demands the sustainable waste management (Kaur et al., 2021). In the other scenario, as reported by International Energy Agency (IEA) the energy crisis is increasing with the expected global energy demand of 17 billion toe by the year 2035 (Ly et al., 2023; Malav et al., 2020). These issues of tremendous waste generation and rising energy demand can be rectified by implementation of WTE technology that involves extracting energy sources from waste (Beyene et al., 2018; Zhang et al., 2015). WTE has the potential of reducing greenhouse gases by reducing the use of fossil fuels, finally contributing towards sustainable waste management and economic growth as well (Beyene et al., 2018; Khan & Tanveer, 2011; Lino et al., 2023; Singh et al., 2023a, 2023b; Zaman & Lehmann, 2013). WTE is also included renewable energy sources (The United States Environmental Protection Agency and many countries of the world are using the technology (Kaur et al., 2021). Depending upon the kind of waste produced different number of methods do exist currently that can be adopted for WTE process (Edjabou et al., 2015). But the energetic routes can be physical, biological and thermal. Reduced derived fuel (RDF) or solid recovery fuel is produced from physical technology. However, incineration, gasification and pyrolysis processes involve the thermal treatment of waste (Mata-Lima et al., 2021; Rezania et al., 2023; Teixeira et al., 2014; Tomic et al., 2017). Aerobic digestion is included in biochemical treatment of the MS. Landfill, which generates biogas, can also be adopted for WTE process if properly handled as the gas can be collected and used for heat and power generation (Kaur et al., 2021).

2 Municipal Solid Waste and Its Composition

Solid waste is simply all the useless, unwanted and discarded domestic, residential and commercial material (Khan et al., 2022; Pop et al., 2022). Garbage, broken wood furniture, discarded fabrics and cloth, plastic material, papery, useless electronic appliances, etc., are all included in MSW (Gundupalli et al., 2017). Table1 represents the composition of MSW of the world.

Table 1 Composition of MSW

S. No.	Waste component	Percentage%	References
1	Plastic	10–24	Darmey et al. (2023); Kaur et al. (2021); Singhal et al. (2022)
2	Metal	0–4	
3	Paper (and cardboard)	5–17	
4	Organics	44–61	
5	Glass	0.8–4	
6	Others	14–18	

3 Energy Recovery Potential from MSW and Its Application for WTE Technologies

MSW that is generally considered as waste can be used for energy producing sources (Halder et al., 2014). It is a well-known fact to the stakeholders of solid waste that MSW is a kind of composition that can be the potential source for energy extraction (Koraia et al., 2017). The caloric content of the MSW makes it useful for energy and power generation (Halder et al., 2014). Different methods can be adopted for yielding energy from MSW viz; thermochemical method including incineration, pyrolysis, and gasification, biological processes like biomethanation, and composting etc. (Tozlu et al., 2016). While being the solution to the waste accumulation problem, these technologies also provide the way for energy generation and at the same time contribute towards the improvement of environmental quality (Halder et al., 2014). China is producing about 300 million tonnes of waste per year (World Energy Resources, 2016). However, to tackle this issue the country is adopting advanced incineration plants with the capacity of processing 800 tonnes/d of MSW to generate power (World Energy Resources, 2016; Zhao et al., 2016). On the other hand, the Greece is using the anaerobic digestion plants with the energy recovery of 695 KWh/ tonne of waste (Ibarra-Esparza et al., 2023). Electricity generating potential of a WTE facility in Central Greece is reported to be 0.55 MW per tonne of MSW, with 22% of net conversion efficiency (Perkoulidis et al., 2010). Talking about Brazil, the power producing capacity from MSW landfills is about 660 MW/day (Kumar & Samadder, 2017). In the light of all above facts, it can be concluded that solid waste is not a waste but it is a precious renewable resource (Koraia et al., 2017).

4 Incentives Given by Government to Promote WTE Technology

Many countries of world use MSW for energy producing sources under the investments and other facility benefits provided by various government and private sectors in the (Kaur et al., 2021). Below mentioned are some examples of such countries.

China

The government of China is providing about 80% of total expenditure for the installation of WTE facility (Kaur et al., 2021). In addition to that govt. has also specified rewards for declining the pollutant emissions in environment (Kaur et al., 2021).

India

According to budget 1999–2000, Ministry of Non-Conventional Energy Sources (MNES) is supporting municipalities, technology providers and others associated with WTE projects by offering funds and grants in the same area (Singh et al., 2011). The Ministry of New and Renewable Energy (MNRE) is supporting the WTE process by providing internet costs, supplying garbage, arranging WTE plant sites, training and guidance in the related area, etc. (Kalyani & Pandey, 2014).

Malaysia

Benefits provided Malaysian govt. to promote WTE facility in Malaysia include Prepayment, tax subsidy, transfer technology aids and relax laws to uplift foreign investors (Sharifah et al., 2013).

Germany

Financial as well technical support is provided by Germany government to promote WTE process. Loans with low interest rates and long payback time are offered to the WTE stakeholders, and power purchase agreement is also provided for the encouraging WTE plant installation (EPA, USEPA, 2015; Kaur et al., 2021).

USA

In USA, percentage of market share of municipal waste management is governed by private companies (Kaur et al., 2021). Investment motivation includes different policies that are commendatory to WTE facility like price on CO_2, charges and taxes of landfills, direct subsidies and many more (Kaur et al., 2021).

Turkey

Turkey uses the sale of electricity to encourage investment by private sector in the waste-to-energy technology (Kaur et al., 2021). Govt. has reduced the tax price for the electricity generated in waste to energy plants (Williams & Helm, 2011, Marktscan, 2013).

5 Waste-To-Energy Technologies for the Recycling of Municipal Solid Waste into Energy Producing Sources

Any waste management system intends at restoring the material and energy from waste, as well as disposing off the residues. Meanwhile the sustainability of the environment is also a necessary condition (Hutajulu et al., 2022). Different waste management techniques are being promoted that include sustainable alternatives to make proper use of MSW for generating resources (Cucchiella et al., 2017). Developed and developing nations are widely using recycling and reusing processes followed waste management techniques (Suthar et al., 2016). A lot number of WTE technologies are being utilized currently and are regarded as optimistic sources of renewable energy (Roy et al., 2022). Commonly used waste treatment process includes (a) thermal treatment that can be incineration, pyrolysis, gasification, refuse derived fuel or RDF, (b) biochemical treatment which in turn includes composting, vermicomposting, anaerobic digestion/biomethanation and (c) chemical treatment like transesterification (Annepu, 2012; Singh et al., 2011).

5.1 *Thermal Conversion Technologies*

These technologies are usually applicable for waste with low moisture content and significant portion of non-biodegradable organic matter and involves the conversion of organic matter in MSW to fuel oil, gas or heat (Kumar & Samadder, 2017).

5.1.1 Incineration

This is one of the most commonly used processes for waste treatment with the capacity of about 70% mass reduction and 90% volume reduction (Kalyani and Panday, 2014). The product of the waste incineration process is energy in the form of electricity (Annepu, 2012; Singh et al., 2011). Here the waste is fed into a chamber and is combusted at high temperature 800 °C to 1000 °C (Havukainen et al., 2017). Produced heat is utilized for boiling the water fed into the boiler of incineration system which generates steam that in turn rotates a turbine generator to produce electricity (Kaur et al., 2021). Pollutant gases produced from the process of incineration are managed by a control system that is fixed with the combustion chamber (Havukainen et al., 2017).

5.1.2 Pyrolysis

Pyrolysis involves burning of the waste in absence of is oxygen (Kalyani and Panday, 2014). The process is often used for the waste with high organic matter (Basu, 2010). In this process, three major products are formed in the process of pyrolysis.

(1) Vaporous part that under the process condensation is converted into oil and water,
(2) Char and ash residue with high carbon content that can be utilized for heating,
(3) CO, CO_2 and CH_4 (Okafor, 2011; Rajasekhar et al., 2015).

5.1.3 Gasification

This process involves incomplete combustion of biomass with gas and char as basic products CO_2 and H_2O formed are converted to CO and H_2 (Sivakumar et al., 2023). Finally, the feedstock is transformed into gas under many heterogeneous reactions with the help of a gasification agent (Bozorgi et al., 2023; Zachl, et al. 2024). The composition of the product gas is carbon monoxide, carbon dioxide, methane, hydrogen, water, inert gasses, small amount of higher hydrocarbons and some char and ash (US, DENREL, 1996). Sometimes an external energy source is required where the pyrolysis does not take place normally under oxidizing agent. This is called Indirect Gasification (Faheem et al., 2024; Li et al., 2023), and the commonly used energy source is steam that has potential to raise hydrogen content of the product gas (Faheem et al., 2024).

5.2 Biological/Biochemical Conversion

Biochemical conversion is comparatively more eco-friendly technique than other existing methods of waste to energy conversion which involves the production of energy from waste with the help of microorganisms. The technique commonly uses the methods of anaerobic digestion and composting (Kalyani and Pandey, 2014).

In anaerobic digestion (AD), organic waste is treated in absence of oxygen under action of microorganisms (Franceschi et al., 2023; Elango et al., 2007; Kumar & Samadder, 2020; Sáez & Río, 2024; Saha & Sumi, 2023). The product of anaerobic digestion is biogas that is used to generate combined heat and power (CHP) and can be used as a transport fuel also. The remnants of the feedstock are incinerated or gasified (Kalyani and Pandey, 2014).

The digested materials can also be converted to compost that supports nutrient cycling (Mathioudakis et al., 2022). Since the organic portion MSW is rich in nitrogen, phosphorus, etc., therefore after segregating this organic portion properly and with the help of vermicomposting method (Joshi et al., 2016) manual composting (Trapani et al., 2023), and the use of microbial cultures (Saha et al., 2010), the nutrients can be recycled to the environment (Saha et al., 2010). The process of composting reduces the waste burden on landfills and is also an important means for the nutrient enhancement in crops (Bhardwaj et al., 2023).

5.3 Landfilling

Landfills are the waterproofed areas for receiving large amount of waste in which the base is sealed in a way to prohibit the ground water pollution from leachate (Kaur et al., 2021). Here, the waste received is compressed in layers within subunits of the landfill called cells (Firmansyah et al.; 2023). The dimensions of a cell in landfill are 50 ft(L), 50 ft.(W) and 14 ft.(H) with the capacity to hold trash of 2500 tons and 1500 lb/cubic yard with the aid of a heavy weight equipment that rolls over the trash many times (Kaur et al., 2021). The cell once completely filled is sealed with around 6 inch soil layer and is compacted again (Okafor, 2011). This technique protects the landfill site from being the hub for scavengers and also prevents the emission of foul odours (Kaur et al., 2021).

5.3.1 Making Landfill Practice Environmentally Sound Compared to Traditional Dumping of Waste into Land

The end-products of the microbial decomposition of the waste in the landfill include CO_2, methane, VOCs and leachate. Besides preventing groundwater contamination from landfill leachate, attention is now being given to the prevention of global warming gas (methane) into the environment, as well as its utilization to generate

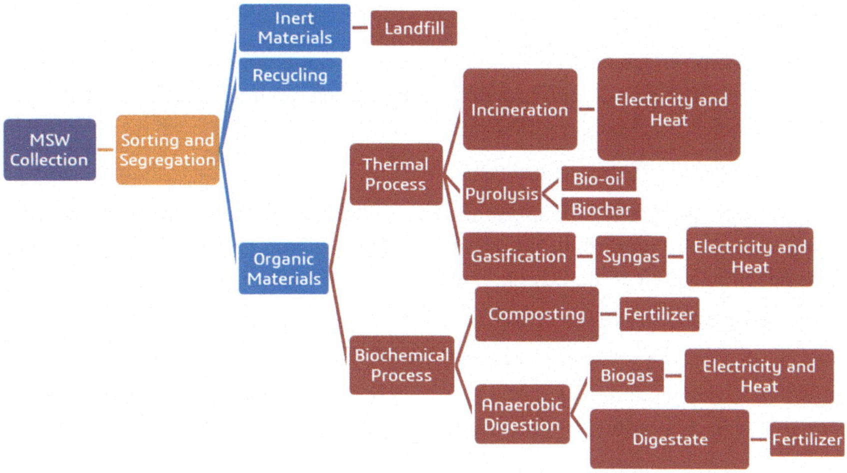

Fig. 1 Summary of different MSW management techniques and their ultimate products (Roy et al., 2022)

power generation, and the ability of landfill to store the carbon for long time (Masters and Ela, 2015). Raw landfill gas is corrosive and needs treatment that usually involves moisture and contaminant removal, while CO_2 is left as such. This ensures the conversion of landfill gas to a medium energy content which can be utilized in steam or gas turbines for generating on-site electricity. This content can be further treated by eliminating CO_2 from it to form a pipeline-quality gas for off-site use (Masters and Ela, 2015).

5.3.2 Refuse Derived Fuels (RDFs)

RDF is a homogeneous product with great combustion properties that is made from MSW (Masters & Ela, 2015). At waste treatment plants, inert fraction, plastic and metals in the collected MSW are segregated manually from combustible fraction which is passed on to a mechanical separation unit. Then the burnable fraction is cut and grinded to form a flocculent material called RDF fluff that can be utilized as a fuel in a boiler for electricity generation. RDF fluff has also demand in many industries as coal substitute (Kaur et al., 2021). The various steps and techniques involved in zero waste management are presented in Fig. 1.

5.4 Benefits from Waste-To-Energy Technology

Major benefits from WTE technology are as under:

1. *Energy Generation*

Energy produced from MSW in developed countries contributes about 5% of total energy generated there which if properly exploited can minimize the demand of fossil fuels (Brunner & Rechberger, 2014).

2. *Environmentally Safe Technology*

Apart from production of large amount of energy WTE is eco-friendly technology and encourages Environmental protection which is always counted in its major benefits (Singh et al., 2011).

3. *Reduction of Greenhouse Gases*

According to study, 52 countries of the world have successfully reduced the level of GHG by implementation of 350 WTE projects which encourages sustainable waste treatment as well as generation of energy without harmful emissions (Islam, 2018).

4. *Means of Renewable Energy*

MSW is counted in the group of biomass fuel of renewable energy sources by the US DOE as it is composed of non-fossil material significantly (Rezania et al., 2023).

5. *Encourages Recycling*

Successful implementation of WTE technologies, e.g. biochemical method depends on the waste composition (Braber, 1995). This places waste segregation and categorization at first step of WTE process to collect the suitable feedstock for treatment which in turn supports waste recycling (Munir et al., 2021).

6. *Saving the Land*

Once installed WTE plants last for years under proper maintenance and demands comparatively very small land than dumping sites hence saves a large land area (Rezania et al., 2023).

7. *Prevention from Traditional Landfill*

Aprox. 70% MSW worldwide undergoes open dumping with the release of harmful emissions and leachate (Ferronato & Torretta, 2019). In contrast to that WTE methods encourage sustainable waste management with the capacity to reduce the number of traditional landfills and prevent the environment from pollution (Rezania et al., 2023).

8. *Maintaining Sanitation*

WTE technology offers sanitation through the destruction of pathogens in waste using thermal treatment methods like incineration, pyrolysis, etc., which is otherwise impossible in other waste management processes (Brunner & Rechberger, 2014).

9. *Mineralization and Immobilization of Hazardous Substances*

Production of large amount hazardous organic materials each year demands safe disposal to which incineration method of WTE technology provides best solution by offering efficient pathway for immobilization of such substances and transforming these into non-toxic end-products (Brunner & Rechberger, 2014).

5.5 Challenges in WTE Projects

Although a lot of benefits are generated by WTE technologies, but these does face certain minor challenges also such as

1. *Improper Management of Waste*

   The composition of MSW has an essential role for efficient performance of WTE technologies (Adipah & Kwame, 2018), e.g. thermal conversion technology demands dry waste, e.g. paper, rubber, etc. (Kumar & Samadder, 2017) while as biological method kind of waste with high moisture, e.g. food waste (Edwards et al., 2015). This makes pre-processing and proper segregation of the MSW an essential condition which is unfortunately a challenge in many (Roy et al., 2022).

2. *Health Hazard*

   Handling of the MSW is always associated health hazard of the workmen as seen in many developing countries. While being at significant risk there is no proper provision for the protection of waste handlers in such countries (Roy et al., 2022).

3. *Cost-Intensive Technology*

   Another challenge of most WTE projects is that they are highly cost-intensive. In developing nations implementation of WTE technology like pyrolysis and gasification remains a challenge due to high Capital and Operating Costs (Alam & Qiao, 2020).

4. *Release of Some Harmful Emissions*

   Although the WTE technologies are comparatively eco-friendly than other waste management practices, but these still release certain harmful emissions unlike other renewable methods of energy generation, e.g. wind and solar where no unwanted emissions are found (Roy et al., 2022). For instance, improper management of incinerators may result in incomplete combustion that in turn may release CO, soot and toxic fumes (Ziraba et al., 2016).

5.6 Conclusion

Solid waste is one of the big challenges if not handled properly, as it has potential to cause great threat to the environmental as well as human health. Development of science and technology has come up with the solution to this problem where solid waste is not merely considered as waste but a resource. Many waste treatment technologies are being used which are feasible for recycling and material recovery from MSW. Along with the reduction of waste load which is the main goal of waste management, waste-to-energy technologies (WTEs) are providing best opportunities for resource recovery in environmentally sound way. Few challenges are faced by WTE technology in some developing countries, which are beaten by abundance of benefits it generates. Hence, WTE technology can be considered as economic and sustainable solid waste management practice.

References

Adipah, S., & Kwame, O. (2018). A novel introduction of municipal solid waste management. *Journal of Environment Science Public Health, 03*, 147–157.

Alam, O., & Qiao, X. (2020). An in-depth review on municipal solid waste management, treatment and disposal in Bangladesh. *Sustainable Cities and Society, 52*, 101775.

Annepu, R. K. (2012). *Sustainable solid waste management in India.* New York: Columbia university in the city of New York, Department of Earth and Environmental Engineering.

Basu, P. (2010). *Biomass gasification and pyrolysis: Practical design and theory* (p. 2010). Academic Press.

Beyene, H. D., Werkneh, A. A., & Ambaye, T. G. (2018). Current updates on waste to energy (WtE) technologies: A review. *Renewable Energy Focus, 24*(2018), 1–11. https://doi.org/10.1016/j.ref.2017.11.001

Bhardwaj, P., Sharma, R. K., Chauhan, A., Ranjan, A., Rajput, V. D., Minkina, T., Mandzhieva, S. S., Mina, U., Wadhwa, S., Bobde, P., & Tripathi, A. (2023). Assessment of heavy metal distribution and health risk of vegetable crops grown on soils amended with municipal solid waste compost for sustainable urban agriculture. *Water, 15*(2), 228. https://doi.org/10.3390/w15020228

Bozorgi, M., Dutta, A., Mahmud, S., & Tasnim, S. H. (2023). Identifying optimal geographic locations for hybrid concentrated solar biomass (HCSB) power plants in Alberta and Ontario, Canada. *Energy Sources, Part b: Economics, Planning, and Policy, 18*(1), 2229320.

Brunner, P. H., & Rechberger, H. (2014). Waste to energy—Key element for sustainable waste management. *Waste Management*, xxx(2014), xxx–xxx

Cucchiella, F., D'Adamo, I., & Gastaldi, M. (2017). Sustainable waste management: Waste to energy plant as an alternative to landfill. *Energy Conversion Management, 131*, 18–31.

Darmey, J., Ahiekpor, J. C., Narra, S., Achaw, O.-W., & Ansah, H. F. (2023). Municipal solid waste generation trend and bioenergy recovery potential: A review. *Energies, 16*(23), 7753. https://doi.org/10.3390/en16237753

Edjabou, M. E., Jensen, M. B., Götze, R., Pivnenko, K., Petersen, C., Scheutz, C., & Astrup, T. F. (2015). Municipal solid waste composition: Sampling methodology, statistical analyses, and case study evaluation. *Waste Management, 36*, 12–23. https://doi.org/10.1016/j.wasman.2014.11.009

Edwards, J., Othman, M., & Burn, S. (2015). A review of policy drivers and barriers for the use of anaerobic digestion in Europe, the United States and Australia. *Renewable and Sustainable Energy Reviews, 52*, 815–828.

Elango, D., Pulikesi, M., Bakaralingam, P., Ramamurthi, V., & Sivanesan, S. (2007). Production of biogas from municipal solid waste with domestic sewage. *Journal of Hazardous Materials, 141*, 301–304.

EPA, USEPA. (2015). *US, advancing sustainable materials management: Facts and figures* (pp. 1–186).

Faheem, H. H., Britt, B., Rocha, M., Zhou, S., Li, C., Cai, W., & Fan, L. (2024). Sensitivity analysis and process optimization for biomass processing in an integrated gasifier-solid oxide fuel cell system. *Fuel, 356*, 129529.

Ferronato, N., & Torretta, V. (2019). Waste mismanagement in developing countries: A review of global issues. *International Journal of Environmental Research and Public Health, 16*, 1060.

Firmansyah, F., Park, I., Corona, M., Aphale, O., Ahuja, A., Johnston, M., Krista, L., Hewitt, E., Tonjes, D. J. (2023). *Variation in municipal solid waste generation and management across time and space*. Available at SSRN: https://ssrn.com/abstract=4626949 or https://doi.org/10.2139/ssrn.4626949

Franceschi, F. F., Castillo, J. S., Cherni, J. A., Acosta-González, A., & Gómez, M. F. (2023). ADPMDesign: The use of a participatory methodology to design a dry anaerobic digestion power plant for municipal solid waste treatment. *Energy for Sustainable Development, 74*, 173–184. https://doi.org/10.1016/j.esd.2023.03.017

Gundupalli, S. P., Hait, S., & Thakur, A. (2017). A review on automated sorting of source-separated municipal solid waste for recycling. *Waste Management, 60*, 56–74.

Halder, P. K., Paul, N., Hoque, M. E., Hoque, A. S. M., Parvez, M. S., Rahman, M. H., & Ali, M. (2014). Municipal solid waste and its management in Rajshahi City, Bangladesh: A source of energy. *International Journal of Renewable Energy Research N, 4*(1), 168–175.

Havukainen, J., Zhan, M., Dong, J., Liikanen, M., Deviatkin, I., & Li, X. (2017). Horttanain, environmental impact assessment of municipal solid waste management incorporating mechanical treatment of waste and incineration in Hangzhou, China. *Journal of Cleaner Production*, 453–461.

Hutajulu, S. M., Marsaulina, I., Siregar, F. A., Indirawati, S. M. (2022). Solid medical waste management strategy in hospitals, Indonesia. *The Open Public Health Journal*, 15(1). https://doi.org/10.2174/18749445-v15-e221223-2022-152

Ibarra-Esparza, F. E., González-López, M. E., Ibarra-Esparza, J., Lara-Topete, G. O., Senés-Guerrero, C., Cansdale, A., Forrester, S., Chong, J. P. J., Gradilla-Hernández, M. S. (2023). Implementation of anaerobic digestion for valorizing the organic fraction of municipal solid waste in developing countries: Technical insights from a systematic review. *Journal of Environmental Management, 347*, 118993.

Islam, K. M. N. (2018). Municipal solid waste to energy generation: An approach for enhancing climate co-benefits in the urban areas of Bangladesh. *Renewable and Sustainable Energy Reviews, 81*, 2472–2486.

Joshi, R., Ahmed, S., & Ng, C. (2016). Status and challenges of municipal solid waste management in India: A review. *Cogent Environmental Science, 2016*(2), 1139434.

Kalyani, K. A., & Pandey, K. K. (2014). Waste to energy status in India: A short review. *Renewable and Sustainable Energy Reviews, 31*, 113–120.

Kaur, A., Bharti, R., & Sharma, R. (2021). Municipal solid waste as a source of energy. *Materials Today: Proceedings*. https://doi.org/10.1016/j.matpr.2021.04.286

Khan, A. H., Sharholy, M., Alam, P., Al-Mansour, A. I., Ahmad, K., Kamal, M. A., Alam, S., Pervez, M. N., & Naddeo, V. (2022). Evaluation of cost benefit analysis of municipal solid waste management systems. *Journal of King Saud University Science, 34*, 101997.

Khan, M., Tanveer, H. (September 28–30, 2012) Production of thermoelectric power from solid waste of urban Lahore. In *Proceedings of the 2011 international conference & utility exhibition on power and energy systems: Issues and prospects for Asia (ICUE), Pattaya City, Thailand*.

Koraia, M. S., Maharb, R. B., & Uqailic, M. A. (2017). The feasibility of municipal solid waste for energy generation and its existing management practices in Pakistan. *Renewable and Sustainable Energy Reviews, 72*, 338–353.

Kumar, A., & Samadder, S. (2020). Performance evaluation of anaerobic digestion technology for energy recovery from organic fraction of municipal solid waste: A review. *Energy, 2020*(197), 117253.

Kumar, A., & Samadder, S. R. (2017). A review on technological options of waste to energy for effective management of municipal solid waste. *Waste Management, 2017*(69), 407–422.

Li, H., Li, M., Wang, H., Tan, M., Zhang, G., Huang, Z., & Yuan, X. (2023). A review on migration and transformation of nitrogen during sewage sludge thermochemical treatment: Focusing on pyrolysis, gasification and combustion. *Fuel Processing Technology, 240*(2023), 107562.

Lino, F. A. M., Ismail, K. A. R., & Castañeda-Ayarza, J. A. (2023). Municipal solid waste treatment in Brazil: A comprehensive review. *Energy Nexus*, 100232.

Ly, N. H., Joo, S. W., Choo, J., Vasseghian, Y., Cho, J., & Rezania, S. (2023). Sustainable cutting-edge techniques for gold valorization from electronic wastes. *Chemical Engineering Journal, 471*, 144324.

Malav, L. C., Yadav, K. K., Gupta, N., Kumar, S., Sharma, G. K., Krishnan, Rezania, S., Kamyab, H., Pham, Q. B., Yadav, S., Bhattacharyya, S., Yadav, V. K., & Bach, Q. V. (2020). A review on municipal solid waste as a renewable source for waste-toenergy project in India: Current practices, challenges, and future opportunities. *Journal of Cleaner Production*, 277, 123227. https://doi.org/10.1016/j.jclepro.2020.123227

Malinauskaite, J., Jouhara, H., Czajczyńska, D., Stanchev, P., Katsou, E., Rostkowski, P., Thorne, R. J., Colon, J., Ponsá, S., Al-Mansour, F., Anguilano, L., Krzyżyńska, R., Lopez, I. C., Vlasopoulos, A., & Spencer, N. (2017). Municipal solid waste management and waste-to-energy in the context of a circular economy and energy recycling in Europe. *Energy*, 141, 2013e2044.

Marktscan, N. (2013). *Development of sanitary landfills and waste-to-energy from landfill gas in Turkey*.

Masters, G. M., & Ela, W. P. (2015). *Introduction to environmental science and engineering and science*. Noida 201301, U.P, India. Pearson Education India.

Mata-Lima, H., Silva, D. W., Nardi, D. C., Klering, S. A., de Oliveira, T. C. F., & Morgado-Dias, F. (2021). Waste-to-energy: An opportunity to increase renewable energy share and reduce ecological footprint in small island developing states (SIDS). *Energies, 14*(22), 7586. https://doi.org/10.3390/en14227586

Mathioudakis, D., Karageorgis, P., Papadopoulou, K., Astrup, T. F., & Lyberatos, G. (2022). Environmental and economic assessment of alternative food waste management scenarios. *Sustainability, 14*, 9634.

Munir, M. T., Mohaddespour, A., Nasr, A. T., & Carter, S. (2021). Municipal solid waste-to-energy processing for a circular economy in New Zealand. *Renewable and Sustainable Energy Reviews, 145*, 111080.

Okafor, N. (2011). *The disposal of municipal solid waste*. Springer Science+Business Media B.V, Environmental Microbiology of Aquatic and Waste System.

Perkoulidis, G., Papageorgiou, A., Karagiannidis, A., & Kalogirou, S. (2010). Integrated assessment of a new waste-to-energy facility in central Greece in the context of regional perspectives. *Waste Management, 30*(7), 1395–1406.

Pop, M., Bucur, I., Zoldan, D., Imre, K., Nichita, I., Cristina, G., Tîrziu, A., & Tîrziu, E. (2022). Chemical and microbiological air quality in a municipal solid waste landfill and its surroundings, in South-Eastern Romania. *Sustainability, 14*, 156.

Rajasekhar, M., Rao, N. V., Rao, G. C., Priyadarshini, G., & Kumar, N. J. (2015). Energy generation from municipal solid waste by innovative technologies—Plasma gasification. *Procedia Materials Science, 10*, 513–518. https://doi.org/10.1016/j.mspro.2015.06.094

Rezania, S., Oryani, B., Nasrollahi, V. R., Darajeh, N., Lotfi Ghahroud, M., & Mehranzamir, K. (2023). Review on waste-to-energy approaches toward a circular economy in developed and developing countries. *Processes, 11*, 2566. https://doi.org/10.3390/pr11092566

Roy, H., Alam, S. R., Bin-Masud, R., Prantika, T. R., Pervez, M. N., Islam, M. S., & Naddeo, V. (2022). A review on characteristics, techniques, and waste-to-energy aspects of municipal solid waste management: Bangladesh perspective. *Sustainability, 14*, 10265. https://doi.org/10.3390/su141610265

Sáez, V. P., & del Río, M. M. (2024). Procedure for waste prevention and management to implement best practices in the design and construction of a building. In: L. Bragança, M. Cvetkovska, R. Askar, & V. Ungureanu, (Eds.), *Creating a roadmap towards circularity in the built environment*. Springer Tracts in Civil Engineering. Springer. https://doi.org/10.1007/978-3-031-45980-1_2

Saha, J. K., Panwar, N., & Singh, M. V. (2010). An assessment of municipal solid waste compost quality produced in different cities of India in the perspective of developing quality control indices. *Waste Management, 2010*(30), 192–201.

Siddiqua, A., Hahladakis, J. N., & Al-Attiya, W. A. K. A. (2022). An overview of the environmental pollution and health effects associated with waste landfilling and open dumping. *Environmental Science and Pollution Research, 29*, 58514–58536.

Singh, P., Boora, A., & Gupta, A. K. (2023a). A review on utilizing municipal solid waste incineration (MSWIA) in construction activates. *IOP Conference Series: Earth and Environmental Science, 1110*(1). IOP Publishing. https://doi.org/10.1088/1755-1315/1110/1/012042

Singh, P., Amardeep, B., & Gupta, A. K. (2023b). A review on utilizing municipal solid waste incineration (MSWIA) in construction activates. *IOP Conference Series: Earth and Environmental Science, 1110*(1). IOP Publishing.https://doi.org/10.1088/1755-1315/1110/1/012042

Singh, R. P., Tyagi, V. V., Allen, T., Ibrahim, M. H., & Kothari, R. (2011). An overview for exploring the possibilities of energy generation from municipal solid waste (MSW) in Indian scenario. *Renewable and Sustainable Energy Reviews, 15*(9), 4797–4808. https://doi.org/10.1016/j.rser.2011.07.071

Suthar, S., Rayal, P., & Ahada, C. P. S. (2016). Role of different stakeholders in trading of reusable/recyclable urban solid waste materials: A case study. *Sustainable Cities and Society, 2016*(22), 104–115.

Teixeira, S., Monteiro, E., Silva, V., & Rouboa, A. (2014). Prospective application of municipal solid wastes for energy production in Portugal. *Energy Policy, 71*, 159–168. https://doi.org/10.1016/j.enpol.2014.04.002

Tomic, T., Dominkovic, D. F., Pfeifer, A., Schneider, D. R., Pedersen, A. S., & Duic, N. (2017). Waste to energy plant operation under the influence of market and legislation conditioned changes. *Energy, 137*, 1119–1129. https://doi.org/10.1016/j.energy.2017.04.080

Tozlu, A., Özahi, E., & Abuşoğlu, A. (2016). Waste to energy technologies for municipal solid waste management in Gaziantep. *Renewable Sustainable Energy Reviews, 54*, 809e815.

Trapani, D. D., Alida, C., Marines, F. D., & Viviani, G. (2023). Biological stability assessment of MSW organic fractions by means of respirometric and germination tests. *Journal of Material Cycles and Waste Management, 25*, 1142–1152.

US DENREL (Department of Energy's National Renewable Energy Laboratory). (1996). *Evaluation of gasification and novel processes for the treatment of municipal solid waste.*

Williams, M., & Helm, A. (2011). *Waste-to-energy success factors in Sweden and the United States.* Analyzing the tranferability of the Swedish waste-to-energy model to the United States.

Zachl, A., Buchmayr, M., Gruber, J., Anca-Couce, A., Scharler, R., & Hochenauer, C. (2024). Evaluation of the influence of the fuel fines content on the stratified downdraft gasification of wood chips. *Fuel, 357*(2024), 129926.

Zaman, A. U., & Lehmann, S. (2013). The zero waste index: A performance measurement tool for waste management systems in a zero waste city. *Journal of Cleaner Production, 2013*(50), 123–132.

Zhang, D., Huang, G., Xu, Y., & Gong, Q. (2015). Waste to energy in China: Key challenges and opportunities. *Energies, 2015*(8), 14182–14196.

Zhao, X. G., Jiang, G. W., Li, A., & Wang, L. (2016). Economic analysis of waste-to-energy industry in China. *Waste Management, 48*, 604–618.

Ziraba, A. K., Haregu, T. N., & Mberu, B. (2016). A review and framework for understanding the potential impact of poor solid waste management on health in developing countries. *Archives Public Health, 2016*(74), 55.

Chapter 9
Modulated Vermibiotechnology for the Management of Solid Waste

Manmeet Kaur and Rouf Ahmad Bhat

Abstract Nowadays, unmanaged solid waste pose a major ecological threat. The waste generation rate due to rapid industrialization, demographic change and urbanization add huge quantity of waste into the environment. Huge amount of waste are disposed-off into landfills without prior treatment. Huge efforts have been put worldwide to sort a better technology for management and generation of revenue from wastes. The investment costs for waste treatment techniques such as gasification, pyrolysis and incineration are very high. Composting and vermicomposting are two methods of treating solid organic waste that are considered sustainable and profitable. Unfortunately, pollutants such as heavy metals are often found in trash, which are harmful to the bacteria that break down the waste. Due to its simplicity, biodegradability, high efficacy, portability, and environmental friendliness, vermicompost is an incredible asset for farmers. In light of the fact that arable land currently needs to be fertilized with biomaterials, vermi biotechnology is an appropriate strategy for sustainable development. The main objective of this book chapter is to comprehend how vermicomposting works as a management strategy to reduce solid organic waste.

Keywords Solid waste · Agricultural waste · Kitchen waste · Vermibiotechnology · Sustainable development

1 Introduction

Solid waste management is a significant ecological issue because of the quickening industrialization, population expansion, and urbanization. Massive amounts of mixed solid and industrial trash are being deposited without any pre-treatment because

M. Kaur
Department of Microbiology, Punjab Agricultural University, Ludhiana 141001, India

R. A. Bhat (✉)
Department of School Education, Jammu and Kashmir, India
e-mail: rufi.bhat@gmail.com

R. A. Bhat et al. (eds.), *Zero Waste Management Technologies*,
https://doi.org/10.1007/978-3-031-57275-3_9

recycling and sorting are still in their infancy (Ugwu et al., 2020). For communities to survive and function properly, proper solid waste management is essential and requires immediate action (Bui et al., 2020). Poor waste management is polluting the biosphere, including the oceans, rivers, and seas. This leads to flooding, drain blockages, disease transmission through vector breeding, increase in respiratory problems due to airborne particles created by waste burning, animal harm from unintentional waste consumption, and a significant decline in economic development (Ferronato & Torretta, 2019). However, garbage frequently includes toxins that are hazardous to decomposers, like heavy metals. As a result, heavy metals pose a problem for effective biological treatments (Sharma et al., 2020). If not managed and used efficiently, manure and other organic or green waste (items such as food, plant, or animal waste) can cause problems for farms. This waste is an important organic resource that can be recovered to produce soil conditioners and nutrient-rich fertilizers (Sayara et al., 2020). In addition, interest in organic farming has grown due to growing knowledge of the harmful effects of agrochemicals on human health (Calabi-Floody et al., 2018). Since organic farming uses natural resources wisely, it also promotes environmental protection (Kareem et al., 2022). Solid waste disposal techniques that involve incineration, pyrolysis, or gasification require significant upfront investments (Hoang et al., 2022). Solid organic waste composting and vermicomposting are considered economically feasible and environmentally sound waste management techniques (Kosamkar et al., 2017). Vermicomposting reduces organic waste faster than traditional composting. With this method, organic waste is decomposed by the action of bacteria and earthworms. When used properly, the resulting material (vermicompost) can be a very effective fertilizer or soil conditioner (Hajam et al., 2023). Vermicomposting, often called wormery, is a fast and efficient way to dispose of organic waste while recycling important resources such as nutrients (Kaur, 2020). Similar to traditional composting, the vermicomposting method involves the biological breakdown of organic waste in an aerobic atmosphere to provide stabilized organic fertilizer (Kauser & Khwairakpam, 2022). Vermicomposting, on the other hand, uses earthworms and is an ecobiotechnological process that, unlike composting, converts high-energy and complex organic materials into a stabilized, humus-like product (Lim et al., 2015). Vermicomposting is critical because it uses low-cost, environmentally friendly alternatives and the activities of the earthworm to turn waste into wealth (Cao et al., 2021). An important environmentally friendly technique for monetizing waste is vermibiotechnology. However, this discovery has caused people around the world to pay much attention to the use of earthworms for environmental and health reasons (Yatoo et al., 2022). Earthworms are used in a scientific approach to waste management that has earned the moniker "Cinderella of organic farming". The biological activity of earthworms creates vermicompost, which is the perfect substrate for preserving and restoring the natural properties of the soil (Akhila & Entoori, 2022). On the other hand, the widespread use of cutting-edge technology has accelerated the development of rural areas and broadened employment prospects. It is crucial that the country's scientific community make significant efforts to improve and standardize worm technology nationwide (Singh & Bhartiya, 2021a, 2021b).

2 Vermicompost

Earthworms devour a variety of organic waste, resulting in a 40–60% reduction in volume. The different types of wastes used in vermitechnology are shown in Fig. 1. Each earthworm weighs between 0.5 and 0.6 g, consumes waste equal to its body weight and excretes feces, which is approximately 50% of the waste it ingests daily (Ahmed & Al-Mutairi, 2022). Castings typically contain between 32 and 66% moisture and have a pH of 7.0. (Reddy et al., 1998). Compared to garden compost, vermicompost has a higher proportion (almost double) of macro and micronutrients. In addition, vermicompost increases nutrient uptake by plants and provides all the nutrients in an easily accessible form (Sreenivas et al., 2000). Earthworms eat various organic wastes, reducing the volume by 40 to 60%. Each earthworm weighs between 0.5 and 0.6 g, ingests waste equal to its body weight and excretes feces, which make up around 50% of its daily waste intake (Al-Maliki et al., 2021). Castings typically have a pH (7.0) and a moisture content (32–66%). Vermicompost has a greater proportion (almost twice as much) of macro- and micronutrients than garden compost (Sofo et al., 2020). Vermicompost also improves plant nutrient uptake, providing all nutrients in an easily digestible form (Sreenivas et al., 2000).

The use of earthworms to break down organic waste into substances similar to finely ground peat is called vermicomposting. Vermicomposting can be done without earthworms by using any composting method after the garbage is partially decomposed (Ahmad et al., 2021). Vermicompost is a type of organic fertilizer enriched with all beneficial soil bacteria and contains all the nutrients that plants absolutely need, including N, P, and K (Singh et al., 2020). In conventionally made

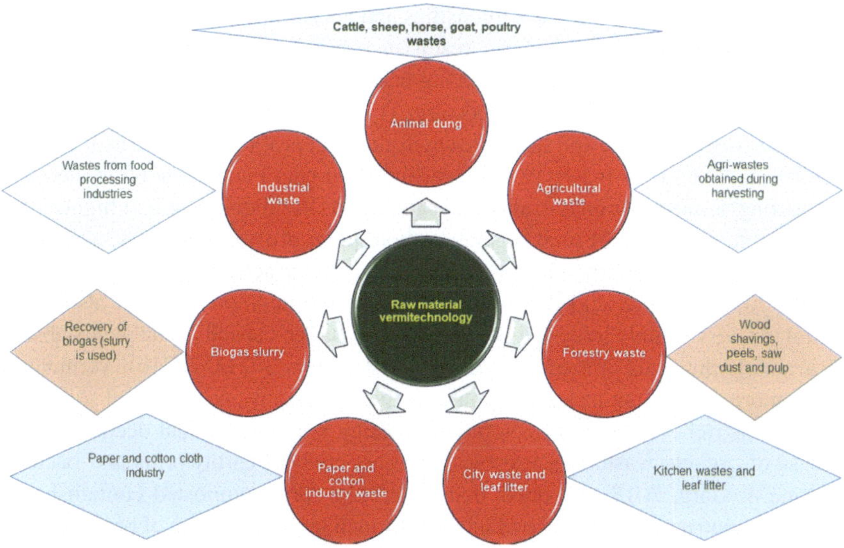

Fig. 1 Raw materials used in vermitechnology

vermicompost, standard concentrations of total nitrogen, phosphorus, and potassium are present at 1.94%, 0.47%, and 0.70%, respectively (Ramos et al., 2022). Numerous epigeal earthworms can be reared in organic waste and consume waste for food. Earthworm species such as *Eudrilus eugeniae*, *Perionyx excavatus*, and *Eisenia fetida* have become major competitors in the organic waste recycling market. Through aeration, bioconversion, excretions, and qualitative or quantitative effects on telluric microflora, earthworms accelerate the process of converting organic waste into elemental form (Kaviraj & Sharma, 2003). Vermicomposting can be an effective method of managing organic waste and converting it into valuable products using earthworms (Chauhan & Singh, 2014).

After eating various types of solid organic waste, earthworms produce vermicompost that resembles peat. Due to its high degree of humification and decomposition, vermicompost is more fragmented, porous and susceptible to microbial activity than the original waste (Raza et al., 2022). Vermicomposting is a quick and effective way to recycle organic waste and turn it into nutrient-rich compost through the action of earthworms and the presence of large populations of actinomycetes and beneficial bacteria (Vuković et al., 2021). It increases the capacity and porosity of drainage water, reducing the amount of water required for plant irrigation. It makes nutrients more accessible and can be used to create complicated fertilizer granules. According to Baghel et al. (2018) the application of vermicompost increased the yield by 9, 14, and 12 q/ha and the germination percentage of green chickpeas, garbanzo beans, and field peas by 92, 90, and 93, respectively. By using vermicompost the amount of harmful pathogens becomes clear reduced. It is generally believed that the thermophilic stage of the composting process occurs when the pathogenic organisms are killed. However, composting with earthworms increases and accelerates the rate of nitrogen mineralization. During the maturation phase of the vermicomposting process, the humification process intensifies and accelerates.

Additionally, this technique can result in a greater reduction in bioavailable heavy metals compared to the composting process. As a method of waste disposal, composting is inferior to vermicomposting. This is one of the most important steps to stabilize organic waste and solve the environmental problem (Mishra et al., 2014). Vermicomposting uses materials that are similar to peat but are finely divided and have strong aeration, drainage, and water retention. Vermicompost had higher than average concentrations of total organic carbon, plant growth regulators, accessible NPK, and other micronutrients, microbial and enzymatic activity, and organic matter (Malik, 2020). Humic acid is more common in vermicompost, which also has a neutral pH, less soluble ions, and better cation exchange capacity. The plants absorb the available nutrients immediately (Singh & Bhartiya, 2021a, 2021b). Vermicompost is more stable than its starting material, has better physico-chemical soil properties and is richer in nutrients. Vermicomposting is a biochemical decomposition process of organic waste that involves the interaction of earthworms and bacteria (Sharma & Garg, 2018). Products that have been vermicomposted contain larger amounts of nutrients available to plants and a much larger spectrum of probiotics for agriculture and aquaculture. Therefore, vermicompost is considered a type of microbial fertilizer that is mainly applied to agricultural soils (Sethi et al., 2023). Vermicompost can be made from animal manure as a substrate. Vermicompost encourages the growth of soil characteristics such as granularity, fine debris, efficient aeration, simple roots, and greater water holding capacity.

Vermicompost is nutrient-rich, physically and chemically superior to its starting material and more stable than soil. Earthworms and bacteria interact in the biochemical breakdown of organic waste, known as vermicomposting (Sharma & Garg, 2018). Vermicomposted products offer higher concentrations of plant-available nutrients and a broader spectrum of probiotics for agriculture and aquaculture. Vermicompost is therefore considered a type of microbial fertilizer that is mainly used on agricultural soils. Animal manure can be used as a substrate for making vermicompost. Vermicompost encourages the development of beneficial soil properties such as granularity, fine grain, effective aeration, easy roots, and increased water holding capacity (Lirikum et al., 2022). Some of the properties that organic fertilizer can improve are aggregation, bulk density, water retention and transfer, and the amount of organic matter in the soil. Vermicompost has been found to affect the growth and productivity of a wide range of crops, including cereals, legumes, sugarcane, ornamental and floral plants, and field crops. However, due to their high salt content, larger doses of vermicompost do not promote plant development. They also showed that the plant only responded more when vermicompost was treated at between 10 and 40 percent of the growing medium (Bansal & Kapoor, 2000).

The production of compost from organic waste by several earthworm species is compared in Table 1. The amount of nutrients such as phosphorous, zinc, potassium, calcium, and nitrogen in the final product was higher with combinations of earthworms and microorganisms compared to conventional compost (Busato et al., 2012).

3 Methods of Vermicomposting

Various vermicomposting methods (Beds or Bin, windrows, flow through reactors, tower, pit or trench, and troughs) can be described under 2 types of system (Pilli et al., 2019).

3.1 Batch System

By using a batch method, the bedding and feed are combined, the worms are added, and the process is left alone to finish.

Table 1 Chemical properties of different worms in vermicomposting

Composting material	Worms	N	K	P	C/P	C/N	References
Kitchen waste	*Lampito mauritti*	3.32 fold increase	1.13 fold increase	1.61 fold increase	1.35 fold decrease	2.79 fold decrease	Tripathi and Bhardwaj (2004)
Food waste	*Eisenia foetida*	3.6 fold increase	40–50%	34–54%	–	11–13 decrease	Yadav and Garg (2011)
Paddy straw	*Bacillus polymixa*	52–72% increase	40–80% increase	34–80% increase	–	Lowest from initial	Das et al. (2016)
Cowpea	*Pontoscolex corethrurus*	62% increase	38% increase	20% increase	–	11	Das et al. (2016)
Sugarcane trash	*Eudrillus eugineae*	63.31% increase	–	2–11 fold increase	–	9.6	Sonowal et al. (2014)
Vegetable waste	*Eudrillus eugineae*	63.75% increase	42–55% increase	31–58% increase	–	–	Biruntha et al. (2020)

3.2 *Continuous Flow Systems*

In continuous flow systems, worms are placed in bedding and then routinely introduced in small amounts of food and fresh bedding.

3.3 *Windrows Vermicomposting*

Most commercial vermicomposting in agriculture occurs through windows, which are wide rows of cow manure. Farmers often stack the manure in rows that are 100 feet or longer, 3 feet high, and 2 feet wide. Farmers sow worms in the swaths while ensuring moisture is maintained in the rows. Adding extra manure to the ends of the existing rows to trap the worms will speed up the process (Enebe & Erasmus, 2023). Various methods of vermicomposting in windrows are listed below.

(a) **Static Pile Windrows (Batch Type)**: Static piles are piles of mixed bedding and forage that have been injected with worms and remain in place until processing is complete (or bedding with forage piled on top). Typically, these posts are long and rickety-shaped, but they can also be square or rectangular. The maximum allowed size (before weaning) is one meter. In order for the worms to live in a healthy environment, the type and amount of bedding must be carefully selected.

(b) **Top-Fed Windrows (Continuous Flow)**: Swaths fed from above are not combined and stacked, but created as a continuous operation. This indicates that the feed is applied in several thin layers (less than 10 cm) after deworming and before bedding. Normally, upon contact with the food, the worms will perch, consume it and then dump their waste to the bottom of the swath. The heap is constructed with the finished product at the bottom, the partially used bedding in the middle and the fresher feed at the top. Regular application of new layers is necessary to replenish the bedding material, which the worms gradually destroy (Kiyasudeen et al., 2020).

(c) **Wedges (Continuous Flow)**: The vermicomposting wedge is an interesting twist on the top-feed windrow. A first batch of worms in the bedding is housed in a makeshift enclosure no more than a meter tall. The side walls of the enclosure can be made of concrete, wood, or even hay or straw bales. Fresh material is regularly fed in through the open side, usually with a shovel loader. The worms will eventually move away from the processed foods and toward the fresh foods. Once the material reaches the open end of the fence, the backing of the fence is removed and the finished product is harvested by scooping it out with a loader.

(d) **Beds or Bin Vermicomposting**: The easiest way to make vermicompost is in a container made of plastic or untreated, non-aromatic wood. To fill the container and provide the worms with a place to burrow, mix a few handfuls of soil with some type of bedding, such as shredded paper, composted animal manure, or dead leaves. Bedding also needs water to maintain moisture and allow the worms to breathe. Tea bags, coffee grounds, and other organic food scraps, including leftover fruit and vegetables, are good for the worms. Egg shells provide the worms with calcium and reduce the acidity of the barrel (Li et al., 2020).

(e) **Top-Fed Beds (Continuous Flow)**: A top-fed bed is an analog of a top-fed windrow. The main difference between a bed and windrow is that a bed is surrounded by four walls and typically a floor. In winter, bales of straw can be used to insulate the edges of the bed. The only insulation required may be a top layer or insulation pad if the containers are large enough, protected from wind and rain, and the feedstock contains sufficient nitrogen. This can be something as simple as straw bales or sacks.

(f) **Stacked Bins (Batch or Continuous Flow)**: Stacked bins offer a solution to the limited space problem by incorporating a vertical component into vermicomposting. The bins must be small enough to be lifted manually or with a forklift, even when filled with wet material. Continuous feeding is possible but requires routine care. Since the material is pre-mixed, worms are placed in the container, it is stacked for a period of time and then emptied, and it is more cost-effective to use a batch process.

3.4 Flow-Through Reactors Method

Stacked bins offer a solution to the limited space problem by incorporating a vertical component into vermicomposting. The bins must be small enough to be lifted manually or with a forklift, even when filled with wet material. Continuous feeding is possible but requires routine grooming. Since the material is pre-mixed, worms are placed in the container, it is stacked for a period of time and then emptied, and it is more cost-effective to use a batch process. Material often falls through after being loosened by a set of hydraulically powered breaker bars as they move down the bottom grid.

3.5 Trough Vermicomposting

Cement troughs can also be used for vermicomposting. Usually, only liquid manure that has matured for at least a week is put into the troughs. In this composting process, only a few centimeters of manure are initially placed on the bottom of the trough. Farmers put in worms and give them a few days to feast on the manure before applying another coat. Additional layers of manure are added every 10 days until the vermicompost has reached the top of the trough.

3.6 Tower Method of Vermicomposting

This process is quick and easy. Once in place, there is no need to clean it or remove the cast. Create multiple holes in a PVC pipe. Restock it after planting it permanently vertically in the center of the garden. Put some worms in there. Fill it with kitchen scraps once or twice a week. Ditch or pit some farmers dig a large hole and bury worms and organic waste in it to create vermicomposting pits. Of course, farmers must line the pit before adding worms and bedding to prevent worms from entering the surrounding soil. A good liner that keeps worms out while still allowing adequate water drainage is canvas feeder bags. Farmers place organic produce such as straw, grass clippings, and manure in the walled pit before covering it with soil. After about a week, farmers place worms in the pit and water it during this time to preserve moisture. The vermicomposting process begins as soon as the worms enter the pit (Rastegari et al., 2023).

3.7 Vermiwash

It is a liquid that accumulates after the water passes through a column created by worm movement and is great for spraying the leaves. It is a mixture of the mucus secreted by earthworms and their excretory secretions, as well as micronutrients derived from organic molecules in the soil. In the natural ecosystem, they are transferred to the leaves, shoots, and other parts of the plant (Sulaiman & Mohamad, 2020). When properly collected, vermiwash is a clear, translucent liquid with a pale yellow tint. In addition to acting as a biopesticide, it has been reported to have great growth-promoting effects. According to Zambare et al. (2008), vermiwash is an aqueous vermicompost extract extracted in the presence of a large population of earthworms. It contains a variety of enzymes, plant growth hormones, vitamins, micro and macronutrients that increase plants' resistance to various diseases and promote their growth and productivity. The sustainability of our agro-ecosystem is seriously threatened by the increasing use of inorganic fertilizers, pesticides and herbicides in the current farming system, as well as the use of limited water resources. In such a case, it is crucial to look for environmentally friendly and efficient solutions. The basic idea behind vermiwash preparation is that soils treated by earthworms have burrows made by them.

Bacteria living in these cavities, also called drilospheres, are abundant. When water flows through these tunnels, nutrients are flushed from the tunnels to the roots, where they are taken up by the plants. This idea is put into practice in the production of Vermiwash. Vermiwash takes 40 to 50 days to prepare to produce the clear, brown liquid that collects at the bottom of the barrel (Gudeta et al., 2021). Shukla and Singh (2010) found that seed germination and legume seedling growth are significantly influenced by nitrogen in the form of mucilage, nitrogenous excretory chemicals, growth, stimulating hormones, and enzymes in vermifuge. The growth and productivity of okra plants are strongly influenced by worm-composed weeds and their aqueous extract. Vermiwash is a natural plant growth supplement for horticultural crops such as tea, coconut, and coconut oil. Vermiwash is an organic fertilizer source containing inorganic N and K. According to Mishra et al. (2014), foliar fertilization is significantly more efficient than soil treatment under dry conditions. It was reported that spraying several tomato varieties with vermiwash significantly increased the plant's growth and fruit production. Foliar sprays with nutrients can help compensate for the reduced ability of roots to take up nutrients early in the reproductive phase due to competition with sinks for carbohydrates. Vermiwash is an excellent foliar spray and insecticide when mixed with diluted cow piss (1 liter vermiwash, 1 liter cow urine, and 8 liters water). Growth and yield of the marigold plants increased significantly through the use of vermifuge. Vermiwash was used weekly, resulting in a 7.3% increase in radish production. It can also be used as a liquid fertilizer (20% to 30% dilution) and prevents harmful fungi from forming their mycelium. It is evident that vermiwash is an effective source of promoting plant growth and controlling a range of microbial diseases. Although much research has been done on vermiwash, there is still no study on the effects of individual waste products and their binary combinations on plant development and agricultural yield. Various animal manure (cow, buffalo, goat, sheep, and horse) mixed with agricultural and kitchen waste has significant impacts on several vegetable crops (Chauhan & Singh, 2014).

4 Biological Wastes

If suitable management and disposal methods are not available, biological wastes produced by biological systems pose a risk to the environment and have a negative impact on both domesticated animals and human life (Alshehrei & Ameen, 2021a, 2021b). Large-scale fly ash production is a significant environmental issue. When coal is burned in power plants, fly ash, a fine residue in exhaust, results, posing severe waste management challenges. It changed the pH and sandy soil's ability to retain water. Numerous hazardous gases were created as a result of the garbage' microbial breakdown. Each year, India produces millions of tons of kitchen and agricultural trash. These wastes contributed to a number of odor and environmental issues in the neighborhood. Industrial waste from gaur gum released toxic gases that had major negative effects on the environment (Nath & Singh, 2016).

There are toxic problems with both household waste and municipal waste. The biggest waste problem in the field is the problem of post-harvest residues of various native crops, such as wheat, pearl millet, sorghum, and vegetable waste. The problems of waste disposal have increased due to the high level of industrialization (Ducasse et al., 2022). These problems range from agricultural waste, kitchen waste, animal waste, etc., to industrial and municipal-level solid waste from textile and sugar factories, the wine industry, dairy sludge, municipal solid waste, etc. Through microbial degradation, these wastes contribute to a number of diseases, odor problems, and pollution problems (Singh & Bhartiya, 2021a, 2021b).

Since the composting of organic matter and its activity when digging are considered promising for tillage, a straightforward, suitable biotechnology for vermicultivation has been developed, which can largely solve the problems of waste processing and management. Vermicompost can be made by using earthworms to break down a range of organic residues including sewage sludge, animal waste, crop residues, kitchen waste, and industrial waste (Gonzalez-Moreno et al., 2022). In fact, the worms increase microbial diversity and activity, which accelerates waste breakdown and nutrient recovery. The waste industry is growing at an annual rate of 2.8%. In addition to the bacteria present in the garbage, the worms' gut microbiota and gut enzymes are also involved in the decomposition (Albasha et al., 2015).

4.1 Municipal Solid Waste (MSW)

Composting and microbiological stabilization are the two most commonly used techniques for disposing of biodegradable waste. Due to the high organic content of

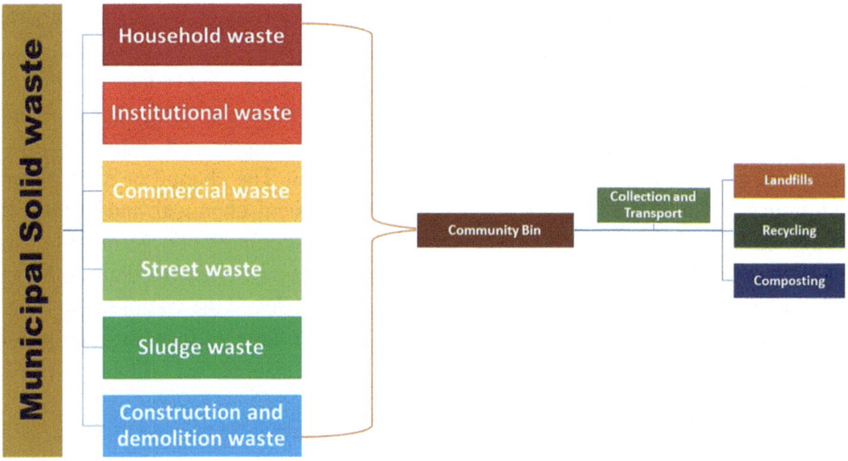

Fig. 2 Schematic flow diagram of municipal solid waste management

MSW, composting can be one of the best agricultural waste management strategies for MSW treatment (Ayilara et al., 2020). It also creates a layer of soil called a conditioner that has beneficial agronomic benefits, making it an effective and financially viable method of balancing organic waste. In addition, it reduces the amount of waste that needs to be disposed of, dramatically reducing the amount of leftover waste that can go to landfill (Srivastava et al., 2020). Collection, pollutant separation, size and mixture, biological decomposition, and other tasks are all part of the existing MSW composting system. The flow chart illustrates the unique method of managing municipal solid waste from source to final disposal as shown in Fig. 2.

4.2 *Agricultural Waste*

The amount of nutrients in vermicompost is mainly influenced by the type and composition of the agro-industrial waste. Nutrients are abundant in agricultural waste. Vermicompost from agro-industrial waste has a higher concentration of micro and macronutrients than the original feed mixture. Besides increasing plant growth and productivity by providing nutrients, vermicomposting is widely considered to be the most promising organic fertilizer available today. It is also inexpensive and environmentally friendly. The degradation activity of the earthworms promotes the mineralization of nutrients, which increases plant productivity. Vermicompost made from agricultural waste promotes soil health and growth, crop quality and yield, and pollution control. In order to solve the problem of agricultural waste disposal, vermicomposting production facilities are being created because this method does not produce hazardous effluents such as pesticide residues, weed seeds, heavy metals, termites or wax, plant root diseases. This is also possible with vermicompost, which can be used at any time for all agricultural, horticultural, ornamental, and vegetable crops. Therefore, disposing of fruit and vegetable waste with vermicompost will soon minimize the need for additional land, improve ecosystems, and reduce ecological risk (Mane & Raskar, 2012). Vermicompost has been shown to lower heavy metal levels in polluted soils in the Thailand region. Vermicompost and cadmium chloride were used in different amounts in the experiment. The change or decrease in cadmium content was calculated by comparing the physical, chemical, and biological properties of the soil before and after treatment with compost and vermicompost. There is encouraging evidence that vermicompost is more effective than compost alone at removing cadmium from sludge waste. This reduced the cadmium content. The amount of soil organic carbon and cadmium load decreased in vermicompost while soil pH and the availability of P, K, Na, Mg, and Ca increased (Nuntawut et al., 2010).

Another study examined the different integrated waste management strategies that can be used in conjunction with different composting systems. The solid waste was composted for 22 days before being subjected to the vermicomposting process. Samples were routinely analyzed for changes in carbon, pH, moisture, nitrogen, and temperature concentrations to determine the quality of the composting process. In addition, the relative amounts of carbon and nitrogen decreased and the moisture content dropped to about 32%. Compost made from cow manure and produced using the epigeal earthworm *E. fetida* has a higher concentration of nitrogen, phosphorus, and potassium nutrients than other types of waste (Wani et al., 2013). Ngakou et al. (2014) found that cow manure compost contained more N, P, and K than kitchen waste compost. By fertilizing maize with bacteria and manure compost fertilizer alone or in combination, researchers aimed to remediate degraded soils (Zhen et al., 2014). They found that using compost manure increased the number of microorganisms due to improved microbial activity and the diversity of degraded irrigation zones. Anwar et al. (2015) found that compost made by combining milk manure, wheat straw, and sawdust increased plant biomass. The compost made from rice straw and cow manure, on the other hand, had a high C/N ratio and high total nitrogen content, making it an excellent soil improver.

4.3 Kitchen Waste

Bioconversion of waste by vermicomposting produces earthworm biomass and vermicompost. Both the pH and the organic carbon content of kitchen waste are high. However, only very small amounts of some nutrients were found, such as total nitrogen-available phosphorus and exchangeable potassium. Vermicomposting has drastically changed the physical and chemical properties of kitchen waste, which could be a useful resource for organic farming. Kitchen waste contains large amounts of organic carbon (7.25%), organic matter (12.49%), and pH (9.32) (Bharadwaj, 2010). Other nutrients such as exchangeable potassium (0.086%), accessible phosphorus (0.11%), and total nitrogen (0.214%) were also present, albeit in incredibly low concentrations. The vermicomposting process has drastically changed the physical and chemical properties of the kitchen waste material, which could be a useful tool for organic farming. Earthworm biotechnology significantly reduces waste while increasing biomass, which is converted into a nutrient-rich form that can be used in agricultural production. Furthermore, vermicomposting process of kitchen waste reduce 61.94% mass of waste (Bharadwaj, 2010).

5 Organic Farming

One of the best practices for sustainable farming is organic farming, which makes waste control easier. This significantly supports the increase in food production, income opportunities, the preservation of soil quality, and environmental protection. The use of organic fertilizer on agricultural land is beneficial for sustainable agriculture. The promotion of organic farming and the use of organic waste were priorities of the tenth five-year plan. India has a long tradition of using organic farming methods. Some examples of fossil fuel-based inputs used in modern agriculture are chemical fertilizers, pesticides, herbicides, and labor-saving but energy-intensive farming equipment. Organic farming is a production system that favors the maximum utilization of organic matter and the use of synthetically manufactured agricultural products to maintain soil productivity, fertility, and pest control under the conditions of sustainable natural resources and a healthy environment (Ranganathan, 2006). In organic farming, vermicomposting is an excellent method to improve soil quality and provide a place for earthworms to dispose of waste. As it contributes to a broader connection between food, environmental quality, and human and animal health safety, this is one of the most exciting features. The use of earthworms for waste disposal is preferable to the use of microbes. A useful by-product of waste is vermicompost, which is used in agriculture as fertilizer to increase food production. Recycling through vermicomposting with earthworms helped solve the waste problem (Garg et al., 2006).

6 Earthworms

Other important tasks performed by earthworms include increasing soil fertility and stabilizing inorganic plant nutrients in organic forms. The worms doubled the inorganic nutrients and added vitamins and some plant growth hormones before depositing their precipitate in the compost. They also help in the disposal of solid organic waste that has rotted and been disposed of in areas where it has harmed human health, the environment, the air, and caused odor nuisance (Edwards & Arancon, 2022). According to Charles Darwin, worms appear to be the main scavengers of vegetation, perforating and shedding the soil, letting through rain and the fibers of the plant, while attracting straw, leaf stalks, and twigs. Most importantly, they continually excrete worm droppings, which are excellent fertilizers for grass and crops (Mondal et al., 2017). Although the worms are omnivores and grazers, they often have fastidious palates. They fed on a wide variety of organic matter, including living and dead animals, diatoms, algae, protozoa, nematodes, and living and dead bacteria and fungi. The garbage was completely shredded and stabilized into organic materials. Worms like soil that is sufficiently moist, warm, and has plenty of decaying organic debris, since they are nocturnal by nature. The bowels of the earth are how Aristotle described earthworms (Ranganathan, 2006). Dubbed the "Farmers Friend," the earthworm serves as a biological indicator of soil fertility. Earthworms support a healthy population of bacteria, fungi, actinomycetes, protozoa, insects, spiders, and centipedes, ensuring a healthy soil. Soil fertility results from maintaining soil quality. The breakdown process is active enough to release the nutrients plants need for optimal growth. Earthworms have developed ingenious systems for temperature regulation because all of their biological functions are isothermal bioreactors and are therefore sensitive to high temperatures (Suthar, 2007).

Soil distribution of earthworms is affected by a variety of factors including soil composition, aeration, temperature, moisture, pH, organic salts, organic matter, manure, and reproductive ability. According to Aalok et al. (2008), pH has a significant impact on diapause conditions in earthworms. They are the most common burrowing invertebrates and their activity has increased the soil's ability to hold water. This also creates ideal conditions for the aerobic growth of bacteria and plant roots. Earthworms could act as helpful biological sensors for soil contamination with harmful metals (Khyade & Pawar, 2016) Earthworms are an essential part of the food chain and can collect hazardous substances from the soil. However, earthworm population is mainly controlled by soil nutrient resources and external biotic parameters. By decomposing organic waste, which is known to accumulate pesticides and heavy metals such as lead, cadmium and nickel in their tissues, earthworms could reduce the pollution risks resulting from these processes (Lang et al., 2023).

Aalok et al. (2008) found that pH significantly affected earthworm diapause conditions. They are the most widespread invertebrates and help the soil retain water while promoting the growth of aerobic bacteria and plant roots. According to Khyade and Pawar (2016), earthworms can serve as useful biological indicators of hazardous metal contamination in soil. Earthworms can scavenge dangerous compounds from the soil and are an important part of the food chain. Both external biotic factors and soil nutrient supply play important roles in regulating earthworm populations. The decomposition of organic waste, which is known to accumulate pesticides and heavy metals such as lead, nickel, and cadmium and mercury poses pollution hazards that can be mitigated by earthworms (Samadhiya et al., 2013).

According to Aalok et al. (2008) the ability of earthworms to digest organic waste increases the surface area for microbial feces to attack. They help maintain a constant pH in the soil through their gut, which has a number of enzymes including lipases, proteases, cellulose, amylase, lichenases, and chitenases. All enzymes are particularly active in a narrow pH range (6.3 to 7.3). Earthworms can extract oxygen from the air and thus supply the intestinal flora. This kills soil pathogens and promotes a range of aerobic waste stabilization measures. They increase soil N-fixing activity by providing an ideal environment of food moisture and air for N-fixing bacteria (Lalandera et al., 2015).

6.1 Earthworms Suitable for Vermicomposting

Together, earthworms and their associated microorganisms controlled the biological waste produced by their metabolism and ensured that nutrients were retained within the worm. In addition to exotic species such as *D. vaneta, L. rubellus,* and *P. excavatus*, the two species *E. fetida* (Savigny) and *E. eugineae* (Kinberg) are mainly used in vermiculture in India. Repeated annual treatments of animal waste and poultry droppings result in higher concentrations of nutrient metals in the soil (Suthar, 2007). Others suggested the use of anecic and endogeic species that live underground. Some vermitechnologists have recommended the use of surface-dwelling epigeal species for vermicomposting.

Earthworms are important in the bioaccumulation of several heavy metals from animal faeces, including cobalt, chromium, cadmium, nickel, lead, and arsenic (Bhartiya & Singh, 2012a, 2012b). While there are more soil earthworms in temperate regions, tropical nations are home to more humus food worms than organic food worms. External biotic conditions and a lack of nutrition resources are the key factors limiting earthworm populations. Earthworms like the meal's nitrogen content because it encourages growth and reproduction. In soils with a moisture content of 12 to 45%, worms were most frequently observed. Their population growth is mostly influenced by the availability of biotic factors outside the soil. Worm-digested animal waste is used in horticultural composts in place of peat where clay is absent (Bo et al., 2021).

Earthworms are important for the bioaccumulation of several heavy metals from animal feces, including cobalt, chromium, cadmium, nickel, lead, and arsenic (Bhartiya & Singh, 2012a, 2012b). While there are more ground earthworms in temperate regions, there are more hummus foodworms than organic foodworms in tropical countries. External biotic conditions and a lack of food resources are the key factors limiting earthworm populations. Earthworms like the nitrogen content of the meal because it promotes growth and reproduction. Worms were observed most frequently in soils with a moisture content of 12 to 45%. Their population growth is mainly influenced by the availability of biotic factors outside the soil. Worm-digested animal waste is used in horticultural compost in place of peat where clay is absent (Cai et al., 2020). Repeated annual treatment of animal waste and poultry litter leads to increased soil nutrient metal concentrations. Worm culture is a revolutionary biotechnology that involves the reproduction and growth of the earthworm *Eisenia fetida* and the use of its feces as an important method of waste recycling and conversion into vermicompost. The earthworms are used as low-cost and sustainable bioreactors for waste disposal. When comparing the various combinations of fly ash and cow manure with inoculation of *Eudrilus eugeniae*, nutrient availability was significantly better with the 1:3 treatments (fly ash to cow manure). Substantial amounts of phosphate-solubilizing bacteria and nitrogen-fixing cellulolytic microorganisms were present in the earthworm strain *E. eugeniae*. *L. mauritii* recommended the use of activated sludge treatment to stop pollution and expressed that the species prefers sites with a high C/N ratio. *E. eugeniae* population and growth are directly affected by time and space. Earthworms are an essential biotic component of soil. They support soil health and comply with environmental regulations by breaking down decaying organic residues (Samadhiya et al., 2013).

7 Solid Waste Pollutants

Depending on the components that make it up, municipal solid waste is divided into many types. In a typical garbage audit, garbage samples from final disposal sites or generators are collected, grouped, and then weighed to determine the toxins in the garbage (Grasserova et al., 2020). The five main categories of municipal waste (MSW) are food waste, metal, paper, plastic, and glass. The largest type of waste in the world, which can account for up to 44% of all waste, is green waste, which includes food waste. Another significant contribution is the 38 percent share of dry recyclables, which include paper, metal, plastic, cardboard, and glass (Deus et al., 2020). Another important factor in pollutant waste is income. With increasing income, the proportion of organic matter decreases. Compared to low-income countries, high-income countries produce more plastic, paper, wood, and rubber waste. Additionally, high-income areas generate 51% more dry, recyclable waste (paper, plastic, metal, cardboard, and glass) than low-income areas (32% less green and food waste and 32% more dry, recyclable waste, respectively). While low- and middle-income countries generate 56% and 53% food waste, respectively, (Alsheehrei & Ameen, 2021a, 2021b).

8 Waste Management Techniques

A large amount of greenhouse gas emissions, mainly CH_4, CO_2, and N_2O, are generated by waste management practices such as incineration, landfill, pyrolysis, vermicomposting, anaerobic digestion, and composting as shown in Table 2 (Rahman & Khondaker, 2012). Vermicomposting is superior to all other waste management methods because it allows year-round composting and can be done both indoors and outdoors. This process enables the rapid production of organic nutrient sources for crops that are more potent in terms of nutrition, physical performance, and biochemistry than other composts (Yadav et al., 2010). A well-known, cost-effective technical method for treating organic waste is worm composting. Vermicomposting is a

Table 2 List of technologies utilized and their impact on environment and economy

S. No.	Waste management technologies	Effect on environment	Effect on economy
1	Incineration	Air emissions	Thermal energy
2	Waste dumping	Odor created due to water pollution	High operational cost
3	Pyrolysis	Degrade plastic, air emissions	Crude oil
4	Vermicomposting	Degrade organic waste using earthworms, reduce heavy metal	Low operational cost
5	Anaerobic digestion	Degrade organic waste, green houses gases emission decreased	Bioenergy
6	Composting	Biodegrade organic waste only	Organic fertilizers

technique that allows solid waste to be turned into invaluable compost that can effectively replace chemical fertilizers while reducing pollution (Monroy et al., 2009). It is therefore imperative to replace these widespread waste management methods with state-of-the-art vermicomposting and waste incineration processes (Khoshnevisan et al., 2021).

9 Detoxification of Pollutants

9.1 Role of Biochar

The effectiveness of the compost and the condition of the soil are directly influenced by the use of biochar. Due to its stable structure, biochar often has a porous, light and carbon-rich component (Pudełko et al., 2021). Despite its high ash content and low nutritional value, it can strengthen soil's physical structure and increase yield. Biochar has been shown in numerous studies to promote plant productivity by sequestering carbon and improving soil physical properties (Rehman et al., 2020). During composting, organic solid waste is converted into humus. Earthworms are present during the humification (decomposition) phase of vermicomposting, which occurs under non-thermophilic conditions (Fernando & Arunakumara, 2021). Through the vermicomposting process, the organic solid waste is stabilized into a soil-enhancing compost that smells dark, nutritious, and earthy. According to Abul-Soud et al. (2009) it also contains significant amounts of major and micronutrients. It is important to emphasize that biochar accelerates the breakdown of heavy metals during vermicomposting and lowers heavy metal levels by absorbing their soluble form (Paul et al., 2020). Biochar is known to improve free air space when airflow increases during composting, improving the environment for aerobic bacteria to thrive and speeding up decomposition (Antonangelo et al., 2020). The porous structure of the biochar absorbs the additional water produced during composting and thus prevents the formation of seepage water. Biochar also acts as a biocatalyst during the decomposition process, accelerating the breakdown of organic materials and producing high-quality compost more quickly (Siedt et al., 2020). The immobilization of heavy metals by biochar and the simultaneous reduction of their bioavailability minimizes the impact of metals (contained in household waste) on bacterial growth and metabolism (Boostani et al., 2021).

9.2 Role of Eisenia Fetida

The Indian subcontinent received the European species Eisenia fetida. It reproduces faster than other earthworm species such as *E. eugeniae* and *P. excavates* (Kumar et al., 2020). It is known that during the vermicomposting process, earthworms make compost by eating complex organic matter and excreting simpler organic matter (Van Groenigen et al., 2019). E. fetida is extremely effective at converting organic waste products such as food scraps, yard waste, agricultural waste, household waste, and more into high-quality compost (Yuvaraj et al., 2020). Due to the biomineralization process, heavy metals are better bound and have a lower bioavailability. Earthworm skin tissues produce extracellular polymeric substance (EPS) that binds heavy metals, reducing their mobility and keeping them attached to the skin during metal exposure (Khan et al., 2019). This EPS acts as an enzyme to promote bacterial growth and the breakdown of waste. In addition, EPS, which consists mainly of proteins and carbohydrates, has been found to increase the nutrient content of compost (Guhra et al., 2020). Because they produce high-quality compost at the end of the worm composting process, earthworms are now of great importance (Thakur et al., 2021).

10 Vermicomposting Process

10.1 Role of Microalgae

It is well-known that microbes play an important role in the biodegradation of complicated chemical compounds into less complex ones. Higher levels of galactosidase, dehydrogenases, proteases, and phosphatases were found in worm composting of duckweed (Gusain & Suthar, 2020). It has also been found that adding nanocarbon or biochar to a dietary supplement significantly increases microbial activity, which in turn increases degradation rate and enzymatic activity (Cao et al., 2020). Leachate was composted and the nutrients from the leachate were successfully converted into algal biomass, which promoted the growth of bacteria and algae in the co-culture

(Tossavainen et al., 2017). In addition, it was found that the spread of algal biomass had little impact on bacterial development. Algae cultivation in composting can improve the nutrient content of the final product and act as a soil conditioner, both of which result in a higher yield (Alobwede et al., 2019). Algal biomass can be used to remove nitrogen and phosphorus from wastewater and improve final compost performance (Cole et al., 2017). Algae can therefore act as a source of nitrogen and phosphorus throughout the composting process in order to improve the N, P, and K content of the finished compost and produce a nitrogen-rich end product, which is necessary for plant growth. Research on the microbial community and its biodynamics is needed to better understand the role of the microbial community (especially microalgae) in the degradation process that contributes to the production of high-quality compost (Lu & Xiao, 2022).

10.2 Bioavailability of Heavy Metals

Heavy metals, such as cadmium, nickel, copper, are known to hinder vermicomposting by suppressing bacterial growth. Metal stress, however, also promotes the reproduction of earthworms. However, earthworms have a unique strategy to mitigate their harmful effects (Wang et al., 2013). When earthworms are exposed to metal stress, they produce a mucus-like substance that helps reduce metal bioavailability through the formation of organometallic complexes and the bioaccumulation of metals on the surface of their tissues (Swati & Hait, 2017). EPS, produced by earthworms and the microbial communities involved in vermicomposting, has also been shown to slow down metal mobility through enzymatic action in gut and gypsum-related processes (Sun et al., 2020). The organic matter excreted by earthworms has functional groups that bind to the metals and form organometallic complexes that reduce metal toxicity due to the lower metal load associated with worm composting (Bai et al., 2020). The bioavailability of metals is therefore reduced by the bioconversion of metals during the composting process (Karimi et al., 2020).

11 Recommendations for Solid Waste Management

The enormous amount of solid waste produced is a testament to its size and is considered one of the biggest problems for neighboring communities (Koval & Mikhno, 2019). To overcome this significant obstacle, the government and local authorities are taking action by issuing a number of plans and policies (Vaverkova et al., 2020). The main obstacles to effective municipal waste management include a lack of competent private operators, poor administrative and institutional arrangements, a skilled workforce, and inconsistent research and data collection (Astrup et al., 2015). As a result, the following suggestions and recommendations for sustainable waste management were made:

11.1 Solid Waste Management Plans

There should first be a political strategy and planning for effective waste management. The achievement of targets and the application of policies are supported by these regional and local waste management plans (Al-Ajlan et al., 2006). The primary objective of these Waste Disposal Guidelines is to educate readers about the types, amounts, and disposal methods of waste. These methods also help in locating areas where specific methods are used to reduce technological waste (Memon, 2010). It also allows to understand the costs related to the disposal and treatment of solid waste.

11.2 Zero-Waste Policy

Government and local authorities could encourage the use of reusable items with a longer shelf life to support a zero-waste policy (Vaverkova et al., 2020). When creating a product, maximum durability and component recycling should be considered. So that one person's waste can be used as raw material for another person and pollution is prevented, zero-waste rules help to integrate sectors, companies, and communities (Ouda et al., 2018). To ensure that the parts of electronic devices can be reused in the future, toxic elements should be disposed of properly and not incinerated (Panda et al., 2010). As incineration causes greenhouse gas emissions, it is not the only solution for solid waste going to landfill. Application of this technique will help eliminate the major threats to human, animal, and plant health posed by air, water, and land pollution (Khan & Kaneesamkandi, 2013).

11.3 Waste Sorting

It is possible to create a sustainable environment by separating solid waste individually (Nilsson, 2010). The three-bin system should be implemented by local governments (Sharma & McBean, 2009). Due to the different labeling or coloring, the solid waste can be disposed of separately (Lebersorger & Schneider, 2011). The recyclable solid waste saves time and labor and can then be sent directly to recycling companies for processing (Sharma & McBean, 2009).

11.4 Recycling

According to Alruqaie and Alharbi (2012), the 3Rs system that promotes energy recovery, recycling, and reuse is still in its infancy and needs to be implemented consistently. Solid waste recycling requires a lot of labor and is carried out by the unorganized sector (Nilsson, 2010). In order to meet the goals of sustainable waste management, recycling should be more frequent.

11.5 Enforcing Law

In order to create a sustainable waste management system, the government must properly enforce the law against people who break the law by disposing of garbage irresponsibly. According to local authorities, the pay-as-you-toss scheme should be introduced (Lebersorger & Schneider, 2011). Second, a law should be passed that requires the payment of an excise tax in addition to product fees when a consumer purchases a specific product. The funds collected are then used to cover the cost of waste disposal.

11.6 Composting

Composting is a waste management technique to reduce the amount of trash to be disposed of. This compost contains a significant amount of organic matter that can be distributed to farmers to improve soil fertility and crop productivity (Adhikari et al., 2008). The development of state-of-the-art technology that can help compost solid waste and accept it as a soil amendment product should be supported by government (Turki et al., 2016).

12 Conclusion

The use of epigeal earthworms reduces the risk of pollution from the decomposition of organic waste, as shown by the above results. A good way to turn trash into nutrient-rich organic fertilizer is vermicomposting. Vermicompost and vermifuge will help the farmland as the plants will thrive and produce more. Recycling various wastes is beneficial when introducing earthworms into cultivated soil. We can conclude that vermibiotechnology is an effective strategy for sustainable development due to the extensive use of vermicompost, vermifuge, and earthworms in the fields.

References

Aalok, A., Tripathi, A. K., & Soni, P. (2008). Vermicomposting: A better option for organic solid waste management. *Journal of Human Ecology, 24*, 54–64.

Abul-Soud, M., Hassanein, M. K., Ablmaaty, S. M., Medany, M., & Abu-Hadid, A. F. (2009). Vermiculture and vermicomposting technologies use in sustainable agriculture in Egypt. *Journal of Agricultural Research, 87*, 389–401.

Adhikari, B. K., Barrington, S., Martinez, J, King S. (2008). Characterization of food waste and bulking agents for composting. *Waste Management, 28*(5), 795–804.

Ahmad, A., Aslam, Z., Bellitürk, K., Iqbal, N., Idrees, M., Nawaz, M., & Aziz, M. M. (2021). Earth worms and vermicomposting: A review on the story of black gold. *Journal of Innovative Sciences, 7*(1), 167–173.

Ahmed, N., & Al-Mutairi, K. A. (2022). Earthworms effect on microbial population and soil fertility as well as their interaction with agriculture practices. *Sustainability, 14*(13), 7803.

Akhila, A., & Entoori, K. (2022). Role of earthworms in soil fertility and its impact on agriculture: A review. *International Journal of Fauna and Biological Studies, 9*, 55–63.

Al-Ajlan, S. A., Al-Ibrahim, A. M., Abdulkhaleq, M., & Alghamdi, F. (2006). Developing sustainable energy policies for electrical energy conservation in Saudi Arabia. *Energ Policy, 34*, 1556–1565.

Albasha, M. O., Gupta, P., & Ramteke, P. W. (2015) Management of kitchen waste by vermicomposting using earthworm, *Eudrilus Eugeniae. International Conference on Advances in Agricultural, Biological & Environmental Sciences (AABES-2015) July 22–23, 2015 London (UK).*

Al-Maliki, S., Al-Taey, D. K., & Al-Mammori, H. Z. (2021). Earthworms and eco-consequences: Considerations to soil biological indicators and plant function: A review. *Acta Ecologica Sinica, 41*(6), 512–523.

Alobwede, E., Leake, J. R., & Pandhal, J. (2019). Circular economy fertilization: Testing micro and macro algal species as soil improvers and nutrient sources for crop production in greenhouse and field conditions. *Geoderma, 334*, 113–123.

Alruqaie, I. M., & Alharbi, B. H. (2012). Environmental advantage assessment of recycling food waste in Riyadh Saudi Arabia. *Research Journal Environment Science, 6*, 230–237.

Alshehrei, F., & Ameen, F. (2021a). Vermicomposting: A management tool to mitigate solid waste. *Saudi Journal of Biological Science, 28*, 84–93.

Alshehrei, F., & Ameen, F. (2021b). Vermicomposting: A management tool to mitigate solid waste. *Saudi Journal of Biological Sciences, 28*(6), 3284–3293.

Antonangelo, J. A., Sun, X., & Zhang, H. (2020). The roles of co-composted biochar (COMBI) in improving soil quality, crop productivity, and toxic metal amelioration. *Journal of Environmental Management, 277*, 111443.

Anwar, Z., Irshad, M., Fareed, I., & Saleem, A. (2015). Characterization and recycling of organic waste after co-composting-a review. *Journal of Agricultural Science, 7*, 68–79.

Astrup, T. F., Tonini, D., Turconi, R., & Boldrin, A. (2015). Life cycle assessment of thermal waste-to-energy technologies: Review and recommendations. *Waste Management, 37*, 104–115.

Ayilara, M. S., Olanrewaju, O. S., Babalola, O. O., & Odeyemi, O. (2020). Waste management through composting: Challenges and potentials. *Sustainable, 12*, 44–56.

Baghel, B., Sahu, R., & Pandey, D. (2018). Vermicomposting an economical enterprise for nutrient and waste management for rural agriculture. *International Journal Current Microbiology Applications Science, 7*, 54–58.

Bai, H., Luo, M., Wei, S., Jiang, Z., & He, M. (2020). The vital function of humic acid with different molecular weight in controlling Cd and Pb bioavailability and toxicity to earthworm (Eisenia fetida) in soil. *Environmental Pollution, 261*, 114222.

Bansal, S., & Kapoor, K. K. (2000). Vermicomposting of crop residues and cattle dung with *Eisenia fetida. Bioresource Technology, 73*, 95–98.

Bharadwaj, A. (2010). Management of kitchen waste material through vermicomposting. *Asian Journal Experimental Biology Science, 1*, 175–177.

Bhartiya, D. K., & Singh, K. (2012a). Heavy metals accumulation from municipal solid wastes with different animal dung through vermicomposting by earthworm *Eisenia fetida. World Applied Sciences Journal, 17*, 33–39.

Bhartiya, D. K., & Singh, K. (2012b). Heavy metals remediation from maize (Zea mays) crop by the use of vermicomposts through vermicomposting by *Eisenia fetida. American-Eurasian Journal Agriculture Environmental Science, 12*, 15–22.

Biruntha, M., Karmegam, N., Archana, J., Karunai Selvi, B., John Paul, J. A., Balamuralikrishnan, B., Chang, S. W., & Ravindran, B. (2020). Vermiconversion of biowastes with low-to-high C/N ratio into value added vermicompost. *Bioresource Technology, 297*, 122398.

Bo, Z. H. O. U., Yiyong, C. H. E. N., Zhang, C., Jianlong, L. I., Hao, T. A. N. G., Jiayu, L. I. U., Jun, D. A. I., & Jinchi, T. A. N. G. (2021). Earthworm biomass and population structure are negatively associated with changes in organic residue nitrogen concentration during vermicomposting. *Pedosphere, 31*(3), 433–439.

Boostani, H. R., Hardie, A. G., Najafi-Ghiri, M., & Khalili, D. (2021). The effect of soil moisture regime and biochar application on lead (Pb) stabilization in a contaminated soil. *Ecotoxicology and Environmental Safety, 208*, 111626.

Bui, T. D., Tsai, F. M., Tseng, M.-L., & Ali, M. H. (2020). Identifying sustainable solid waste management barriers in practice using the fuzzy Delphi method. *Resources, Conservation and Recycling, 154*, 104625.

Busato, J. G., Lima, L. S., Aguiar, N. O., Canellas, L. P., & Olivares, F. L. (2012). Changes in labile phosphorus forms during maturation of vermicompost enriched with phosphorus-solubilizing and diazotrophic bacteria. *Bioresource Technology, 110*, 390–395.

Cai, S., Wang, J., Lv, W., Xu, S., & Zhu, H. (2020). Nitrogen fertilization alters the effects of earthworms on soil physicochemical properties and bacterial community structure. *Applied Soil Ecology, 150*, 103478.

Calabi-Floody, M., Medina, J., Rumpel, C., Condron, L. M., Hernandez, M., Dumont, M., & de La Luz Mora, M. (2018). Smart fertilizers as a strategy for sustainable agriculture. *Advances in Agronomy, 147*, 119–157.

Cao, Y., Tian, Y., Wu, Q., Li, J., & Zhu, H. (2021). Vermicomposting of livestock manure as affected by carbon-rich additives (straw, biochar and nanocarbon): A comprehensive evaluation of earthworm performance, microbial activities, metabolic functions and vermicompost quality. *Bioresource Technology, 320*, 124404.

Chauhan, H. K., & Singh, K. (2014). Potency of vermiwash with *Azadirachta indica* A. Juss on yield of gram (*Cicer arietinum*) and infestation of Helicoverpa armigera (Hübner)'. *America-Eurasian Journal Toxicology Science, 6*, 87–93.

Cole, A. J., Paul, N. A., De Nys, R., & Roberts, D. A. (2017). Good for sewage treatment and good for agriculture: Algal based compost and biochar. *Journal of Environmental Management, 200*, 105–113.

Das, D., Bhattacharyya, P., Ghosh, B., & Banik, P. (2016). Bioconversion and biodynamics of *Eisenia foetida* in different organic wastes through microbially enriched vermiconversion technologies. *Ecological Engineering, 86*, 154–161.

Deus, R. M., Mele, F. D., Bezerra, B. S., & Battistelle, R. A. G. (2020). A municipal solid waste indicator for environmental impact: Assessment and identification of best management practices. *Journal of Cleaner Production, 242*, 118433.

Ducasse, V., Capowiez, Y., & Peigné, J. (2022). Vermicomposting of municipal solid waste as a possible lever for the development of sustainable agriculture. A review. *Agronomy for Sustainable Development, 42*(5), 89.

Edwards, C. A., & Arancon, N. Q. (2022). The role of earthworms in organic matter and nutrient cycles. In *Biology and Ecology of Earthworms* (pp. 233–274). Springer US.

Enebe, M. C., & Erasmus, M. (2023). Vermicomposting technology-A perspective on vermicompost production technologies, limitations and prospects. *Journal of Environmental Management, 345*, 118585.

Fernando, K. M. C., & Arunakumara, K. K. I. U. (2021). Sustainable organic waste management and nutrients replenishment in the soil by vermicompost: A review.

Ferronato, N., & Torretta, V. (2019). Waste mismanagement in developing countries: A review of global issues. *International Journal of Environmental Research and Public Health, 16*, 1060.

Garg, P., Gupta, A., & Satya, S. (2006). Vermicomposting of different types of waste using Eisenia fetida: A comparative study. *Bioresource Technology, 97*, 91–95.

Gonzalez-Moreno, M. A., García Gracianteparaluceta, B., Marcelino Sádaba, S., Prieto Cobo, E., & Seco Meneses, A. (2022). Vermicomposting of lavender waste: A biological laboratory investigation. *Agronomy, 12*(12), 2957.

Grasserova, A., Hanc, A., Innemanová, P., & Cajthaml, T. (2020). Composting and vermicomposting used to break down and remove pollutants from organic waste: A mini review. *European Journal of Environmental Sciences, 10*(1), 9–14.

Gudeta, K., Julka, J. M., Kumar, A., Bhagat, A., & Kumari, A. (2021). Vermiwash: An agent of disease and pest control in soil, a review. *Heliyon, 7*(3).

Guhra, T., Stolze, K., Schweizer, S., & Totsche, K. U. (2020). Earthworm mucus contributes to the formation of organo-mineral associations in soil. *Soil Biology and Biochemistry, 145*, 107785.

Gusain, R., & Suthar, S. (2020). Vermicomposting of duckweed (*Spirodela polyrhiza*) by employing *Eisenia fetida*: Changes in nutrient contents, microbial enzyme activities and earthworm biodynamics. *Bioresource Technology, 12*, 123585.

Hajam, Y. A., Kumar, R., & Kumar, A. (2023). Environmental waste management strategies and Vermi transformation for sustainable development. *Environmental Challenges*, 100747.

Hoang, A. T., Varbanov, P. S., Nižetić, S., Sirohi, R., Pandey, A., Luque, R., & Ng, K. H. (2022). Perspective review on municipal solid waste-to-energy route: Characteristics, management strategy, and role in circular economy. *Journal of Cleaner Production, 359*, 131897.

Kareem, A., Farooqi, Z. U. R., Kalsom, A., Mohy-Ud-Din, W., Hussain, M. M., Raza, M., & Khursheed, M. M. (2022). Organic farming for sustainable soil use, management, food production and climate change mitigation. *Sustainable Agriculture: Technical Progressions and Transitions*, 39–59.

Karimi, F., Rahimi, G., & Kolahchi, Z. (2020). Interaction effects of salinity, sewage sludge, and earthworms on the fractionations of Zn and Cu, and the metals uptake by the earthworms in a Zn-and Cu-contaminated calcareous soil. *Environmental Science and Pollution Research, 27*, 10565–10580.

Kaur, T. (2020). Vermicomposting: An effective option for recycling organic wastes. In S. K. Das, Organic Agriculture. https://doi.org/10.5772/intechopen.91892

Kauser, H., & Khwairakpam, M. (2022). Organic waste management by two-stage composting process to decrease the time required for vermicomposting. *Environmental Technology and Innovation, 25*, 102193.

Kaviraj, & Sharma, S. (2003). Municipal solid waste management through vermicomposting employing exotic and local species of earthworms. *Bioresource Technology, 90*, 69–73.

Khan, M. B., Cui, X., Jilani, G., Lazzat, U., Zehra, A., Hamid, Y., Hussain, B., Tang, L., Yang, X., & He, Z. (2019). *Eisenia fetida* and biochar synergistically alleviate the heavy metals content during valorization of biosolids via enhancing vermicompost quality. *Science of the Total Environment, 684*, 597–609.

Khan, M. S. M., & Kaneesamkandi, Z. (2013). Biodegradable waste to biogas: Renewable energy option for the Kingdom of Saudi Arabia. *International Journal of Innovation and Applied Studies, 4*, 101–113.

Khoshnevisan, B., Duan, N., Tsapekos, P., Awasthi, M. K., Liu, Z., Mohammadi, A., Angelidaki, I., Tsang, D. C., Zhang, Z., Pan, J., Ma, L., & Liu, H. (2021). A critical review on livestock manure biorefinery technologies: Sustainability, challenges, and future perspectives. *Renewable and Sustainable Energy Reviews, 135*, 110033.

Khyade, V. B., & Pawar, S. R. (2016). Physical, nutritional and biochemical status of vermiwash produced by two earthworm species *Lampito mauritii* (L) and *Eudrillus eugeniae* (L)'. *World Science News, 42*, 28–55.

Kiyasudeen, K., Ibrahim, M. H., & Ismail, S. A. (2020). Vermicomposting of organic wastes and the production of vermicompost. In *Biovalorisation of Wastes to Renewable Chemicals and Biofuels* (pp. 277–285). Elsevier.

Kosamkar, V., Chavan, F.I., & Hussain, M. (2017) Solid waste management by vermicomposting. *International Conference Proceeding ICGTETM Dec 2017|ISSN: 2320–2882.*

Koval, V., & Mikhno, I. (2019). Ecological sustainability preservation of national economy by waste management methods. *Economics Ecological Socium, 3*, 30–40.

Kumar, R., Sharma, P., Gupta, R. K., Kumar, S., Sharma, M. M. M., Singh, S., & Pradhan, G. (2020). Earthworms for eco-friendly resource efficient agriculture. *Resources use Efficiency in Agriculture*, 47–84.

Lalandera, C. H., Komakecha, A. J., & Vinnerasa, B. (2015). Vermicomposting as manure management strategy for urban small-holder animal farms—Kampala case study. *Waste Management, 39*, 96–103.

Lang, B., Betancur-Corredor, B., & Russell, D. J. (2023). Earthworms increase mineral soil nitrogen content–a meta-analysis. *Soil Organisms, 95*(1), 1–16.

Lebersorger, S., & Schneider, F. (2011). Discussion on the methodology for determining food waste in household waste composition studies. *Waste Management, 31*, 24–33.

Li, W., Bhat, S. A., Li, J., Cui, G., Wei, Y., Yamada, T., & Li, F. (2020). Effect of excess activated sludge on vermicomposting of fruit and vegetable waste by using novel vermireactor. *Bioresource Technology, 302*, 122816.

Lim, L. S., Wu, T. Y., Lim, P. N., & Shak, K. P. Y. (2015). The use of vermicompost in organic farming: Overview, effects on soil and economics. *Journal of the Science of Food and Agriculture, 95*, 43–56.

Lirikum, Kakati, L. N., Thyug, L., & Mozhui, L. (2022). Vermicomposting: an eco-friendly approach for waste management and nutrient enhancement. *Tropical Ecology, 63*(3), 325–337.

Lu, Q., & Xiao, Y. (2022). From manure to high-value fertilizer: The employment of microalgae as a nutrient carrier for sustainable agriculture. *Algal Research, 67*, 102855.

Malik, J. A. (2020). Urban solid waste management techniques with special reference to vermicomposting. In *Waste Management Techniques for Improved Environmental and Public Health: Emerging Research and Opportunities* (pp. 53–79). IGI Global.

Mane, T. T., & Raskar, S. S. (2012). Management of agriculture waste from market yard through vermicomposting. *Research Journal of Record Science, 15*(2277), 2502.

Memon, M. A. (2010). Integrated solid waste management based on the 3R approach. *Journal Materials Cycles Waste Management, 12*, 30–40.

Mishra, K., Singh, K., & Tripathi, C. P. M. (2014). Management of infestation of pod borer (*Lucinodes orbonalis Guenee*) and productivity enhancement of brinjal (*Solanum melogena*) through vermiwash with biopesticide. *International Journal of Advanced Research, 2*, 80–89.

Mondal, T., Datta, J. K., & Mondal, N. K. (2017). Chemical fertilizer in conjunction with biofertilizer and vermicompost induced changes in morpho-physiological and biochemical traits of mustard crop. *Journal of the Saudi Society of Agricultural Sciences, 16*, 35–44.

Monroy, F., Aira, M., & Domínguez, J. (2009). Reduction of total coliform numbers during vermicomposting is caused by short-term direct effects of earthworms on microorganisms and depends on the dose of application of pig slurry. *Science of the Total Environment, 407*, 11–16.

Nath, S., & Singh, K. (2016). Analysis of different nutrients status of liquid bio-fertilizer of different combinations of buffalo dung with gram bran and water hyacinth through vermicomposting by *Eisenia fetida. Environmental Development Sustainable, 18*, 45–56.

Ngakou, A., Koehler, H., & Ngueliaha, H. C. (2014). The role of cow dung and kitchen manure composts and their non-aerated compost teas in reducing the incidence of foliar diseases of *Lycopersicon esculentum* (Mill). *International Journal of Agriculture Research Innovation Technology, 4*, 88–97.

Nilsson, P. (2010). Waste collection: Equipment and vehicles. *Solid Waste Technology Management, 1*, 51–76.

Nuntawut, C.-N., Chuleemas, B. I., & Mongkon, T.-O. (2010). Vermicompost: Tool for agro-industrial waste management and sustainable agriculture. *International Journal of Environmental Rural Development, 1*, 38–43.

Ouda, O., Peterson, H., Rehan, M., Sadef, Y., Alghazo, J., & Nizami, A. (2018). A case study of sustainable construction waste management in Saudi Arabia. *Waste Biomass Valorizat., 9*, 41–55.

Panda, A. K., Singh, R. K., & Mishra, D. K. (2010). Thermolysis of waste plastics to liquid fuel: A suitable method for plastic waste management and manufacture of value added products—A world prospective. *Renewable and Sustainable Energy Reviews, 14*, 233–248.

Patel, P., Rajput, S. (2018). Vermicompost, its uses and benefit in sustainable agriculture. *Accent Journal of Economics Ecology & Engineering, 5*(1), 172–174.

Paul, S., Kauser, H., Jain, M. S., Khwairakpam, M., & Kalamdhad, A. S. (2020). Biogenic stabilization and heavy metal immobilization during vermicomposting of vegetable waste with biochar amendment. *Journal of Hazardous Materials, 390*, 121366.

Pudełko, A., Postawa, P., Stachowiak, T., Malinska, K., & Drózd, D. (2021). Waste_derived biochar as an alternative filler in biocomposites-mechanical, thermal and morphological properties of biochar added biocomposites. *Journal of Cleaner Production, 278*, 123850.

Rahman, S. M., & Khondaker, A. N. (2012). Mitigation measures to reduce greenhouse gas emissions and enhance carbon capture and storage in Saudi Arabia. *Renewable and Sustainable Energy Reviews, 16*, 46–60.

Ramos, R. F., Santana, N. A., de Andrade, N., Romagna, I. S., Tirloni, B., de Oliveira Silveira, A., Domínguez, J., & Jacques, R. J. S. (2022). Vermicomposting of cow manure: Effect of time on earthworm biomass and chemical, physical, and biological properties of vermicompost. *Bioresource Technology, 345*, 126572.

Ranganathan, L. S. (2006). Vermicomposting enhances humification, mineralization and chelation. *Journal of Annamalai University Science, 42*, 1–14.

Rastegari, H., Nooripoor, M., Sharifzadeh, M., & Petrescu, D. C. (2023). Drivers and barriers in farmers adoption of vermicomposting as keys for sustainable agricultural waste management. *International Journal of Agricultural Sustainability, 21*(1), 2230826.

Raza, S. T., Wu, J., Rene, E. R., Ali, Z., & Chen, Z. (2022). Reuse of agricultural wastes, manure, and biochar as an organic amendment: A review on its implications for vermicomposting technology. *Journal of Cleaner Production, 360*, 132200.

Reddy, R., Reddy, M. A. N., Reddy, Y. T. N., Reddy, N. S., Anjanappa, N., & Reddy, R. (1998). Effect of organic and inorganic sources of NPK on growth and yield of pea [*Pisum sativum* (L)]. *Legume Research, 21*, 57–60.

Rehman, A., Nawaz, S., Alghamdi, H. A., Alrumman, S., Yan, W., & Nawaz, M. Z. (2020). Effects of manure-based biochar on uptake of nutrients and water holding capacity of different types of soils. *Case Studies Chemistry Environment Engineering, 2*, 100036.

Samadhiya, H., Dandotiya, P., Chaturvedi, J., & Agrawal, O. P. (2013). Effect of vermiwash on growth and development of leaves and stem of tomato plants. *International Journal of Current Research, 5*, 20–23.

Sayara, T., Basheer-Salimia, R., Hawamde, F., & Sánchez, A. (2020). Recycling of organic wastes through composting: Process performance and compost application in agriculture. *Agronomy, 10*, 1838.

Sethi, D., Kusumavathi, K., Ravindran, B., Panda, N., Padhan, K., Dash, S., Sahoo, T. R., Mangaraj, S., Dhal, A., Swain, S. K., & Sarkar, S. (2023). Bioconversion of organic wastes into wealth by vermitechnology: A review. *Recent Trends in Solid Waste Management, 1*, 27–53.

Sharma, H. B., Vanapalli, K. R., Cheela, V. R. S., Ranjan, V. P., Jaglan, A. K., Dubey, B., Goel, S., & Bhattacharya, J. (2020). Challenges, opportunities, and innovations for effective solid waste management during and post COVID-19 pandemic. *Resources, Conservation and Recycling, 162*, 105052.

Sharma, K., & Garg, V. K. (2018). Comparative analysis of vermicompost quality produced from rice straw and paper waste employing earthworm *Eisenia fetida* (Sav.). *Bioresource Technology, 250*, 8–15.

Sharma, M., & McBean, E. (2009). Strategy for use of alternative waste sort sizes for characterizing solid waste composition. *Waste Management Research, 27*, 38–45.

Shukla, R. C., & Singh, K. (2010). Effect of vermicompost of different organic wastes germination, growth and productivity of certain crops. *Journal of Solid Wastes Technology Management, 36*, 70–81.

Siedt, M., Schäffer, A., Smith, K. E. C., Nabel, M., Roß-Nickoll, M., & van Dongen, J. T. (2020). Comparing straw, compost, and biochar regarding their suitability as agricultural soil amendments to affect soil structure, nutrient leaching, microbial communities, and the fate of pesticides. *Science of the Total Environment, 20*, 141607.

Singh, K., & Bhartiya, D. K. (2021a). A review article on vermibiotechnology and waste management. *World Reviews Science Technology Sustainable Development, 17*, 81–97.

Singh, K., & Bhartiya, D. K. (2021b). A review article on vermibiotechnology and waste management. *World Review of Science, Technology and Sustainable Development, 17*(1), 81–97.

Singh, S., Singh, J., Kandoria, A., Quadar, J., Bhat, S. A., Chowdhary, A. B., & Vig, A. P. (2020). Bioconversion of different organic waste into fortified vermicompost with the help of earthworm: A comprehensive review. *International Journal of Recycling Organic Waste in Agriculture, 9*(4), 423–439.

Sofo, A., Mininni, A. N., & Ricciuti, P. (2020). Soil macrofauna: A key factor for increasing soil fertility and promoting sustainable soil use in fruit orchard agrosystems. *Agronomy, 10*(4), 456.

Sonowal, P., Khwairakpam, M., & Kalamdhad, A. S. (2014). Vermicomposting of solid pulp and paper mill sludge (SPPMS) using *Eudrilus Eugeniae* earthworm. *International Journal of Environmental Sciences, 5*, 502.

Sreenivas, C., Muralidhar, S., & Rao, M. S. (2000). Vermicompost: A viable component of IPNSS in nitrogen nutrition of ridge gourd. *Annals Agriculture Research, 21*, 108–113.

Srivastava, V., Vaish, B., Singh, R. P., & Singh, P. (2020). An insight to municipal solid waste management of Varanasi city, India, and appraisal of vermicomposting as its efficient management approach. *Environmental Monitoring and Assessment, 192*, 1–23.

Sulaiman, I. S. C., & Mohamad, A. (2020). The use of vermiwash and vermicompost extract in plant disease and pest control. In *Natural Remedies for Pest, Disease and Weed Control* (pp. 187–201). Academic Press.

Sun, F.-S., Yu, G.-H., Zhao, X.-Y., Polizzotto, M. L., Shen, Y.-J., Zhou, H.-B., Zhang, X., Zhang, J.-C., & He, X.-S. (2020). Mechanisms of potentially toxic metal removal from biogas residues via vermicomposting revealed by synchrotron radiationbased spectromicroscopies. *Waste Management, 113*, 80–87.

Suthar, S. (2007). Nutrient changes and biodynamics of epigeic earthworm *Perionyx excavates* (Perrier) during recycling of some agriculture wastes. *Bioresource Technology, 98*, 8–14.

Swati, A., & Hait, S. (2017). Fate and bioavailability of heavy metals during vermicomposting of various organic wastes—a review. *Process Safety and Environment Protection, 109*, 30–45.

Thakur, A. N. J. A. N. A., Kumar, A. D. E. S. H., Kumar, C. V., Kiran, B. S., Kumar, S. U. S. H. A. N. T., & Athokpam, V. A. R. U. N. (2021). A review on vermicomposting: By-products and its importance. *Plantarum Cell Biotechnology Molecules Biology, 22*, 156–164.

Tossavainen, M., Nykanen, A., Valkonen, K., Ojala, A., Kostia, S., & Romantschuk, M. (2017). Culturing of Selenastrum on diluted composting fluids; Conversion of waste to valuable algal biomass in presence of bacteria. *Bioresource Technology, 238*, 205–213.

Tripathi, G., & Bhardwaj, P. (2004). Comparative studies on biomass production, life cycles and composting efficiency of *Eisenia fetida* (Savigny) and *Lampito mauritii* (Kinberg). *Bioresource Technology, 92*, 275–283.

Turki, A., Al-Hadeedi, Y., & Al-Romian, F. (2016). Chemical properties of locally composts produced in Saudi Arabia and the need for regulations. *Recent Trends PGPR Research Sustainable Crop Product, 148*, 23–31.

Ugwu, C. O., Ozoegwu, C. G., & Ozor, P. A. (2020). Solid waste quantification and characterization in university of Nigeria, Nsukka campus, and recommendations for sustainable management. *Heliyon, 6*, e04255.

Van Groenigen, J. W., Van Groenigen, K. J., Koopmans, G. F., Stokkermans, L., Vos, H. M. J., & Lubbers, I. M. (2019). How fertile are earthworm casts? A meta-analysis. *Geoderma, 338*, 525–535.

Vaverkova, M. D., Paleologos, E. K., Dominijanni, A., Koda, E., Tang, C.-S., Małgorzata, W., Li, Q., Guarena, N., Mohamed, A.-M.O., & Vieira, C. S. (2020). Municipal solid waste management under COVID-19: Challenges and recommendations. *Environmental Geotechnical, 40*, 1–16.

Vuković, A., Velki, M., Ečimović, S., Vuković, R., Štolfa Čamagajevac, I., & Lončarić, Z. (2021). Vermicomposting—Facts, benefits and knowledge gaps. *Agronomy, 11*(10), 1952.

Wang, L., Zhang, Y., Lian, J., Chao, J., Gao, Y., Yang, F., & Zhang, L. (2013). Impact of fly ash and phosphatic rock on metal stabilization and bioavailability during sewage sludge vermicomposting. *Bioresource Technology, 136*, 281–287.

Wani, K. A., Mamta, W. K. A., & Rao, R. J. (2013). Bioconversion of garden waste, kitchen waste and cow dung into value-added products using earthworm *Eisenia fetida. Saudi Journal of Biology Science, 20*, 149–153.

Yadav, A., & Garg, V. (2011). Recycling of organic wastes by employing *Eisenia fetida. Bioresource Technology, 102*, 74–80.

Yadav, K. D., Tare, V., & Ahammed, M. M. (2010). Vermicomposting of source-separated human faeces for nutrient recycling. *Waste Management, 30*, 50–56.

Yatoo, A. M., Bhat, S. A., Ali, M. N., Baba, Z. A., & Zaheen, Z. (2022). Production of nutrient-enriched vermicompost from aquatic macrophytes supplemented with kitchen waste: Assessment of nutrient changes, phytotoxicity, and earthworm biodynamics. *Agronomy, 12*, 1303.

Yuvaraj, A., Thangaraj, R., Ravindran, B., Chang, S. W., & Karmegam, N. (2020). Centrality of cattle solid wastes in vermicomposting technology–A cleaner resource recovery and biowaste recycling option for agricultural and environmental sustainability. *Environmental Pollution, 12*, 115688.

Zambare, V. P., Padul, M. V., Yadav, A. A., & Shete, T. B. (2008). Vermiwash: Biologically and microbiological approach as ecofriendly soil conditioner ARPN. *Journal of Agriculture Biology Science, 3*, 1–5.

Zhen, Z., Liu, H., Wang, N., Guo, L., & Meng, J. (2014). Effects of manure compost application on soil microbial community diversity and soil microenvironments in a temperate cropland in China. *PLoS ONE, 9*, e108555.

Chapter 10
Crop Residue Management Practices for Sustaining Soil Health

J. A. Ruley

Abstract In the twenty-first century, several innovations have been developed that are focused on realization of sustainability. Zero waste management is a prime example and is practiced with goal of ensuring that the environment is kept clean by ridding it of waste that has in the past, caused deleterious effects. These have manifested especially where man has illicitly dumped or disposed them of as waste materials leading to both direct and indirect effects on ecosystems. In this chapter, priority is given to crop residues management as a form of zero waste management. Emphasis is on sustainable ways and means of disposing off crop waste as opposed to widely applied traditional means that have caused insurmountable havoc on the environment and the constituents of entire ecosystems. The praxis of the chapter is to showcase pitfalls of traditionally used methods of disposing off crop residues while angling for more sustainable means that lead to cradle-to-cradle models of protecting environment. The chapter also provides key interventions that ought to be undertaken if the 'from soil to soil' zero waste management technology is to be adopted and its sustainable benefits are to become a common place in all farming zones worldwide.

Keywords Crop residues · Zero waste management · Recycling · Environment · Waste materials

1 Introduction

Man is a prime ecosystem engineer. On the skin of earth habited by man, several socio-economic activities have been established as sources of livelihood. In their presence, man's survival is guaranteed. One of these is agriculture which provides man with food, a physiological need that is indispensable if man is to live as professed by Abraham Maslow while theorizing the Needs-hierarchy theory in 1943 (Maslow, 1943). It is not by surprise that as demographers are anticipating a bulge in world

J. A. Ruley (✉)
Department of Agricultural Sciences, University of Juba, P.O. Box 82, Juba, South Sudan
e-mail: janenajeb@yahoo.com

R. A. Bhat et al. (eds.), *Zero Waste Management Technologies*,
https://doi.org/10.1007/978-3-031-57275-3_10

population to clock 10 billion by eve of 2050, efforts for boosting productivity of farms are considered as a real deal to curtail what might culminate into a human disaster. For this reason, soil health is a base factor that must be enhanced if purported levels of agricultural productivity are to materialize. One of the latest developments that has become a catch phrase for development practitioners is investment of reasonable amounts of money and time to ensure that most activities focus on zero waste management, aware that soil ecosystems are final resting points for waste. Defined, zero waste denotes ways and means resigned to ensure effective management of waste products and processes with the aim of systematically avoiding and eliminating volume of waste produced while also riding it of toxic content before dumping (Awasthi et al., 2021; United States Environmental Protection Agency [USEPA], 2022).

The main objective of zero waste management is to conserve and recover all resources whilst refraining from burning or burying them (USEPA, 2022). This innovation therefore does with re-designing of the resource life cycles that materials generated along crop production points or phases are reused with minimal and or no waste left behind (USEPA, 2022). This philosophy is alternatively called cradle-to-cradle model or closed loop system (Zero Waste International Alliance [ZWIA], 2020). By cradle-to-cradle, we mean a circular model which engenders minimization of waste while keeping resources in use for quite a long period of time (ZWIA, 2020). This model calls for promotion of sustainability and zero waste management by among others reducing the bulkiness and quantities of waste generated, reusing and recycling them. Cradle-to-cradle (C2C) circular model is an antithesis of the cradle-to-grave (C2G) philosophy. C2G is a wasteful linear model where resources are extracted and moved along a chain till the end, a point where the collected waste is dumped at landfills. For this reason, C2G is an open loop system that results in substantial losses (ZWIA, 2020). According to ZWIA (2020), sustainable natural cycles in farming and composting provide a clear and lively example of C2C. This accounts for why this chapter is devoted to crop residue management practices and settling for sustainable ones that promote the "from soil to soil" idealities.

2 Crop Residues

According to Meena et al. (2022), crop residues are leftovers. Essentially, they are the non-food components of plants. These include stalks, leaves, straw, and husks. Generally, they are parts of the plants that are abandoned in the gardens after crops have been harvested and threshed (Meena et al., 2020). This conception however provides a narrow view of what actually, crop residues are. Collectively, they are called lignocellulosic biomass (Kaur & Singh, 2022; Meena et al., 2022). By degree of importance, the largest proportion of crop residues originate from cereal production at 74% followed by sugar crops (sugar cane and sugar beet) at 10%. Legumes, tubers and oil seeds contribute the remaining 16% (Lu, 2020). This implies that recycling of crop residues should be encouraged among the farming households and large scale farm. In the twenty-first century and beyond, massive production of crop residues is a necessary evil (Srivastava et al., 2023). This rise in tonnage or metric tons of crop residues is largely explained by the changes in farming trends. To mention for example, farmers are increasingly adopting modern agricultural technology, applying hybrid seeds, fertilizers and using of irrigation and pesticides which has increased on crop productivity (Srivastava et al., 2023). This development has caused an exponential rise in quantities of crop residues at both farm and market levels.

A broader description is provided by British Columbia Ministry of Agriculture (2014) which denotes crop residues as leftovers that are left on the field at two stages; during crop growth and after harvesting or processing. During crop growth, activities such as thinning and pruning lead to production of a heap of the fresh parts of the plants such as leaves and branches or an entire plant in case of thinning. At harvest, the residues such as shoot systems for tubers and root systems for plants with above ground harvestable parts lead to generation of residues. For crops such as coffee, rice and cocoa, husks are produced when the crops are undergoing processing while for others such as sugarcane, bagasse is produced. The crop residues are supposed to be disposed of as some are known of producing nuisance odors and become havens for municipal waste houseflies which are vectors. Therefore, much as some crop residues are applied and left to decompose on their own, for others, effective recycling is necessary to realize the tripartite benefits of prevention of erosion, control of pests and preservation or improvement of soil fertility (British Columbia Ministry of Agriculture, 2014).

3 Rationale of Recycling Crop Residues

Meena et al. (2020) enlisted the following importances of recycling crop residues;

1. Recycling crop residues is encouraged because the materials act as reservoirs for plant nutrients such as organic carbon and essential nutrients such as P and N. Besides, the crop residues when recycled help to prevent the losses of these essential nutrients from the soil.
2. The crop residues also have an effect on the availability of soil nutrients through chelation.
3. The crop residues also in buffering the capacity of the soil leading to stabilization of the soil PH and CEC.
4. Soil with crop residues have an improved structure compared to the ones without.
5. The crop residues also help in transmission of heat in the soil and help further in harmonizing it.
6. The soil organic carbon enriched in the soil by the crop residues provides rich source of food for the heterotrophs such as autochthonous microorganisms leading to rhizobcateria abundance and richness.

7. The crop residues affect soil chemistry by influencing the efficiency of chemicals, their degradability and toxicity.
8. The crop residues increase abundance of plant growth substances in the soil environment.
9. The crop residues also have a chelating effect on the toxicities of pollutants in the soul such as heavy metals.
10. Crop residues especially mulches help in conserving the soil and water hence checking against possibilities of capillary loss, soil aridity, and prevention of surface run offs that erode soil on bare grounds.
11. Crop residues provide a cost effective way of maintaining soil fertility and therefore, they help in bridging the fertilizer gap.
12. Crop residues help in sequestration of carbon which helps in minimization of environmental pollution.

The suite of benefits above provides a clear manifestation of the discoveries made in the era of zero waste management. According to Meena et al. (2020), Pili et al. (2019) and British Columbia Ministry of Agriculture (2014), for quite a long period of time, crop residues have been regarded as waste materials. Indeed, to many people, even in present time, crop residues are a waste material. Characteristically, as is the case with the other waste materials, they must be disposed of (Meena et al., 2020). Before drawing our attention to zero waste management of crop residues, we also need to appreciate the efforts that have been undertaken under the traditional lens to discard crop residues as waste. The inherent loopholes introduce us to the rationale of encouraging zero waste management of crop residues now and in the future.

4 Traditional/Unsustainable Methods of Recycling Crop Residues by Farming Communities

Despite progressive increase in quantity of crop residues, avenues for their proper management are still limited. According to Srivastava et al. (2023), in several ecological environments, management of crop residues is a teething problem to farmers who are faced with lack of dumping lands, shortages of better means of transporting them to landfills, and lack of planned utilization of the waste through recycling, among others.

The most commonly used traditional means of disposing off crop residues is by incineration in the gardens or open burning (Adejumo & Adebiyi, 2020; Lu, 2020; Pili et al., 2019). The commonest method employed to dispose of crop waste in several agroecosystems is through burning (Sachdeva et al., 2023; Sarkar et al., 2020; Srivastava et al., 2023). Several reasons are advanced for the adoption of burning as the preferred means of recycling crop residues;

1. The cost of hiring labor to collect and recycle residues by others means is high in relative and absolute terms. Besides, there is a high net cost for procuring implements to help in the mechanical collection of the residues leading to adoption of the short cut route of burning the waste materials.
2. The approach provides the fastest way of clearing the gardens or cropped farms of the residual biomass which eases the two activities of land preparation and planting.
3. Burning crop residues is preferred because it helps in controlling the growth of weeds that rid the soil of nutrients while also helping in eliminating or altering the natural habitats of pathogens and insect pests from the soil ecosystem.
4. Burning crop residues to some extent improves soil fertility by increasing the short-term.
5. Availability of some essential nutrients such as phosphorus and potassium. As well, soil acidity is believed to reduce when crop residues are burnt in situ.
6. Aruya et al. (2016) established that farmers are compelled to opt for burning after carrying out cost benefit analyses in comparison with other means of recycling that may even be available and practices by other farmers in their areas. Many farmers consider residue utilization as a wasting activity given the low financial returns they are capable of realizing from adopting the costlier recycling approaches. They therefore find it easier and cheaper to torch them in situ.

While the above cases may convincingly induce farmers to continue with the practice, burning crop residues in situ is disastrous (See Fig. 1). Much as it rids crop gardens of residual leftovers, the method posts deleterious effects to the environment (Sarkar et al., 2020). Studies (such as Adejumo & Adebiyi, 2020; Lu, 2020; Meena et al., 2022; Sarkar et al., 2020) have reported the following negative impacts of recycling crop residues through burning;

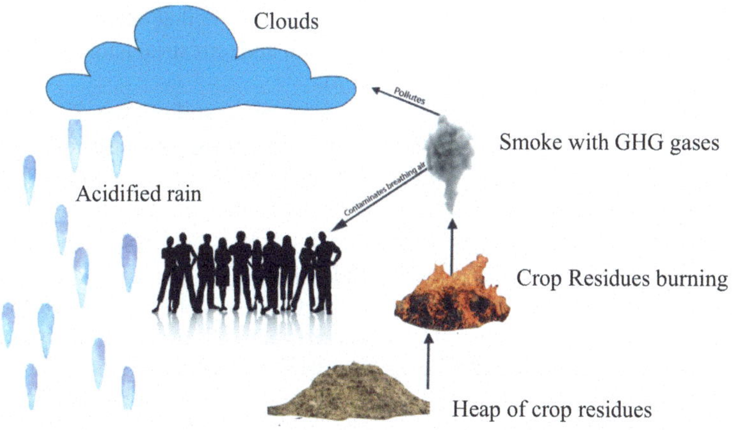

Fig. 1 Deleterious effects of burning crop residues on human and environmental health

1. Burning crop residues produces climate-relevant emissions leading to air pollution as harmful gases, smoke, and dust are released in the atmosphere. Whenever crop residues are openly burnt, they emit methane (CH_4) and nitrous oxide (N_2O) gases. Both CH_4 and N_2O are greenhouse gases (GHGs) (Sarkar et al., 2020). Pili et al. (2019) cohere with the position when they argue that whenever one ton of crop residues is burned, about 3 kg of methane are emitted into the atmosphere. The figures are worrying for intensive cropped lands such as rice and what produced under green revolution practices. This implies that burning of crop residues increases on the global methane budget. Notable effects are observable in drastic reduction in the milk yield of animals while in humans, the high risk of air pollution may induce sudden break down of chest diseases such as asthma due to difficulty in breathing.
2. Burning large quantities of crop residues also interrupts the balance of nature. The flames have the potential of scaring birds and insects away from the sites while fire fearing animals such as hippopotami and cowardly animals such as hyenas are likely to be scared away from the environments leading to imbalance of nature, all of which affects man in one way or the other. For example, when bats are scared away from the environment, the numbers of mosquitoes and houseflies grows manifold. These duo are disease carrying vectors.
3. Burning generates man-made smog. This refers to a lot of smoke and when it is severe, it may obstruct visibility in a given area leading to traffic accidents. The smog may be less severe when the crop residues are burnt after they have dried. However, when they are laden with water or partially dry, this encourages smoldering which prolongs the smog in a given area.
4. Burning of crop residues is equally harmful to the water ecosystems. Meena et al. (2020) established that when a ton of stubble is burnt along with 199 kg of ash, particulate matter, carbon monoxide, carbon dioxide, and sulphur dioxide are released in the atmosphere. These and other chemically and radiative important trace gases and aerosols affect the quality of rain water that enters ground water sources. This has a deleterious effect on human and terrestrial animal life as well as the aquatic organisms, macro and micro.
5. The burning of crop residues over a large area affects soil physics. For example, it causes an increase in the soil temperature. As a result, it causes death of soil microbiota and yet these are conditioning agents in the soil ecosystem.
6. Soil essential nutrients such as N and C and also decimated when the crop residues are recycled by burning. Open air burning leads to loss of soil C stock, loss of soil moisture. This renders the burnt soil infertile.
7. Recycling crop residues by burning affects the physical properties of the soil. Khouzani and Ghahfarokhi (2022) observed that soil on which plant residues has been burnt gets affected in terms of strength, specific gravity, granulation composition. Furthermore, the soil becomes compact, pores are filled and soil aeration processes are impeded. Large clods are formed which cause a drop in the available soil organic matter.

Apart from burning, the other least commonly used methods of recycling crop residues is through deep burying and dumping. According to Pili et al. (2019), deep burying involves use of excavation pits where crop residues are poured and left to decompose. The possibility of farmers drawing the compost and applying it in the gardens may not be guaranteed. The same applies to dumping in compost pits or garbage collection sites. By applying the two methods, the farmers are resigned to losing the crop residues as they highly regard them as waste materials. These approaches together with burning as discussed above portray gross mismanagement options undertaken by farmers to recycle crop residues. Any farmer who opts to use these methods has totally departed from the virtues of zero waste management which entails reuse and minimal or no generation of waste for dumping. To maximize the economic benefits and environmental standards listed in the foregoing Sect. 4, sustainable ways of recycling crop residues must be explored and disseminated among the farming communities so that zero waste management of crop residues is realized. Below, is an account of the sound crop residues management practices that resonate with the proposition of the chapter, "from soil to soil".

5 Modern/Sustainable Ways of Recycling Crop Residues

At present, 2445.2 metric tons of crop residues are produced worldwide (Fu et al., 2021). Incidentally, a large percentage of these residues are disposed of through open burning (Fu et al., 2021) yet they are rich in nutrients and are amenable to decomposition by soil living microorganisms. Burning is not a zero waste management practice since it impacts the environment negatively. Given this higher tonnage of crop residues generated and the anticipated increase at a future date, efforts must be taken to ensure that the crop residues are returned to the soil ecosystem so that their benefits are not forfeited or foregone. This calls for proper disposal methods so that soils are kept in a healthy condition (Fu et al., 2021) (Fig. 2).

5.1 Application as a Surface Mulch

As earlier noted, crop residues from cereals such as straw provides an excellent source of mulches. While the application of mulches over large chunks of land might prove laborious unless mechanization is applied, mulching reduces loss of top soil layers compared to bare land. In environments with saline soils, mulching controls the rate of soil evapotranspiration. Supporting evidence of this effect is provided by Fan et al. (1993) who reported that a significantly higher decrease in the level of soil salinity from 0.44% to 0.07%. Comporting evidence is also traceable in the work of Yang et al. (2006) who established that in mulches lowered the levels of salinity in the soil environment compared to the control. One other way through which returning of crop residues protects the soil ecosystems from environmental stressors is through

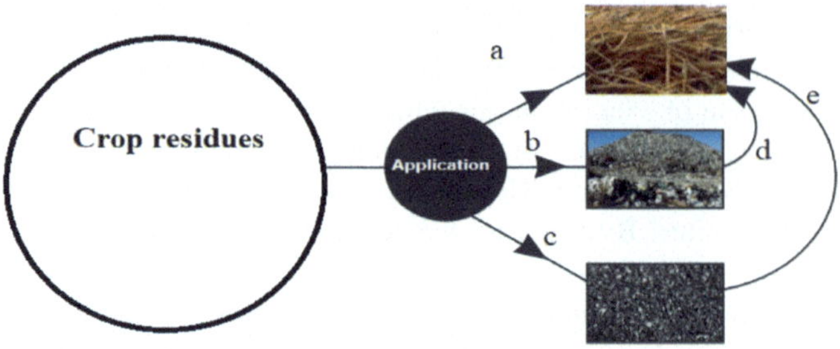

A= Residues such as straw applied directly as mulches; b=Crop residues composted; c= Crop residues converted into biochar; d and e= composted manure and biochar applied to gardens.

Fig. 2 Modern ways of disposing of crop residues. **a** Residues such as straw applied directly as mulches; **b** Crop residues composted; **c** Crop residues converted into biochar; **d** and **e** Composted manure and biochar applied to gardens

mulching to retain soil moisture thereby decreasing the possibilities of soil aridity. Crop residue mulches control soil moisture levels, regulate the soil temperature and minimize the levels of soil evaporation. By doing so, the soil ecosystem is protected from becoming arid. Even in the so called dry lands, mulches help in the conservation of soil moisture by lowering the levels of capillary water loss.

5.2 *In Situ Incorporation*

Crop residues are also beneficial when they are applied in the soil before planting successive crops in the same field. The crop residues are added in order to improve the physicochemical properties of the soil (such as structure, porosity, water retention and availability, nutrient cycling, CEC) and biological properties (such as carbon sequestration, microbial abundance, and diversity). Candidates for this method should be quick decomposing residues.

5.3 *Recycling Crop Residues by Composting*

Pili et al. (2019) define composting as a process of speeding up the natural decay of organic materials under controlled conditions. This among others includes the provision of ideal conditions for detritus-eating organisms to thrive. Their abundance and biodegradation activities enhance fast decomposition of the organic materials. The ideal conditions for quick decomposition of plant materials is provided by adding livestock manure to the compost pits. The manure helps to balance off the excess carbon that slows the decomposition process while and excess nitrogen that increases the acidity of the pile leading to decimation of some soil living microorganisms.

Compositing can either be Indore, Coimbatore, Bangalore, vermicomposting, or NADEP. Despite providing the classification, the operating mechanism of either method is outside the scope of this chapter.

5.4 Recycling Crop Residues into Biochar

Crop residues constitute the most commonest and abundant raw materials for making biochar (Afraz et al., 2023; Breulmann et al., 2015; Len et al., 2022). Recently, production of biochar using crop residues has been identified as a priority environmental clean-up strategy in cropped land where accumulation of crop residues especially after harvest had become a nuisance (Billa et al., 2019; Gupta et al., 2020). Biochar nourishes crops with essential nutrients such as K, Mg, K, Ca, and P (Mona et al., 2020; Zheng et al., 2022). The minerals are abundant in biochar because they form when crop residues undergo pyrolysis. Besides providing crop plants with essential nutrients, biochar also provides excellent conditions for soil living microorganisms such as rhizobcateria (Zheng et al., 2022). Biochar posts several benefits for environmental clean-up. Production of biochar is one of the zero waste strategies for managing the crop residues (Sachdeva et al., 2023). When biochar is produced, the release of greenhouse gases to the environment, and consequently to the soil is mitigated. More on, the chances of introducing toxic compounds to the environment through open burning are minimized when the residues are collected and transformed into biochar by pyrolysis (Sachdeva et al., 2023). Biochar rids soil of heavy metals, improves the SOC stocks in the soil and regulates the release of carbon into the atmosphere (Enaime & Lübken, 2021; Khawkomol et al., 2021; Zhao, 2020). This corrects the deficits that induce sudden climate change, improves the productivity of crop enterprises, hence boosting agricultural development, plays a key role in wastewater treatment, and equally, it causes a tremendous improvement in soil health. On account of the above, Sachdeva et al. (2023) surmised that the crop residues make best raw materials for production of biochar since they are rich in fibrous cellulose that improves on nitrogen and other nutrient uptake by the plants from soil. Biochar is a soil conditioner and therefore, a priority zero waste management strategy that improves the physicochemical and biological properties of soil while also improving the stocks of carbon in the soil. Collectively, these attributes have caused a massive effect on mitigation of climate change while also improving the livelihoods through higher food production and consequently, food security (Haider et al., 2022; Guo et al., 2020; Taraqqi-A-Kamal et al., 2021). Biochar also rids the soil of toxins while also trapping excess nitrogen in the soil consequently protecting the quality of the ground water as well as other open water sources that are liable to infiltration by runoff erosion. This improves the healthiness of soil.

6 Composite Roles of Sustainably Recycled Crop Residues in the Soil Environment

Crop residues serve as one of the soil-based climate change mitigation strategies (Stella et al., 2019). According to Rumpel et al. (2018), use of crop residues forms part of the reconditions under the '4 per 1000' initiative. Launched by the French Government at the COP21 climate change summit in Paris in 2015, the initiative is aimed at boosting the storage of carbon in agricultural soils by 0.4% per annum. The goal of the initiative is to help in mitigating climate change while also causing an increase in the levels of food security (Rumpel et al., 2018).

6.1 *Impacts on Soil Textual Properties*

Direct return of raw or unprocessed crop residues has a conditioning effect on the soil leading to greater improvements in the physical properties of the soil. According to Fu et al. (2021), returning of crop residues such as wheat and rice straw leads to a corresponding increase in soil moisture content, decreased bulk density, soil aggregate stability, and higher levels of soil porosity. The increase in soil moisture content arises from drastic drop in surface run off, impeded evaporation of moisture from the soil, improved saturation of water in the soil, and squared potential of the soil to allow water infiltration. Scientific studies have proven that application of crop residues is a conditioner to the soil. For example, Lou et al. (2011) reported a decrease in the temperature of surface layers of soil and a proportionate increase in soil moisture in a maize field experiment when straw was added compared to the control while a study by Zhao et al. (2019) noted a 19% increase in soil water content in experiments where straw was added compared to control. In the same study, the soil bulk density in the surface soil layer decreased by about 6% after 7 years of application of wheat and maize straw. Comporting views are shared by Xu et al. (2016) whose application of maize straw (4500 kg ha^{-1}) decreased the soil bulk density by 9.5% and 8.6% at 0–20 cm and 20–40 cm soil layers, respectively.

Crop residues have a significant impact on the porosity of soil. Fu et al. (2021) recommended that best results of soil total porosity are observable when crop residues are crushed and mixed with soil during deep ploughing. To this effect, (31) reported that soil porosity can increase up to 30 cm depth when wheat straw is returned to the soil while Zheng et al. (2019) noted that maize straw increased porosity by 21.7%.

6.2 *Impacts on pH and CEC*

In soils with low buffering capacity, application of raw or unprocessed crop residues have an influence on soil pH. Studies (such as Butterly et al., 2012; Pan et al., 2021) reported an increase in soil pH. Fu et al. (2021) attribute the pH conditioning effect to the rise in cation concentration, stabilization of the C and N in the soil. The pH effects are known to persist in the soil for as long as 3 years (Fu et al., 2021). Mulching crop residues and harvested fresh biomass have a greater effect on the increase of CEC in the soil. However, where harvested biomass is to be used, larger quantities should be applied.

6.3 Provision of Limiting Nutrients

In nutrient deficient soils, decomposing crop residues supply the limiting nutrients hence offsetting the deficiencies. Fu et al. (2021) noted that when crop residues are applied and decay in soil, they improve nutrient cycling. More specifically, there is a notable increase in SOC and SOM. For SOC, decomposing crop residues are rich in carbon (about 40%) and once this has been added to soil ecosystem, the soil aggregates are stabilized. For SOM, there is a marked increase in available N, P, and K. The higher the availability of N, the greater will be proteins, amino acids and nucleic acids in soil all of which are essential for metabolic functions of growing plants.

6.4 Maintenance of Carbon Stocks in the Soil

Crop residues are a key component of the carbon cycle. Using simulations, Stella et al. (2019) noted that any possible loss of SOC can be buffered by returning harvest residues to soil and in SOC deficient soil types such as clayey, an increase in SOC can be realized with addition of crop residues. Their removal from soil depletes the SOC pools. Based on this argument, their return to soil is more of a necessity since they provide a natural way of enhancing the stocks of SOC using land-based means. Therefore, any efforts aimed at restoration of SOC and mitigation of C emissions in Agriculture cannot isolate utilization of crop residues. Therefore, from soil, crop residues should be returned to soil (Stella et al., 2019).

7 Conclusion

Soil is a critical natural resource that supports environmental and agricultural productive functions such as filtering pollutants, plant growth, and food production. Therefore, hopes are high among soil scientists and any other agro-based professionals that zero waste management of crop residues holds several benefits that will greatly improve productivity of farmers. Thus, both short term and futuristic strategies ought to be undertaken to ensure that effective management of crop residues as key soil conditioning amendments is promoted. As shown, studies have proven that crop residues that are returned to the garden as mulch result in an increase in organic matter content and available nitrogen content while they also increase crop yields and water use efficiency (WUE). Other desirable effects posted by returning crop residues to the soil include improvement in soil organic matter (SOM), soil physical properties, stabilizes soil structure and causes a reduction in soil bulkiness density. Other desirable effects posted by returning crop residues to the soil include improvement in soil organic matter (SOM), soil physical properties, stabilizes soil structure and causes a reduction in soil bulkiness density. These gains are lost with improper management of crop residues. Therefore, crop residues from soil ought to be returned to soil following standards specified by Zero Waste International Alliance (ZWIA).

8 Recommendations

Characteristically, traditional management practices for crop residues restrains them from posting beneficial effects to soil ecosystem. Thus, efficacy of recycling procedure used ought to be assessed for better zero waste management. This is in tandem with zero waste management procedures which are process efficient. The efficacy of crop residues should be boosted before they are applied or returned to soil. They should be returned to soil in the same fashion as other organic amendments. For example, the aspect of size of residues should be considered keenly. Farmers should be sensitized on chopping and spreading chippings throughout gardens. The best application method is incorporation into the soil as opposed to piling them on the surface. Incorporating them in the soil hastens the rate of decomposition. Further, emphasis should be on compositing the residues. By doing so, their quality as amendments is bolstered. Equally, this aids in controlling of pests and pathogens that are common with direct returning of the residues to gardens. By this, we mean that sensitization and empowerment of farmers and others players in agricultural development such as agricultural extension workers and soil scientists should be empowered.

While it is the case in the parlance of legal frameworks that he/she who generates solid waste must be responsible for its disposal, normally at a fee, crop residues are not much of a priority in the developing countries compared to developed ones. The farmers will always take short cut and burn residues in order to evade likely fees. Thus, it is important to note that the net benefits of zero waste management will accrue when political leadership of different countries especially those running prosperous agrarian economies have established a subsidy that will induce farmers to buy the idea of recycling crop residues by zero waste management approaches.

References

Adejumo, I. O., & Adebiyi, O. A. (2020). Agricultural solid wastes: Causes, effects, and effective management. *Strategies of Sustainable Solid Waste Management, 8.*

Afraz, M., Muhammad, F., Nisar, J., Shah, A., Munir, S., Ali, G., et al. (2023). Production of value added products from biomass waste by pyrolysis: An updated review. *Waste Management Bulletin.*

Aruya, E. I., Yusuf, R. O., & Yusuf, Y. O. (2016). An assessment of crop residue characteristics and factors militating against efficient management in the Ikara local government area of Kaduna state, Nigeria. *Waste Management Environment VIII, 1*, 333–344.

Awasthi, A. K., Cheela, V. S., D'Adamo, I., Iacovidou, E., Islam, M. R., Johnson, M., Miller, T. R., Parajuly, K., Parchomenko, A., Radhakrishan, L., Zhao, M., & Li, J. (2021). Zero waste approach towards a sustainable waste management. *Resources, Environment and Sustainability, 3*, 100014.

Billa, S. F., Angwafo, T. E., & Ngome, A. F. (2019). Agro-environmental characterization of biochar issued from crop wastes in the humid forest zone of Cameroon. *International Journal of Recycling of Organic Waste in Agriculture, 8*, 1–13.

Breulmann, M., van Afferden, M., & Fühner, C. (2015). Biochar: bring on the sewage. *Nature, 518*(7540), 483–483.

British Columbia Ministry of Agriculture. (2014 May). *Crop residue management*. Farm practices crop residue management order no. 870.218-30. Retrieved on 1 January 2023 from https://www2.gov.bc.ca/assets/gov/farming-natural-resources-and-industry/agriculture-and-seafood/agricultural-land-and-environment/strengthening-farming/farm-practices/870218-30_crop_residue_management.pdf

Enaime, G., & Lübken, M. (2021). Agricultural waste-based biochar for agronomic applications. *Applied Sciences, 11*(19), 8914.

Fan, X. W., Chi, B. L., Jiao, X. Y., Li, D. W., & Zhang, Z. P. (1993). Soil improvement and yield increment in salt-alkalined fields by straw mulch. *Agricultural Research in the Arid Areas, 11*(4), 13–18.

Fu, B., Chen, L., Huang, H., Qu, P., & Wei, Z. (2021). Impacts of crop residues on soil health: A review. *Environmental Pollutants and Bioavailability, 33*(1), 164–173.

Guo, X. X., Liu, H. T., & Zhang, J. (2020). The role of biochar in organic waste composting and soil improvement: A review. *Waste Management, 102*, 884–899.

Gupta, R. K., Hussain, A., Sooch, S. S., Kang, J. S., Sharma, S., & Dheri, G. S. (2020). Rice straw biochar improves soil fertility, growth, and yield of rice–wheat system on a sandy loam soil. *Experimental Agriculture, 56*(1), 118–131.

Haider, F. U., Coulter, J. A., Liqun, C. A. I., Hussain, S., Cheema, S. A., Jun, W. U., et al. (2022). An overview on biochar production, its implications, and mechanisms of biochar-induced amelioration of soil and plant characteristics. *Pedosphere, 32*(1), 107–130.

Kaur, K., & Singh, P. (2022). Crop residue burning in India: Potential solutions. In *Agricultural waste-new insights*. IntechOpen.

Khawkomol, S., Neamchan, R., Thongsamer, T., Vinitnantharat, S., Panpradit, B., Sohsalam, P., ... & Mrozik, W. (2021). Potential of biochar derived from agricultural residues for sustainable management. Sustainability, 13(15), 8147.

Khouzani, M. R. Z., & Ghahfarokhi, Z. D. (2022). Evaluation of agricultural waste management mechanism in Iran. *Industrial and Domestic Waste Management, 2*(2), 113–124.

Lou, Y., Xu, M., Wang, W., Sun, X., & Zhao, K. (2011). Return rate of straw residue affects soil organic C sequestration by chemical fertilization. *Soil and Tillage Research, 113*(1), 70–73.

Lu, X. (2020). A meta-analysis of the effects of crop residue return on crop yields and water use efficiency. *PLoS ONE, 15*(4), e0231740.

Maslow, A. H. (1943). A theory of human motivation. *Psychological Review, 50*(4), 370–396.

Meena, H. N., Jat, S. L., Meena, M. S., & Singh, S. K. (2020). Crop residue generation, recycling and its management for agricultural sustainability. *Indian Journal of Fertilisers, 16*(11), 1152–1161.

Meena, H. N., Singh, S. K., Meena, M. S., Narayan, R., & Bheem, S. (2022). *Crop residue: Waste or wealth?* (pp. 1–30). Technical bulletin 2022, published by ICAR-agricultural technology application research institute, Zone-II, Jodhpur.

Pan, X. Y., Shi, R. Y., Hong, Z. N., Jiang, J., He, X., Xu, R. K., & Qian, W. (2021). Characteristics of crop straw-decayed products and their ameliorating effects on an acidic Ultisol. *Archives of Agronomy and Soil Science, 67*(12), 1708–1721.

Pilli, K., Revathi, B., Jaison, M., & Sridhar, D. (2019). Chapter-2 recycling of crop residues for sustainable soil health management. *Current Research in Soil Fertility, 19*, 19.

Rumpel, C., Lehmann, J., & Chabbi, A. (2018). '4 per 1,000'initiative will boost soil carbon for climate and food security. *Nature, 553*(7686).

Sachdeva, S., Kumar, R., Sahoo, P. K., & Nadda, A. K. (2023). Recent advances in biochar amendments for immobilization of heavy metals in an agricultural ecosystem: A systematic review. *Environmental Pollution*, 120937.

Sarkar, S., Skalicky, M., Hossain, A., Brestic, M., Saha, S., Garai, S., Ray, K., & Brahmachari, K. (2020). Management of crop residues for improving input use efficiency and agricultural sustainability. *Sustainability, 12*(23), 9808.

Srivastava, R. K., Sarangi, P. K., Shadangi, K. P., Sasmal, S., Gupta, V. K., Govarthanan, M., Sahoo, U. K., & Subudhi, S. (2023). Biorefineries development from agricultural byproducts: Value addition and circular bioeconomy. *Sustainable Chemistry and Pharmacy, 32*, 100970.

Stella, T., Mouratiadou, I., Gaiser, T., Berg-Mohnicke, M., Wallor, E., Ewert, F., & Nendel, C. (2019). Estimating the contribution of crop residues to soil organic carbon conservation. *Environmental Research Letters, 14*(9), 094008.

Taraqqi-A-Kamal, A., Atkinson, C.J., Khan, A., Zhang, K., Sun, P., Akther, S., Zhang, Y (2021) Biochar remediation of soil: Linking biochar production with function in heavy metal contaminated soils. *Plant Soil Environment, 67*(4), 183–201. https://doi.org/10.17221/544/2020-PSE

United States Environmental Protection Agency. (2022 October 6). How communities have defined zero waste. Retrieved on 20 December 2022 from https://www.epa.gov/transforming-waste-tool/how-communities-have-defined-zero-waste

Xu, J., Zong, M. H., Fu, S. Y., & Li, N. (2016). Correlation between physicochemical properties and enzymatic digestibility of rice straw pretreated with cholinium ionic liquids. *ACS Sustainable Chemistry and Engineering, 4*(8), 4340–4345.

Yang, Y. M., Liu, X. J., Li, W. Q., & Li, C. Z. (2006). Effect of different mulch materials on winter wheat production in desalinized soil in Heilonggang region of North China. *Journal of Zhejiang University Science. B, 7*(11), 858.

Zero Waste International Alliance (ZWIA). (2020 June 23). *What is zero waste?—A guide to resource recovery and conservation.* Retrieved on 4 January 2023 from https://www.zerowaste.com/blog/what-is-zero-waste-a-guide-to-resource-recovery-and-conservation/

Zhao, M., Dai, Y., Zhang, M., Feng, C., Qin, B., Zhang, W., ... & Qiu, R. (2020). Mechanisms of Pb and/or Zn adsorption by different biochars: Biochar characteristics, stability, and binding energies. *Science of the Total Environment, 717*, 136894.

Zhao, X., Yuan, G., Wang, H., Lu, D., Chen, X., & Zhou, J. (2019). Effects of full straw incorporation on soil fertility and crop yield in rice-wheat rotation for silty clay loamy cropland. *Agronomy, 9*(3), 133.

Zheng, Y., Han, X., Li, Y., Yang, J., Li, N., & An, N. (2019). Effects of biochar and straw application on the physicochemical and biological properties of paddy soils in Northeast China. *Science and Reports, 9*, 16531.

Zheng, X., Xu, W., Dong, J., Yang, T., Shangguan, Z., Qu, J., ... & Tan, X. (2022). The effects of biochar and its applications in the microbial remediation of contaminated soil: A review. *Journal of Hazardous Materials, 438*, 129557.

Chapter 11
Biotreatability of Industrial Hazardous Wastes

Neha Salaria, Amit Kumar, Sukhvinder Singh, Abhilasha Rajput, Mohee Shukla, and Sharad Kumar Tripathi

Abstract Hazardous waste contamination due to industrial activity is now identified as a potential risk to both the environment and serious health problems. This chapter explores the methods of biotreatments of industrial pollutants. Industrial pollutants can be stabilized by their hydrolysis, oxidation, reduction, precipitation, volatilization, and solubilization. The electrons in oxidation–reduction are donated by (a) organic compounds, (b) reduced inorganic substances such as NH_4^+, Fe^{2+}, H_2, H_2S, S, (c) in oxygenic photosynthesis H_2O act as an electron donor. In oxidation-reduction reactions, electron acceptors differ based on the specific metabolic pathway. During fermentation, a fraction of an organic molecule functions as the electron acceptor, facilitating the conversion of substrates into products. Conversely, in anaerobic respiration, electron acceptors include organic compounds like SO_4^{2-}, CO_2, Fe^{3+} NO_3^-, NO_2^-. Of particular interest is co-metabolism, a phenomenon where microorganisms utilize hazardous substances, both organic and inorganic, as a supplementary energy source. This simultaneous biodegradation process showcases the remarkable adaptability and versatility of microorganisms in their quest for sustenance. The efficacy of biodegradation processes for hazardous waste can be enhanced through mechanical disintegration techniques or by deferring the treatment of hydrophobic substances. These methods aim to increase the available reaction surface area, thereby facilitating the breakdown of pollutants by microorganisms. Additionally, the removal of hazardous substances from wastewater can be achieved through advanced oxidation processes. These techniques employ powerful oxidizing agents to degrade pollutants into harmless byproducts, thus contributing to the remediation of contaminated water sources. By the use of genetic engineering, microbial strains can be used to create an ability to degrade biodegradation of xenobiotics or to amplify this capability by amplification of respective genes. Building multimodal metabolic pathways to encompass the broadest spectrum of biodegradable xenobiotics and the pace of biodegradation is an alternative method for the destruction of industrially dangerous

N. Salaria · A. Kumar · S. Singh · A. Rajput · S. K. Tripathi (✉)
Department of Life Sciences, Arni University, Kangra, Himachal Pradesh 176401, India
e-mail: sharadtripathi92@gmail.com

M. Shukla
Biological Product Laboratory, Department of Botany, University of Allahabad, Prayagraj 211002, India

R. A. Bhat et al. (eds.), *Zero Waste Management Technologies*,
https://doi.org/10.1007/978-3-031-57275-3_11

chemicals. In the biotreatment of hazardous wastes, self-aggregating microbial cells, flocs and granules, and artificially aggregated cells adsorbed on solid particles are frequently utilized.

Keywords Biohazardous wastes · Biological treatment · Aerobic methods · Anaerobic methods · Genetic engineering · Xenobiotics

1 Introduction

Waste management is one of the primary environmental issues of the present. Waste is increasing due to modernization and changes in the lifestyle and consumption of the human population. Garbage is becoming more and more accumulated as a result of rising human activity and urbanization. Waste is produced as a result of manufacturing procedures, industries, and municipal solid waste (Kan, 2009). Containers, packaging, food scraps, and other diverse inorganic wastes from domestic, industrial, as well as commercial sources make up municipal solid trash. The majority of affluent nations have strict regulations governing garbage management. Waste management's main objective is to eliminate dangerous industrial waste from the source. The collection, processing, transportation, but also disposal of waste are all part of the waste management process. Waste management is crucial for maintaining public health and protecting the environment from contamination. Industrial hazards have an effect on the environment, animals, plants, and people's health. Recycling, composting, sewage treatment, incineration, and landfilling are some of the waste management techniques. Recycling is the process of recovering the material from a product after the user has used it. A continuously new method of the recycling has been also developed. It involves energy recovery, biological reprocessing, as well as physical reprocessing. The industrial waste is classified into different types. It includes solid waste, liquid waste, and chemical waste. There are different types of methods for treatment of the liquid, solid, and chemical waste which includes biological treatment. Chemical treatment, rapid carbonation method, coupled electrocoagulation method, and selective oxidation process are some examples of photosynthesis bacterial methods. Biological waste can be transformed into useful energy and other byproducts via biological or thermochemical ways. The production of biowaste to produce value-added products will enhance circular economy which in turn gains sustainability (Rajendran et al., 2021).

1.1 Waste

Any undesired material is considered waste. Waste classification is exceedingly challenging. Commercial wastes, animal wastes, domestic wastes, industrial solid wastes, biomedical wastes, construction wastes, hazardous wastes, biodegradable wastes as well as non-biodegradable wastes are the broad categories into which the garbage is divided. Agricultural trash, paper, wood, and metal scraps are also included in the rubbish (Hajam et al., 2023). Waste in the UK is often divided into two categories: hazardous garbage and non-hazardous waste. In the UK, there are a total of 17 locations for waste disposal, which is strictly regulated (Environment Agency, 2009).

1.2 *Industrial Hazardous Wastes*

Industrial hazardous waste is the term for waste produced by industrial processes. Waste that is solid, liquid, or gaseous might be categorized as hazardous industrial waste. These wastes, either alone or in combination with other wastes, are dangerous to humans. The features of the hazardous waste can be used to identify it. The hazardous waste has a variety of characteristics, including flammability, corrosively, reactivity, or toxicity as shown in Table 1 (Liu et al., 1997).

The industrial hazardous waste is produced by the following procedures:

a. Pyrolyte activities and petrochemical processes.
b. An oil- and gas-related drilling operation production.
c. Cleaning, emptying, and maintaining ships' tanks for petroleum oil.
d. Refining petroleum, treating used oil, and recycling waste oil.
e. Business activities involving hydraulic systems or other applications that use mineral or synthetic oil as lubricant.

Table 1 Characteristics of industrial hazardous waste and their effects

S.No	Characteristics of the industrial hazardous waste	Effects of the industrial hazardous waste
1	Flammable	This type of the waste causes harmful gases at high temperature and pressure and responsible for the damage to the surrounding
2	Oxidizing	This type of waste helps in combustion of the material by yielding oxygen
3	Poisonous	This type of waste is responsible for causing death if inhaled or by coming in contact
4	Infectious	Infectious hazardous waste contain microorganisms which release toxins and responsible for diseases in humans as well as in plants
5	Corrosives	This type of hazardous waste is chemically active in nature and responsible for damage to plants as well as animals
6	Eco-toxic	This kind of hazardous waste causes bioaccumulation, which negatively affects the environment
7	Toxic	These types of hazardous wastes are carcinogenic in nature. If they are inhaled or ingested, results can cause chronic effects

f. The industrial or secondary manufacturing of zinc.
g. Copper secondary production.
h. Production of aluminium, both primary and secondary.
i. Manufacturing of basic ingredients for plastic

1.3 Types of Industrial Hazardous Wastes

The hazardous industrial wastes are poisonous in nature. The proper disposal of the trash is required to minimize the negative effects of these wastes on the environment. The UK Environment Agency divided the waste into two categories: controlled waste and non-controlled waste (Environment Agency, 1999). Controlled waste is the term used to describe the waste produced by industrial, commercial, and residential organizations. The term "non-controlled waste" is used to describe the waste that agriculture mines produce. Solid, liquid, and gaseous types of industrial hazardous waste can all be encountered. The wastewater released from brewing process for beer extraction contains a sufficient amount of H_2 and CH_4 which will be used for production of biogases in concerns with wastewater treatment (Kumar et al., 2022).

1.3.1 Solid Industrial Waste

The solid industrial waste is classified into different types on the basis of their source. Industrial solid waste, institutional solid waste, municipal solid waste, construction solid waste, and household solid waste are examples of solid waste. The industrial solid wastes are corrosive in nature that is responsible for corroding metal containers and highly reactive in nature releasing toxic fumes when heated.

1.3.2 Chemical Industrial Waste

The chemical industrial waste may be defined as any unusual chemical that is present in excess and responsible to cause damage to the human health. The chemical industrial wastes are generated from the chemical industries, refining processes, drilling operation of oil. The chemical industrial wastes include-

a. Arsenic-containing sludge.
b. Sludge made with phosphate.
c. Alkali residues.
d. Sludge containing mercury.
e. Chromium sludge from a water-cooling tower.
f. Distillation leftovers.
g. Vinyl chloride monomer manufacturing leftovers.

Processing waste sludge or residues that contain acid, other hazardous metals, or organic complexes.

1.4 Effects of Hazardous Wastes

1.4.1 Effect of Hazardous Waste on Health

The World Health Organization's (WHO) Member States in Europe are particularly worried about hazardous waste, and the Sixth Ministerial Conference on Environment and Health made this issue a top priority (WHO meeting report, 2015). Everything present in our surrounding has direct or indirect impact on our health. Health issues are directly associated with steps that are taken to be disposed of and handling of hazardous wastes. Exposure of our body to hazardous wastes might be direct (i.e. during recycling and recovery of the industrial waste material, exposure by emissions from incinerators, landfill sites, sewage treatment, etc.) and indirect (i.e. by the consumption of contaminated food, water, and other eatables).

Even though we live in a time of significant technological developments and advanced waste management systems, public acceptability of these systems and facilities is still not very high. It makes hazardous materials more dangerous to the environment and human health. It has been seen in numerous research (prospective cohort, retrospective case–control, and cross-sectional) that exposure to hazardous wastes causes serious illnesses. Disorders that are both acute and chronic in nature can result from human exposure to toxins generated during garbage disposal. Sever (1997) suggested that residents who live close to places where waste has been dumped are more likely to experience birth abnormalities and other malignant illnesses. Heavy metals from industrial wastes, for instance, mercury, nickel, lead, and cadmium, are also released into the environment and contaminate both soil and different water bodies. Peplow (1999) summarized that organisms in an ecosystem that are exposed to these toxins poison their food chain.

These heavy metals have extremely detrimental impacts on health in the form of numerous illnesses, whether exposure is direct or indirect. The following effects on human health are demonstrated by high concentrations of heavy metals: Lead exposure can result in mortality due to loss of brain function, central nervous system damage, and kidney failure (USGAO, 2000). High level of nickel contamination results in worsening of eczema in humans. In several studies, individuals exposed to chromic acid and other waste materials have reported a range of skin-related disorders, notably dermatitis and skin ulceration. These conditions highlight the potential health impacts associated with such exposures. For further details on the effects of these substances on health, please refer to the accompanying Table 2.

Instead of heavy metals, some organic compounds, such as polychlorinated dibenzo-p-dioxins (PCDDs) and polychlorinated dibenzofurans (PCDFs) produced by the incomplete combustion of municipal wastes, medical wastes, household wastes, during the production of chemicals or pesticides, etc., also contribute to an elevated risk of developing sarcomas. (Zambon et al., 2007). Some indirect health issues also seen by the contribution of the greenhouse gases with disposed material become more significant. Additionally, as a result of ozone loss, earth's temperature

Table 2 Types of hazardous waste, its sources, and effects on health

Sr. No.	Hazardous waste	Source of hazardous waste	Effects on health
1	*Heavy metals*		
	Arsenic	From non-anthropogenic geochemical formation	Anaemia and several cancerous conditions
	Cadmium	From fertilizer industries and from battery waste	Responsible for skeletal and pulmonary disorders
	Chromium	From the mining areas	Responsible for respiratory disorders
	Mercury	From health care institutes	Vision problem, effect on brain and CNS
2	*Hydrocarbons*		
	Benzene	From petrochemical industries	Blood cancer and defects in bone marrow
	Vinyl chloride	From plastic	Lungs and liver cancers
3	*Organic chemicals*		
	Dioxins	From herbicides	Skin disorders and cancer
	PCBs	From electronic waste and hydraulic fluids	Skin diseases and Gastrointestinal disorders

is rising day by day, which will have an impact on elderly people who already have cardiovascular and respiratory conditions like asthma.

1.4.2 Effect of Hazardous Waste on Environment

Industrial wastes that have been dumped are contaminating the environment much more quickly than we anticipated. Industries play a significant role to the production of potentially noxious wastes worldwide. The dumping and disposal of waste substances is a major concern due to the detrimental effects of illegal disposal and dumping practices on the environment and human health. Soil, air, and water are all vulnerable to contamination through various pathways such as infiltration, leaching, bioaccumulation, and biological uptake. This contamination can occur through processes like fugitive dust emissions, groundwater seepage, surface water runoff, and soil erosion, affecting both the environment and biota. A significant environmental concern is the gases released by the garbage. Certain bacteria engage in anaerobic respiration which generates methane (CH_4), one of the most powerful greenhouse gases. These bacteria thrive in landfills. At sites with maximum anaerobic decomposition, methane makes up about 50% of the composition of landfill gases (Cointreau-Levine, 1997). Such gases influence the environment and significantly enhance the greenhouse effect.

Even though leather itself degrades very slowly and the tanning process makes it more resistant to processes like microbial degradation, chemical degradation, thermal degradation, etc., the leather industry still produces a significant amount of non-biodegradable solid waste (Han et al., 2003). This will also lead to further consequences in the form of groundwater system degradation and agro-based activity (Mwinyihija, 2012). Significant amounts of volatile organic compounds (VOC) like amines, hydrogen sulphides, volatile hydrocarbons, ammonia, etc. are released during the tanning processes, and if not properly controlled, these emissions could pose a serious threat to the environment. Groundwater that is impure or contaminated due to landfill leachate poses a risk to downstream surface waters and is a concern for the environment related to waste disposal (Christensen et al., 1994). So, to make our environment free from all these effects, it is very necessary to develop and introduce some useful treatment methods or strategies and waste management facilities to make environment safe.

1.4.3 Effect of Hazardous Waste on Soil and Plants

The substances released from different industrial wastes and household waste dumping sites have an influence on the fertility of the soil. Plant growth is supported by the soil. Emissions from discarded materials have an adverse effect on the productivity of the soil as well as on plant characteristics like size, composition, quality, and yield. One of the primary causes of soil pollution is the release of heavy metals from hazardous materials. These heavy metals alter the soil's biochemical and biological characteristics. Clay content, organic matter, pH, and other biological and biochemical characteristics are some of the things that pollutants greatly affect. There are numerous beneficial microbes in the soil that are essential to increasing the fertility of the soil because they interfere with the behaviour of microorganisms; the soil biota is negatively impacted by heavy metals. Plant health is directly correlated with these soil-related factors present in the soil that are essential for increasing the soil's fertility.

Heavy metals like Hg, Se, As, and Cd cannot significantly influence plant growth. However, certain metals, such as Zn, Mn, Co, Ni, and Fe (in their ideal concentrations), are crucial for the normal growth and metabolism of plants; if their concentration exceeds their ideal values, the plants become toxic and suffer damage (Garnier et al., 2006). As heavy metals are absorbed by plants from contaminated soil, they build up along the food chain and endanger both human and animal health. Increased lead (Pb) concentration in soil reduces soil productivity, and even small concentrations of lead can interfere with important plant processes like mitosis, photosynthesis, and water absorption. Heavy metals' toxicity causes abnormalities in plants like chlorosis, the wilting of older leaves, and stunted plant growth.

1.4.4 Effect of Hazardous Waste on Aquatic Life

Due to certain unusual human activities, aquatic life also gets disturbed. Waste products from industries, dumping of medical, and other hazardous substances are thrown directly into the water bodies. This contaminant seriously disrupts the natural equilibrium of the aquatic environment. Results limited diversity of aquatic system with extent of contaminations (Ayandiran et al., 2009). When material containing heavy metals is dumped into an aqueous environment, it settles down and becomes part of the sediment. This settled material continuously upsets the biological activities in the aquatic life as well as photosynthetic process of various algae. Dyes used in various industries, such as the textile and tanning industries, are highly toxic. When these dyes contaminate aquatic systems, they have noxious effects on both aquatic flora and fauna. Some dyes and their breakdown products exhibit mutagenic and carcinogenic properties, posing significant health risks to organisms and disrupting the ecological balance of aquatic environments. As fish ingest these pollutants, their chances of survival are significantly reduced. When these contaminated fish are consumed by carnivores and omnivores, the predators themselves become more susceptible to various carcinogenic conditions, further decreasing their survival rates. This bioaccumulation of toxins not only affects individual species but also has cascading effects throughout the entire ecosystem. This leads to disruptions in the food chain and the food web. Hazardous wastes are highly conspicuous in aquatic ecosystems, including ponds, rivers, and lakes, and significantly degrade their cleanliness and water quality.

In a study, dye containing wastewater (treated and untreated) on freshwater fish named *Gambusia affinis* (common name: western mosquitofish) showed virulent effects on red blood cells (RBCs), reduced their number, and variation in their shape and size (Soni et al., 2006). There are five different ways for pollutants to enter the bodies of fish: through ingested food, non-food particles, oral water consumption, skin, as well as gills. As pollutants get absorbed by fish, they accumulate in the muscles and other tissues. When these contaminated fish are consumed by humans and other organisms, it results in harmful impacts on consumers and imbalances the food chain. In South India, the female crab *Spiralothelphusa hydrodroma* is an important food source. However, contamination in water bodies leads to a loss of nutritive value in these crabs, including reductions in lipids, proteins, and carbohydrates (Sekar et al., 2009). Diatoms living under the water also get affected by the presence of virulent substances or wastes coming from industries. Presence of hazardous waste in water bodies also affects multiple aspects of algae for instance pigment content, growth, protein content, and other valuable contents. For making aquatic environment safe from pollutants, it is very important to introduce some waste management methods or strategies to preserve our aquatic life as well as other linked organisms depending upon aquatic ecosystem for their survival.

2 Biological Treatment Methods

There are many conventional treatments methods that have been used to treatment of hazardous waste treatment plants. These techniques are getting evolved together with evolution of instruments and techniques which results in development of new biological tools. Genetic engineering offers to improve biological processes by the reduction

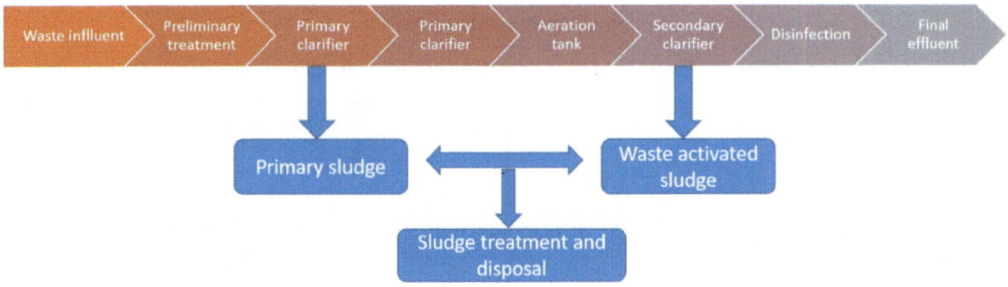

Fig. 1 Flowchart of activated sludge treatment process

of side reactions, higher product concentrations. Biotechnology also substitutes the hazardous material to non-hazardous material in the field of agriculture. In biological treatment, several conventional methods are used, such as the activated sludge treatment process (Fig. 1), which relies on microorganisms to decompose organic matter in wastewater. Another method is bioremediation, where naturally occurring or introduced microorganisms break down pollutants in contaminated environments. These techniques harness the natural capabilities of microorganisms to clean up hazardous substances, making them effective solutions for managing and reducing environmental pollution.

An essential component of the treatment facility that handles both municipal as well as industrial waste is biological treatment (Fig. 2). The biological treatment makes use of several microorganisms, fungi, bacteria, yeasts, and algae. It involves two methods, i.e. aerobic and anaerobic methods. In the aerobic method, degradation of waste material takes place in the oxygen-rich atmosphere. The aerobic treatment requires proper amount of certain temperature and pH regime, and the absence of toxic substances (Hynninen, 2000). The aerobic and anaerobic biological treatment employing microbial treatment alone or in combination with physiochemical management process is used to remediate landfill leachate (LFL) (El-Saadony et al., 2023).

Reactions products of aerobic treatment are carbon dioxide, water, and excess biomass. The paper mills are fully degraded by the aerobic treatment. The main characteristics of aerobic treatment are:

a. High levels of extra sludge are needed.
b. Extremely high energy need.
c. Requirement of high space.
d. Biological materials are fully degraded.

Due to the aeration of the effluent, the disadvantage of aerobic treatment is that it has high operating costs. The anaerobic treatment is used for the highly polluted effluents (Hamm, 2006). Reactions products of anaerobic treatment are carbon dioxide, methane, and excess biomass.

Defining features of anaerobic as well as aerobic wastewater treatment:

a. No requirement of high amount of excess sludge.
b. Production of energy by the utilization of biogas.

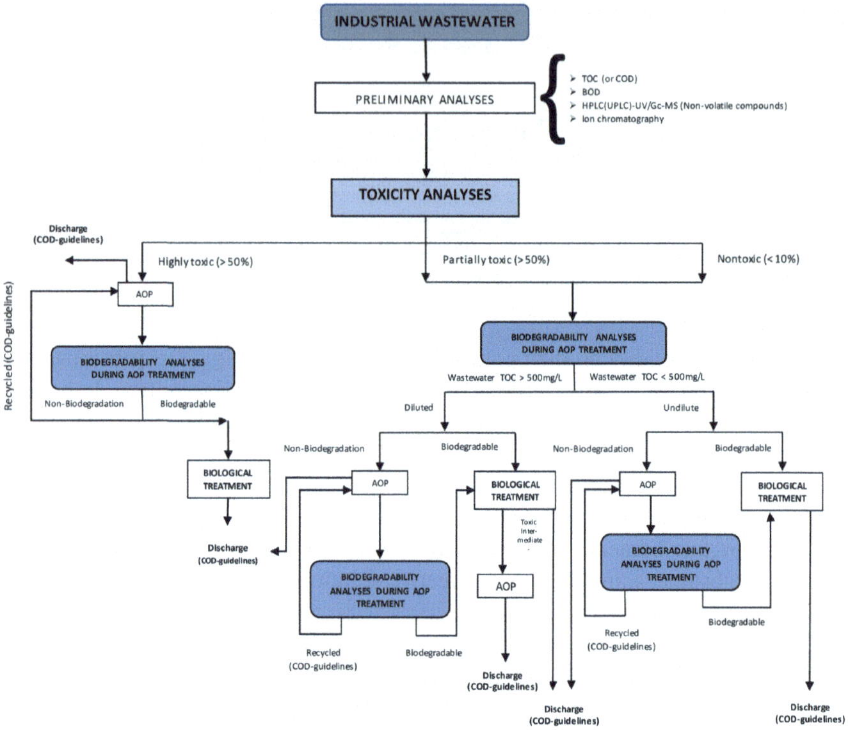

Fig. 2 Common waste management strategies (*Source* Oller et al., 2011)

c. No requirement of the high energy.
d. No requirement of space.
e. Biological materials are not fully degraded.

2.1 Aerobic Methods

Aerobic treatment is very simple, stable, and efficient method of treatment that produces high quality of secondary effluents. In Aerobic method, C*hlorella pyrenoidosa* was able to eliminate over 97% of phenol at phenol concentrations of < 0.6 g/L at pH 8.0 reported by the Dayana Priyadharshini and Bakthavatsalam in (2019). Khan et al. developed sulfolane-degrading aerobic granules (SDAG) in 2019 using two different techniques. The first one involves modifying pre-grown granules for sulfolane-containing environments and collaborating with a bacterial culture for sulfolane-contaminated sites. Aerobic treatment uses oxygen for the breakdown of the organic contaminant. Aerobic treatment is appropriate for a variety of industrial wares such as chemical, food as well as beverages and municipal. Sewage and industrial wastewater treatment can be done using a variety of techniques. These include:

a. **Conventional activated sludge**—The aerobic microorganisms in the aeration tank of conventional activated sludge break down organic material. Sludge is produced as a result of this process, and it is later separated from the cleaned water in a sedimentation tank.
b. **Moving bed biofilm reactor (MBBR)**—In the moving bed biofilm reactor method, biofilms are grown on the plastic carrier and circulated in the aeration tanks.
c. **Membrane bioreactor (MBR)**—Membrane bioreactor is a highly developed method that incorporate membrane filtration with the activated sludge process.

The aerobic treatment has a variety of traits:

a. Requirement of energy very high.
b. High sludge production.
c. Requires 2–4 weak to start up.
d. High nutrient requirements.
e. Low alkalinity requirements.
f. No biogas production.

2.2 *Anaerobic Methods*

One of the most important techniques for treating or degrading waste products from various sources is the anaerobic method. This technique essentially involves a bacterially mediated process that takes place in the lack of oxygen. Use of this biological anaerobic pre-treatment method leads to potential decrease in the sludge production, and furthermore, this is directly associated saving of consumed energy. Biomass is consumed by microbes, which primarily release carbon dioxide (CO_2) as well as methane (CH_4). Mixture of CO_2 and CH_4 and hydrogen sulphide gases are major constituents of biogas released by the anaerobic digestion by anaerobic organisms or methanogens inside an anaerobic digester, bioreactor as well as biodigester. Released biogas can be converted into electricity by burning it and left solids can be degraded by aerobic digestion to produce compost (De Baere, 2006). The anaerobic method's ability to produce biogas depends on a number of variables, including pH, temperature, moisture, and the feedstock's carbon to nitrogen ratio, among others.

In a study, it had been reported that thermophilic conditions are susceptible for the degradation of deca-brominated diphenyl ether (deca-BDE), and pH 7 was the optimal reaction condition (Shi et al., 2019). In another study, Wang et al. (2019) reported that Fe (III) and other matters consisting fenton sludge which get activate anaerobic digestion process in which oxidation of organic matter takes place in the presence of iron-reducing bacteria (IRB) and Fe (III) stimulated dissimilatory iron reduction (DIR). The cultivation of high-performing dechlorination microbes for the anaerobic degradation of 2,4,6 trichlorophenol required fresh aerobically activated sludge, low Eh, and high culture temperature, according to Song et al., (2019a, 2019b).

2.3 Anoxic Methods

In anoxic treatment, no involvement of the free and molecular oxygen but bound oxygen may be present. It is based on the presence or absence of the nitrogen. The anoxic condition requires nitrates and nitrites. Both nitrates and nitrites contain bound oxygen because oxygen is present in their molecular structure. The aim of the anoxic treatment is to reduce the nitrogen and phosphorous from waste effluents before discharge into the water bodies. The release of the waste of the water bodies without anoxic treatment results in eutrophication. The eutrophication is the increase in the concentration of the nitrogen and phosphorous in the water bodies. The removal of the nutrients plays a significant influence in the running of the treatment plant because it increases the thickness of the sludges and, due to a lack of oxygen, causes the growth of filamentous organisms to be reduced. The small amount of sludge generated is nitrated, which is utilized in the aerobic zone's anoxic treatment to reduce biological oxygen demand. The anoxic treatment requires high amount of energy. The anoxic treatment responsible for the release of the greenhouse gases into atmosphere organic nitrogen, ammonia, ammonium ions, nitrate, and nitrites are all forms of nitrogen that can be found in waste. The primary treatment of the wastewater removes only up to 15% phosphorous by the sedimentation method. The traditional techniques for phosphorus removal also involve transferring soluble phosphorous to a sold phase, followed by solid–liquid separation. Chemical phosphorous, biological phosphorous, and a combination of the two phosphorous are all removed using the traditional methods. Chemical precipitation is the common method of removal of the phosphorous. It involves the addition of the salt. The common salts used for the chemical precipitation are used chemical salts are aluminium sulphate or alum [$Al_2(SO_4)_3$], ferric chloride [$FeCl_3$], sodium acuminate [$Na_2Al_2O_4$], polyaluminium chloride (PAC) [$Al_nCl_{(3n-m)}(OH)_m$], ferrous sulphate (Iron 11 sulphate) [Fe_2SO_4] as well as quicklime (CaO), or hydrated lime $Ca(OH)_2$ But phosphorous fraction from the wastewater is removed by the chemical precipitation.

2.4 Phytoremediation

The removal of contaminants from the soil with the help of plants is known as phytoremediation. Contaminants are fixed in the ground and absorbed by plants and accumulated to organ of itself and are released finally to atmosphere (Fig. 3).

The phytoremediation mainly works on the following four basic principles and each has a different mechanism of action (Cobett, 2000; Cobett & Goldsbrough, 2002).

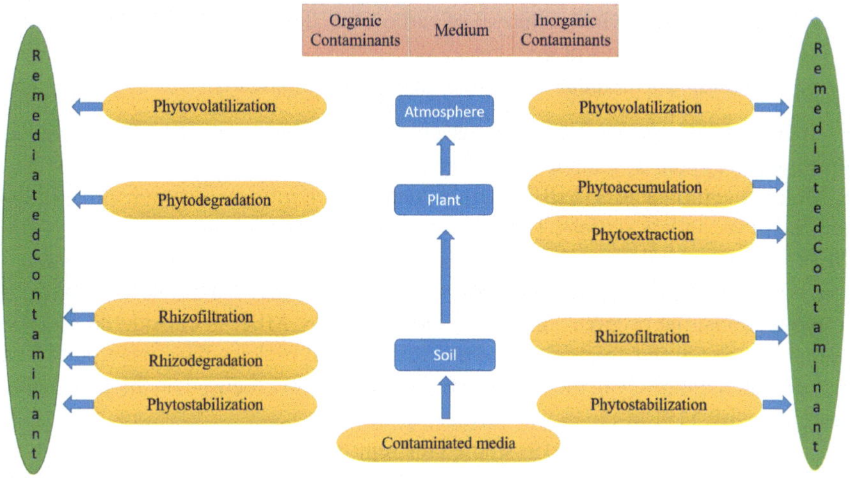

Fig. 3 Uptake mechanism of pollutants and other wastes by phytoremediation process

1. Phytoextraction—It involves the mechanism by which plants absorb metals from the soil and transport to upper part of plant which is harvestable.
2. Phytostabilization—As it is clear from the name, it is stabilized polluted soil with the help of plants.
3. Rhizofiltration—These plants used to filter and remediate the aquatic medium.
4. Phytovolatilization—This includes the extraction of metals by plants through soil and releasing them into environment.

2.5 *Bioreactors*

For disposal purposes, waste materials from various sources pour in an open environment resulting producing harmful impact on biodiversity. For the degradation of these hazardous wastes, it is important to remove the toxicity or noxiousness of these wastes. As technology has advanced, various techniques have emerged to address issues related to the degradation of hazardous wastes, leading to the evolution of bioreactors. Based upon high biomass and high liquid velocities, the reactors can be tall as well as small both are completely closed. This becomes an essential characteristic of bioreactors that it reduces the elimination of harmful emissions. Mainly in anaerobic system where the release of biogas takes place is important to avoid the emission. Due to this region, bioreactors are designed in such a way that it would contain separate sections like distribution section, separation section, reactor section, and gas accumulation section where the biogas is stored and flows to boiler house.

Stirred tank bioreactors are used for aerobic wastewater treatment which is one of the oldest applications of bioreactors in field of technology. Conventional scheme mainly contains two bioreactors or stirred tanks arranged in series:

a. First bioreactor in which removal of carbon, i.e. destruction of organic material and nitrification, takes place.
b. Second tank in which denitrification takes place anoxically.

After denitrification of waste material, the clarification of effluent material is further taking place in sedimentation tank.

Membrane bioreactors (MBR) mainly consist of membrane filtration units for the removal of secondary settlers. There are certain advantages of membrane bioreactors (MBR) which include good effluents quality, reduced footprint and sludge production, good disinfection capability, improved nitrification, etc. With these advantages or applications, the progress in field of biotechnology MBR is considered as good municipal wastewater treatments technology. Several reviews have been published on the MBR technology in the last few years which mainly focused on the wastewater treatment with MBRs.

Razanamahandry et al. (2019) studied the removal of antibiotic sulfamethoxazole (SMX) by anaerobic membrane bioreactor. If the feed concentration was 10–1000, 10,000, or 100,000 g/L, the average SMX separation was 97.1%, 97%, and 88%, respectively. Anaerobic microorganisms' ability to grow was hampered by high SMX loading.

2.6 *Land Farming and Composting*

In a variety of land-based systems, some sludge and solid contaminants with hazardous wastes can be biologically treated, either in the soil or through composting.

2.6.1 Land Farming

Land farming suggests that a crop of microorganisms was grown that used the waste as a source of carbon as well as energy. To dispose of the waste from the refineries like API separator sludge, the petroleum refinery frequently uses these lands for farming. Among the requirements for land farming sites are the following: (1) Proper waste incorporation along with application rates; (2) maintaining the pH of the soil; (3) prevention of soil aggregation; (4) bioaugementation to ensure good microbe health; and (5) proper timing for cycle of growth and nutrition absorption.

2.6.2 Composting

Composting can use a mixed organic/soil matrix, making it similar to land farming. Typically, it is an aerated mixture, well-mixed process that takes place on soft surfaces or in enclosed systems. It is an example of a "dry" anaerobic or aerobic procedure in that the biological growth is supported by particles of cellulosic and lignin-based solids. Blending contaminated material, for instance, sludge or soft, with dry organic material, including straw as well as agricultural runoff, but also allowing the saprophytic microbes to break down the matrix over a period of weeks to months, based on the method, constitutes composting.

2.7 Enzyme Systems

There has been a lot of media coverage of enzyme use in industry. The features of the substrate, enzyme, and process all play a role in how an enzyme system is used. Enzymes can either function outside of the cell or be membrane-bound or dependent on cofactors, or they can function inside the cell and also be excreted into the surrounding media. Although enzyme retrieval can be costly, it may be preferred in commercial treatment if a single-step reaction is found to decontaminate a hazardous material; Munnecke (1979) has identified the application of enzyme extracts for application to pesticide residues of the organophosphorus type. As their composition can be quite expensive, maintaining enzyme stability and activity is crucial. Enzymes may be used either free or immobilized, the latter on carriers with either a fixed or fluidized configuration. The enzymes in their free state are wasted with discharge after becoming a component of the substrate matrix. Immobilization is thus preferred to preserve the enzyme and enable recycling or repeated use of it. If the hydrophobic material cannot be dissolved by emulsification or if enzyme encapsulation is used, free enzyme systems are needed for water-insoluble substrates. Klibanov (1983) lists techniques for immobilizing enzymes. Depending on conformational as well as physical/chemical changes, such methods can either stabilize or destabilize the enzyme. Many immobilization methods work with whole cells that have the required enzyme systems.

2.8 Spill Clean-Up Systems

Toxic wastes can be removed from both land and water using biological technologies. Accidents involving the spilling of organic compounds are also subject to the same principles that apply to land treatment. The main distinction is that spills are frequently connected with loads of a pure substrate at incredibly high rates, which permits the usage of a particular microbial culture. In most cases, it is a very limited area that gets affected. On the other hand, abandoned or unsafe landfills that need clean-up can present very unlike issues from spills, such as soil contamination that is diffuse and unidentified and surface runoff or leaching into aquifers. In these cases, a variety of systems can be designed, ranging from impoundments converted to biooxidation ponds to collect and treat runoff, and aquifer containment with either in situ (Weldon, 1982) or withdrawal and surface treatment.

2.9 Photosynthetic Bacterial Method (Such as S^{2-} or $S_2O_3^{2-}$)

The oldest prokaryotes and those that are most common in the environment, including soil, lakes, and oceans, are photosynthetic bacteria (PSB). They have a simple photoenergy synthesis system (Chen et al., 2020). The most outstanding quality of photosynthetic bacterial is that they have access to light as a source of energy as well as can employ organic carbon substances, sulphides (for instance S^{2-} or $S_2O_3^{2-}$). Rather than through photosynthetic growth, PSB grow phototrophically. The entire process does not release O_2 or need H_2O as a source of electrons (Puyol et al., 2017). Wastewater can be cleaned of COD, nitrogen, phosphorous, and various kinds of heavy metals by photosynthetic bacteria. The current methods for removing nitrogen and phosphorus are complicated, expensive, and energy intensive. For their growth, photosynthetic bacteria require a lot of nitrogen and phosphorus. As a result, using photosynthetic bacteria to remove nutrients from wastewater has become a popular alternative to traditional wastewater treatment methods. In wastewater treatment processes, it is difficult to keep these kinds of bacteria in pure cultures. A group of chemotrophic and photosynthetic bacteria can purge a variety of organic substances from wastewater. Non-photosynthetic bacteria can use the oxygen produced by some strains of photosynthetic bacteria. Additionally, organic compounds that are mineralized by non-photosynthetic bacteria supply vital minerals for the growth of photosynthetic bacteria. Therefore, bacterial consortiums that are both photosynthetic as well as non-photosynthetic have greater efficiency than those that are only made up of photosynthetic bacteria. This is so that they can be used in place of photosynthetic consortium, which is only applicable during industrial wastewater polishing stage of hazardous waste treatment, in both secondary and tertiary treatment stages of waste treatment. The use of photosynthetic bacteria in the treatment of dangerous or resistant pollutants, such as pharmaceuticals, oils, heavy metals, farm chemicals as well as dyes, is also supported by experimental evidence (Talaiekhozani & Rezania, 2017).

2.10 Biochemical Method

By using specific chemicals, industrial harzardous waste is subjected to chemical treatment in order to minimize its potential danger. Chemical treatment helps to recover beneficial byproducts from toxic waste, which reduces the total costs of hazardous waste disposal of industries. Chemical treatments have been extensively employed for a variety of toxic wastes because they are extremely effective including solid waste as well as liquid waste or wastewater (Bustos-Terrones et al., 2021; Reddy et al., 2010). Chemical processing can transform hazardous toxic waste into non-hazardous byproducts, facilitating safe disposal. Oxidation, precipitation, neutralization, coagulation, ion-exchange, electrochemical processes as well as flocculation are the main ones that are involved (Patel et al., 2021).

Hazardous pollutants that are dangerous in both low and high concentrations are treated with chemicals. Any chemical treatment approach is considerably influenced by the environment and the target pollutants. To satisfy the criteria of laws relating to the environment, it must be chosen either alone or in combination with other methods to produce successful outcomes of treatment in a way that is both economical and environmentally friendly. Chemical remediation has been discovered to be extremely efficient for hazardous solid waste, liquid waste, and effluent discharged. However, it is constantly constrained by the massive cost of chemicals and the production of secondary waste that needs to be treated before being properly discarded. The appropriate technology must be used to treat toxic waste due to strict environmental standards, which aids in resource reduction and recovery. Chlorination is an effective chemical treatment, but it produces disinfection byproducts (Song et al., 2021). Numerous recalcitrant pollutants can be effectively treated chemically with ozonation, but this process consumes a lot of energy (Ben Hamida et al., 2018). This approach is more costly due to its limited water solubility, selectivity towards pollutants, as well as reduced mineralization efficiency. When coagulants are used in traditional water treatment methods, a sizable amount of silt is produced. In order to reduce and treat hazardous pollutants, using more chemicals is not a sensible strategy for a long-term fix. Additionally, the Fenton process is efficient at degrading organic toxins; however, it has a constrained pH range and produces iron sludge (Azbar et al., 2004). Sustainable management is needed for the secondary waste produced by the Fenton process. Emerging pollutants can be treated very effectively using catalyst-based methods, and a variety of homogeneous as well as heterogeneous catalysts have been created for this objective (Do Minh et al., 2019). Without an effective system for recovering catalyst, finer catalyst particles may enter either finish up in solid sludge or effluent streams, both of which can have negative environmental impacts. To minimize chemical utilization as well as secondary waste, the development of effective, environmentally responsible chemical treatment techniques is becoming increasingly necessary. In an effort to reduce toxicity of customized catalysts, more emphasis should be placed on the green method of catalyst synthesis (Yadav et al., 2021). Utilizing biobased materials, natural minerals for creating ion-exchange materials, coagulant, or catalyst can make chemical treatment secure and more environmentally friendly (Srivastava, 2021).

2.11 Electrochemical Treatment

The tools provided by electrochemical technology are ideal for addressing environmental issues. Since electron, a clean reagent, is used as the experiment's main reagent, there is no need to add any more (Rajeshwar & Ibanez, 1997). The creations of eco-friendly electrochemical techniques to recover hazardous industrial effluent are receiving increasing amounts of attention. Due to the stringent environmental regulations on effluent wastewater, electrocoagulation (EC) has been the subject of numerous studies (Bhagawan et al., 2017). The coagulating agent's creation in the electrochemical process (EC) is controlled in situ using a sacrificial anode. To carry out the treatment in this instance, it is not necessary to add chemical coagulants or flocculants. Either direct oxidation or indirect oxidation is used to destroy the pollutants in the electrochemical process. The most widely used electrochemical method for treating wastewater today is electro-oxidation. In direct electro-oxidation, pollutants are directly transferred from the adsorbed pollutant to the anode surface and degraded there. Despite being theoretically possible at potentials relatively low than for oxygen evolution, the process is inefficient (Feng et al., 2016). However, due to contaminants that have been adsorbed on the surface, it is likely that electrodes will prevent additional oxidation. In indirect electro-oxidation, contaminants in the bulk solution are detoxified by oxidation/reduction reagents produced anodically, including chlorine as well as hypochlorite, ozone, hydrogen peroxide, persulfate, but also metal mediators. Although indirect electrochemical treatment was effective at treating wastewater of industries, reports of the creation of organic compounds that contain chloride have been made (Chiang et al., 1997; Ribordy et al., 1997). Although electrochemical treatments have been around for a while, they have recently gained popularity due to more stringent laws governing the environment on industrial hazardous waste and improved effectiveness. In order to treat wastewater, a number of new as well as improved innovations, which including advanced electrochemical oxidation processes or novel fusions of currently used methods, have been developed over the past few decades. Among these techniques are electro-Fenton, electro-oxidation, and photo-electrochemical, along with electrochemical processes. Combining these frequently results in improved process economy, faster reaction times, and greater process efficiency. Selected contaminants from industrial hazardous wastewater can now be removed using electrochemical treatments, which have been demonstrated to be a successful strategy (Malinović et al., 2022). Currently with the use of hybrid technology of bioelectrochemical emerges as an alternative energy conversion device for bioelectricity generation with concurrent waste remediation (Chandrsekhar et al., 2020).

2.12 Combined Electrocoagulation and Electro-Oxidation Method

The two processes can be combined to create a useful hybrid process because electro-oxidation (EOx) is a complete but time-consuming process; Electrocoagulation (EC) is a quick but incomplete method. Pollutants can be quickly removed by EC, but not completely. Pollutants can be consistently but slowly eliminated by EOx. The best qualities of both EC and EOx processes are combined with quick and thorough pollution abatement in sequential and simultaneous operations (Özyurt et al., 2017). The disposal of different pollutants has been accomplished with combined treatment through the use of sequential EOx treatment of the electrocoagulated effluent or simultaneous EC and EOx treatment with the proper configuration of electrode. Sequential EC-EOx seeks to combine the positive effects of the two approaches into a novel strategy (Chakchouk et al., 2017). Along with all of the advantages of sequential EC-EOx, simultaneous EC-EOx provides a hybrid system that combines monopolar (MP) EOx and bipolar (BP) EC in a single reactor (Naje et al., 2015). Innumerable EC and EOx studies have examined the treatment of industrial effluents including the production of dairy, pulp, the olive mill, textile, paper, pharmaceutical, tannery, bilge, refinery, slaughterhouse, and potato chips, but also leather. Using EC and EOx techniques, it has reportedly been possible to eliminate BOD, oil, COD, phenol, phosphate, arsenic, colour, lignin, turbidity, chromium, and sulphate from industrial effluents. A brief summary of recent studies revealed that applying combined treatment resulted in removal efficiencies of at least 90.4% colour, 54.16% COD, 37% turbidity, 90% BOD, 94.4% phenol, 72% TOC, 42.9% phosphate, and 97% TSS, and at most 45.28 kWh/m3 SEEC. By utilizing the synergistic effects, merging these two methods results in a useful hybrid. Since EOx is capable of completely mineralizing persistent organic pollutants but is also a slow as well as energy-intensive technique for the elimination of suspended solids, EC is a quick and effective technique for industrial wastewaters; however, it is an ineffective technique for eliminating prolonged solubilized organic pollutants. The EC-EOx procedure is an excellent alternative for managing industrial wastewater in terms of removal of pollutants, but more investigation on applications to these wastewaters is required (Özyurt et al., 2017).

3 Conclusions

The large number of studies and reviews showed different ways to overcome for the management of wastes. In some studies, it was found that chemical coagulation and advanced oxidation processes (AOPs) were used to treat acetate and polyester-dying fibre effluent. AOPs are frequently used as a primary pre-treatment stage for remediating industrial wastewater. Advanced oxidation processes (AOPs) and biological systems were recently combined in experiments to treat a variety of industrial wastewater. It includes a broad range of industrial pollutants as well as model solutions for actual effluents containing different prolonged substances. In addition, there has been an increase in interest in the development of biological reactor configurations that can best treat industrial wastewater. The biological configurations include membrane bioreactors, fixed biomass, aerobic granular biomass systems, and suspended biomass, among others to design a unique biological system for each target industrial wastewater in order to design the consuming steps in the treatment line to increase efficiency. Combining chemical treatment with biological processes enhances biodegradability by breaking down complex hazardous substances into simpler forms that microorganisms can more easily decompose. The biological treatment of industrial wastes was achieved with the help of different microorganisms like fungi, bacteria, and algae via the aerobic and anaerobic methods. *Chlorella pyrenoidosa organisms* are used in treatment of wastewater aerobically and remove more than 97% of phenol from the wastes. Biological treatment may be of the conventional, moving bed biofilm reactor (MMBR) along with membrane bioreactor (MBR). The byproducts were obtained from the treatment of wastes like CO_2, CH_4, and H_2S which are further utilized in the production of electricity and in production of other useful products. Some antibiotic formulation harmful byproducts pose a serious threat to the environment because of their high toxicological effects. Procaine penicillin G, a commercially significant-lactam-type penicillin active ingredient, has been chosen to assess the recalcitrance and toxicity of raw and pre-treated penicillin formulation effluent in order to modify their biodegradability and eliminate their toxicity in the environment. In conclusion, it was found that different aspects of biological treatment can be used to treat the wastes from industries. The performance of comprehensive economic studies that could position this cutting-edge technology as cost-competitive is the aspect of combined AOPs/biological systems for the treatment of scarcely biodegradable specific industrial wastewater that is most lacking, aside from the experimental and modelling work.

References

Ayandiran, T. A., Fawole, O. O., Adewoye, S. O., & Ogundiran, M. A. (2009).

Azbar, N. U. R. I., Yonar, T., & Kestioglu, K. (2004). Comparison of various advanced oxidation processes and chemical treatment methods for COD and color removal from a polyester and acetate fiber dyeing effluent. *Chemosphere, 55*(1), 35–43.

Bakthavatsalam, A. K. (2019). A comparative study on growth and degradation behavior of *C. pyrenoidosa* on synthetic phenol and phenolic wastewater of a coal gasification plant. *Journal of Environmental Chemical Engineering, 7*(3), 103079.

Bhagawan, D., Poodari, S., Chaitanya, N., Ravi, S., Rani, Y. M., Himabindu, V., & Vidyavathi, S. (2017). Industrial solid waste landfill leachate treatment using electrocoagulation and biological methods. *Desalination and Water Treatment, 68*, 137–142.

Bustos-Terrones, Y. A., Hermosillo-Nevárez, J. J., Ramírez-Pereda, B., Vaca, M., Rangel-Peraza, J. G., Bustos-Terrones, V., & Rojas-Valencia, M. N. (2021). Removal of BB9 textile dye by biological, physical, chemical, and electrochemical treatments. *Journal of the Taiwan Institute of Chemical Engineers, 121*, 29–37.

Chakchouk, I., Elloumi, N., Belaid, C., Mseddi, S., Chaari, L., & Kallel, M. (2017). A combined electrocoagulation-electrooxidation treatment for dairy wastewater. *Brazilian Journal of Chemical Engineering, 34*, 109–117.

Chandrasekhar, K., Kumar, G., Mohan, S. V., Pandey, A., Jeon, B. H., Jang, M., & Kim, S. H. (2020). Microbial electro-remediation (MER) of hazardous waste in aid of sustainable energy generation and resource recovery. *Environmental Technology and Innovation, 19*, 100997.

Chen, J., Wei, J., Ma, C., Yang, Z., Li, Z., Yang, X., & Zhang, C. (2020). Photosynthetic bacteria-based technology is a potential alternative to meet sustainable wastewater treatment requirement. *Environment International, 137*, 105417.

Chiang, L. C., Chang, J. E., & Tseng, S. C. (1997). Electrochemical oxidation pretreatment of refractory organic pollutants. *Water Science and Technology, 36*(2–3), 123–130.

Christensen, T. H., Kjeldsen, P., Albrechtsen, H. J. R., Heron, G., Nielsen, P. H., Bjerg, P. L., & Holm, P. E. (1994). Attenuation of landfill leachate pollutants in aquifers. *Critical Reviews in Environmental Science and Technology, 24*(2), 119–202.

Cobbett, C., & Goldsbrough, P. (2002). Phytochelatins and metallothioneins: Roles in heavy metal detoxification and homeostasis. *Annual Review of Plant Biology, 53*, 159–182.

Cobbett, C. S. (2000). Phytochelatins and their roles in heavy metal detoxification. *Plant Physiology, 123*, 825–832.

Cointreau-Levine, S. (1997). *Occupational and environmental health issues of solid waste management: Special emphasis on middle and lower-income countries* (pp. 1–25). Draft, World Bank Report.

De Baere, L. (2006). Will anaerobic digestion of solid waste survive in the future. *Water Science and Technology, 53*(8), 187–194.

Do Minh, T., Ncibi, M. C., Srivastava, V., Thangaraj, S. K., Jänis, J., & Sillanpää, M. (2019). Gingerbread ingredient-derived carbons-assembled CNT foam for the efficient peroxymonosulfate-mediated degradation of emerging pharmaceutical contaminants. *Applied Catalysis b: Environmental, 244*, 367–384.

El-Saadony, M. T., Saad, A. M., El-Wafai, N. A., Abou-Aly, H. E., Salem, H. M., Soliman, S. M., Abd El-Mageed, T.A., Elrys, A.S., Selim, S., Abd El-Hack, M.E. and Kappachery, S., & AbuQamar, S. F. (2023). Hazardous wastes and management strategies of landfill leachates: A comprehensive review. *Environmental Technology and Innovation*, 103150.

Environment Agency. (2009). Merchant hazardous landfills. Available at http://www.evironment-agency.gov.uk

Environment Agency. Waste Statistics for England and Wales 1998–99

Feng, Y., Yang, L., Liu, J., & Logan, B. E. (2016). Electrochemical technologies for wastewater treatment and resource reclamation. *Environmental Science: Water Research and Technology, 2*(5), 800–831.

Garnier, J., Quantin, C., Martins, E. S., & Becquer, T. (2006). Solid speciation and availability of chromium in ultramafic soils from Niquelândia, Brazil. *Journal of Geochemical Exploration, 88*(1–3), 206–209.

Hamida, S. B., Iftekhar, S., Ambat, I., Srivastava, V., Sillanpää, M., Amri, Z., & Ladhari, N. (2018). Dry and wet ozonation of denim: Degradation products, reaction mechanism, toxicity and cytotoxicity assessment. *Chemosphere, 203*, 514–520.

Hamm, U. (2006). Environmental aspects. In H. Holik (Ed.), *Handbook of Paper and Board* (pp. 208–218). Wiley-VCH.

Hajam, Y. A., Kumar, R., & Kumar, A. (2023). Environmental waste management strategies and Vermi transformation for sustainable development. *Environmental Challenges*, 100747

Han, B., Jaurequi, J., Tang, B. W., & Nimni, M. E. (2003). Proanthocyanidin: A natural crosslinking reagent for stabilizing collagen matrices. *Journal of Biomedical Materials Research Part a: An Official Journal of the Society for Biomaterials, the Japanese Society for Biomaterials, and the Australian Society for Biomaterials and the Korean Society for Biomaterials, 65*(1), 118–124.

Hynninen, P. (ed.). (2000). *Papermaking science and technology book 19 environmental control.*

Kan, A. (2009). General characteristics of waste management: A review. *Energy Education Science and Technology Part A—Energy Science and Research, 23.*

Klibanov, A. M. (1983). Immobilized enzymes and cells as practical catalysts. *Science, 219*(4585), 722–727.

Kumar, J. A., Sathish, S., Krithiga, T., Praveenkumar, T. R., Lokesh, S., Prabu, D., Renita, A. A., Prakash, P., & Rajasimman, M. (2022). A comprehensive review on bio-hydrogen production from brewery industrial wastewater and its treatment methodologies. *Fuel, 319*, 123594.

Liu, D. H., & Lipták, B. G. (Eds.). (1997). *Environmental engineers handbook.* CRC Press.

Malinović, B. N., Markelj, J., Žgajnar Gotvajn, A., Kralj Cigić, I., & Prosen, H. (2022). Electrochemical treatment of wastewater to remove contaminants from the production and disposal of plastics: A review. Environmental *Chemistry Letters*, 1–23.

Munnecke, D. M. (1979). Hydrolysis of organophosphate insecticides by an immobilized-enzyme system. *Biotechnology and Bioengineering, 21*(12), 2247–2261.

Mwinyihija, M. (2012). Pollution control and remediation of the tanning effluent. *The Open Environmental Pollution and Toxicology Journal, 3*(1).

Naje, A. S., Chelliapan, S., Zakaria, Z., & Abbas, S. A. (2015). Enhancement of an electrocoagulation process for the treatment of textile wastewater under combined electrical connections using titanium plates. *International Journal of Electrochemical Science, 10*(6), 4495–4512.

Oller, I., Malato, S., & Sánchez-Pérez, J. (2011). Combination of advanced oxidation processes and biological treatments for wastewater decontamination—A review. *Science of the Total Environment, 409*(20), 4141–4166.

Özyurt, B., Camcıoğlu, Ş, & Hapoglu, H. (2017). A consecutive electrocoagulation and electro-oxidation treatment for pulp and paper mill wastewater. *Desalination and Water Treatment, 93*, 214–228.

Patel, A., Arkatkar, A., Singh, S., Rabbani, A., Medina, J. D. S., Ong, E. S., & Mungray, A. K. (2021). Physico-chemical and biological treatment strategies for converting municipal wastewater and its residue to resources. *Chemosphere, 282*, 130881.

Peplow, D. (1999). *Environmental impacts of mining in Eastern Washington*. University of Washington Water Center.

Puyol, D., Barry, E. M., Hülsen, T., & Batstone, D. J. (2017). A mechanistic model for anaerobic phototrophs in domestic wastewater applications: Photo-anaerobic model (PAnM). *Water Research, 116*, 241–253.

Rajendran, K., Awasthi, S. K., Liu, T., Sarsaiya, S., Wainaina, S., Awasthi, M. K., Duan, Y., Jain, A., Sindhu, R., Binod, P., Pandey, A., & Taherzadeh, M. J. (2021). Techno-economics and life-cycle assessment of biological and thermochemical treatment of bio-waste. *Renewable and Sustainable Energy Reviews, 144*, 110837.

Rajeshwar, K., & Ibanez, J. G. (1997). *Environmental electrochemistry: Fundamentals and applications in pollution sensors and abatement*. Elsevier.

Razanamahandry, L. C., Onwordi, C. T., Saban, W., Bashir, A. K. H., Mekuto, L., Malenga, E., Manikandan, E., Fosso-Kankeu, E., Maaza, M., & Ntwampe, S. K. O. (2019). Performance of various cyanide degrading bacteria on the biodegradation of free cyanide in water. *Journal of Hazardous Materials, 380*, 120900.

Reddy, D. A., Khandelwal, S. K., Muthiah, R., Shanmugamani, A. G., Paul, B., Rao, S. V. S., & Sinha, P. K. (2010). Conditioning of sludge produced through chemical treatment of radioactive liquid waste–Operating experiences. *Annals of Nuclear Energy, 37*(7), 934–941.

Ribordy, P., Pulgarin, C., Kiwi, J., & Peringer, P. (1997). Electrochemical versus photochemical pretreatment of industrial wastewaters. *Water Science and Technology, 35*(4), 293–302.

Sekar, P., Prasad, S. H., & Raman, M. D. (2009). Effect of textile dye industry effluent on the nutritive value of fresh water female crab *Spiralothelphusa hydrodroma* (Herbst). *Journal of Applied Sciences Research*, (November), 2041–2048.

Sever, L. E. (1997). Environmental contamination and health effects: What is the evidence. *Toxicology and Industrial Health, 13*(2–3), 145–161.

Shi, C., Hu, Y., Kobayashi, T., Zhang, N., Kuramochi, H., Zhang, Z., & Xu, K. Q. (2019). Anaerobic degradation of deca-brominated diphenyl ether contaminated in products: Effect of temperature on degradation characteristics. *Bioresource Technology, 283*, 28–35.

Song, J., Zhao, Q., Guo, J., Yan, N., Chen, H., Sheng, F., & An, D. (2019a). The microbial community responsible for dechlorination and benzene ring opening during anaerobic degradation of 2, 4, 6-trichlorophenol. *Science of the Total Environment, 651*, 1368–1376.

Song, T., Li, S., Jin, J., Yin, Z., Lu, Y., Bao, M., & Li, Y. (2019b). Enhanced hydrolyzed polyacrylamide removal from water by an aerobic biofilm reactor-ozone reactor-aerobic biofilm reactor hybrid treatment system: Performance, key enzymes and functional microorganisms. *Bioresource Technology, 291*, 121811.

Song, Z. M., Yang, L. L., Lu, Y., Wang, C., Liang, J. K., Du, Y., & Wu, Q. Y. (2021). Characterization of the transformation of natural organic matter and disinfection by products after chlorination, ultraviolet irradiation and ultraviolet irradiation/chlorination treatment. *Chemical Engineering Journal, 426*, 131916.

Soni, P., Sharma, S., Sharma, S., Kumar, S., & Sharma, K. P. (2006). A comparative study on the toxic effects of textile dye wastewaters (untreated and treated) on mortality and RBC of a freshwater fish *Gambusia affinis* (Baird and Gerard). *Journal of Environmental Biology, 27*(4), 623–628.

Srivastava, V. (2021). Grand challenges in chemical treatment of hazardous pollutants. *Frontiers in Environmental Chemistry*, 23.

Talaiekhozani, A., & Rezania, S. (2017). Application of photosynthetic bacteria for removal of heavy metals, macro-pollutants and dye from wastewater: A review. *Journal of Water Process Engineering, 19*, 312–321.

USGAO. (2000). *Health effect of lead in drinking water*. U.S. General Accounting Office reports.

Wang, M., Zhao, Z., & Zhang, Y. (2019). Disposal of Fenton sludge with anaerobic digestion and the roles of humic acids involved in Fenton sludge. *Water Research, 163*, 114900.

Weldon, R. A. (1982). *Personal communication*. SunTech Group.

World Health Organization. (2015). *Waste and human health: Evidence and needs: WHO meeting report 5–6 November 2015: Bonn, Germany* (No. WHO/EURO: 2015–5441–45206–64594). World Health Organization. Regional Office for Europe.

Yadav, N., Garg, V. K., Chhillar, A. K., & Rana, J. S. (2021). Detection and remediation of pollutants to maintain eco-sustainability employing nanotechnology: A review. *Chemosphere, 280*, 130792.

Zambon, P., Ricci, P., Bovo, E., Casula, A., Gattolin, M., Fiore, A. R., & Guzzinati, S. (2007). Sarcoma risk and dioxin emissions from incinerators and industrial plants: A population-based case-control study (Italy). *Environmental Health, 6*(1), 1–10.

Chapter 12
Metamorphosis of Culinary By-Product: Journeying Towards an Ecological and Sustainable Tomorrow Through Biodiesel and Bio-Lubricant Production

Dipti Rawat, Kashama Sharma, Asha Kumari, and Rahul Sharma

Abstract Waste cooking oil (WCO) is produced when edible vegetable oil is used to fry food items. WCO usage has a detrimental impact on both human and animal health, and improper disposal has a bad impact on the environment by clogging drains and contaminating both land and water sources. The current efforts to convert WCO into biotechnologically useful forms and discuss their potential uses as feedstock for biofuel, bisabolene, bio-lubricants, liquid detergents, dishwashing soap, aromatherapy candles, plasticizers, polyurethane foam, surfactants, and asphalt rejuvenator are covered in this review. There are also details about the global scenario for WCO generation, their physicochemical characteristics, and potential uses for them. These applications guarantee the proper use of WCO as priceless consumer goods and industrial products. More research is required before WCO may be used to produce value products and advance the circular economy.

Keywords Waste Cooking Oil (WCO) · Biofuel · Circular economy · Bio-lubricants · Biotechnological conversion

---

D. Rawat
Department of Physics, Dev Bhoomi Uttarakhand University, Manduwala, Naugaon, Uttarakhand 248007, India

K. Sharma · A. Kumari · R. Sharma (✉)
Department of Chemistry, Career Point University, Bhoranj (Tikker—Kharwarian), MDR 35, Hamirpur, Himachal Pradesh 176041, India
e-mail: rahul.che@cpuh.edu.in

Centre for Nano-Science and Technology, CPU, Hamirpur, Himachal Pradesh 176041, India

Center for Green Energy Research, Career Point University-Hamirpur Campus, Hamirpur 176041, India

R. A. Bhat et al. (eds.), *Zero Waste Management Technologies*,
https://doi.org/10.1007/978-3-031-57275-3_12

1 Introduction

Rapid industrialization has occurred in recent years as a result of a mix of factors such as fast population growth, technical improvement, increasing investment encouragement, trade and capital liberalisation, and the development of the financial sector (Ahmad et al., 2011; Foo et al., 2022; Hosseini et al., 2012; Yaakob et al., 2013). Researchers and entrepreneurs must identify affordable and sustainable resources that can support creative manufacturing processes in order to meet resource demand and cater to continuous industrial improvements. Several scientists have turned the waste remediation, which turns trash into useable raw materials and value-added products, to recover resources and optimise resource utilisation. Waste cooking oil (WCO) conversion and transformation into a variety of products is one of the situation's low-hanging fruits. Waste cooking oil (WCO) is regularly created as a by-product while cooking with various edible vegetable oils in residences, hotels, restaurants, and catering firms. "Vegetable oils are a crucial component of a balanced diet because they account for between 15 and 20% of the calories consumed in most cg3 industrialised countries and are predominantly composed of triacylglycerols (88–98%)" (Awogbemi & Komolafe, 2011; Orsavova et al., 2015). Vegetable oil consumption worldwide has increased dramatically, rising from 150 million metric tonnes (MMT) in 2013–14 to 200 MMT in 2020–21. According to projections, the global market for waste cooking oil (WCO) will grow from $5.50 to $8.48 billion by 2027 (Shahbandeh, 2013). When WCOs are consumed, they unfortunately have a detrimental effect on the ecology and the health of both people and animals. Waste cooking oil (WCO) is frequently disposed of incorrectly in metropolitan areas by being dumped into sewers, drains, open spaces, rivers, and woods. This inappropriate disposal pollutes land and marine habitats, releasing foul odours in the process. Foam formation, an increase in the organic load on water sources, a hindrance to wastewater treatment, a decrease in the concentration of dissolved oxygen, and a disturbance in the ecosystem's balance are other effects of improper WCO disposal (Lopes et al., 2020). Nonetheless, some agricultural wastes can be catalysed with WCO to produce biodiesel for several applications (Awogbemi & Komolafe, 2011; Liang et al., 2013; Lu & Wu, 2014). Appropriate strategies for effective management, disposal, reduction, recycling, conversion, utilisation, valorisation, and remediation of waste cooking oil (WCO) have been proposed to limit the detrimental impacts on the environment (Dhanasekaran et al., 2019; Gupta & Rathod, 2018). The amount of academic work published from 2010 onwards was determined in this chapter by using specific keywords and phrases such as "application of discarded vegetable oil", "application of leftover cooking oil", "utilisation of waste cooking oil", "converting of waste cooking oils", "converting waste frying oils", "used consumable vegetable oils", "utilised frying oils", and "biotechnological transformation of waste cooking oils". The quantity of research works has consistently increased, according to several writers. The limits imposed by COVID-19, which slowed down research activity worldwide, are to blame for the decrease in productivity seen in 2020. Although most of them did not mention biotechnology in the article's title or content, they did discuss alternative WCO use methods. Notwithstanding this deluge of study, the crucial question is: The possibilities for biotechnological conversion and use of waste cooking oil (WCO) can greatly expand its role in advancing the circular economy idea. As the amount of WCO produced grows, it is critical to keep researchers, industry participants, and stakeholders up to date on the newest developments in the sector. This question inspired the present intervention, which ensures that stakeholders are up to date on the most recent innovations in using WCO effectively and sustainably (Lopes et al., 2020). In this regard, the current evaluation focuses on the most important cutting-edge methods for transforming WCO into useful goods for domestic and industrial purposes. The objective of this study is to determine the direction for future research in the field and to update the body of existing knowledge.

2 Characteristics and Chemical Compositions of WCO

When frying, vegetable oils are subjected to frequent heating at elevated temperatures, typically between 150 and 200 °C, are contaminated with moisture, and undergo chemical and physical changes to their structure, including the formation of total polar compounds, as well as severe thermal deterioration (acid value, fatty acid composition, and colour, viscosity, and density) (Lopes et al., 2020). Several of the properties and the composition of saturated fatty acids alter as a result of the physical, chemical, and compositional changes brought on by thermal degradation, including oxidation, hydrolysis, polymerization, and others. Use can change properties such as "density, kinematic viscosity, specific heat, iodine value, peroxide value, saponification value, moisture content, flash point, number of single/double bonds, and percentage of mono/polyunsaturated components". The purpose of different studies was to examine the fatty acid (FA) content and thermophysical parameters of purified vegetable oil and waste cooking oil (WCO) obtained from various sources (Awogbemi et al., 2021a, 2021b, 2021c; Pelemo et al., 2021). These studies employed the "American Society for Testing and Materials (ASTM)", the American Oil Chemists' Society (AOCS) standards and other industry-standard methods for testing purpose. According to researchers, factors such as the level of utilization, frying temperature, origin of the original oil, and type of food fried influence the chemical and physical characteristics of waste cooking oil (WCO) and its fatty acid composition.

3 Avenues for Utilisation of WCO

In the past five years, novel approaches and technologies have emerged for the biotechnological reutilisation of WCO, and new methods and technologies have been created over the last five years. Some of the technologies, as shown in Fig. 3, involve the use of raw materials to produce biofuel, bio-lubricants, and other products. To reduce the environmental effect of WCO, "waste minimization techniques such as waste reduction, waste recycling, and waste utilisation are used". The various biotechnological possibilities for the use of WCO result in waste minimization, cost reduction in waste disposal, and waste utilisation for the generation of renewable energy sources (Gupta & Rathod, 2018; Hajam et al., 2023; Kuniyil et al., 2021). These strategies help the circular economical setup by lowering costs, assuring domestic energy supplies, maximising material use, and producing less CO_2. The sustainable use of waste cooking oil (WCO) helps to achieve the United Nations' Sustainable Development Goals, which include ensuring clean water and sanitation, increasing access to affordable and clean energy, conserving aquatic and terrestrial ecosystems, mitigating climate change, and promoting sustainable cities and communities. However, a number of issues, including collection, recovery, yield, conversion efficiency, discrepancies in quality, and productivity, have hindered the industrial implementation of WCO. Researchers have continued to focus on these issues to push the boundaries of resource usage. With positive results, a number of treatments have been used, including the creation of platforms for WCO collecting and adjustment of conversion parameters (Awogbemi et al., 2021a, 2021b, 2021c; Pelemo et al., 2020).

3.1 WCO as Biofuel Feedstock

Cooking oil residue from frying foods is made of plant- or animal-based fats. Cooking oil is a glycerol ester that is exclusively soluble in organic solvents and is made up of many important fatty acids (Suzihaque et al., 2022). Consuming cooking oil waste is bad for the environment and the health of consumers. The toxic chemicals produced when the oil oxidises are thought to cause cancer when fried foods are consumed (Ganesan et al., 2019). WCO is commonly available but frequently discarded after use, especially after frying. The expected production of WCO in the nations of Asia, as well as the location of the oil, is shown in Table 1 (Kheang et al., 2006). Estimates that each year in Malaysia, 540,000 tonnes of WCO made of vegetable and animal fats are thrown without being addressed as wastes. According to Haruhiro et al. (2015), most of the used cooking oil from homes is dumped in the drains and soil. Therefore, using the WCO as the raw material for more affordable, environmentally friendly biodiesel is one suggestion to address these problems. "Due to the increase in WCO production, it may also be used as a feedstock for the manufacturing of biodiesel. The oxidation, hydrolysis, polymerization, and material transfer between the food and the oil that take place during frying are necessary because WCO has different properties from fresh oil" (Suzihaque et al., 2022). In terms of density, kinematic viscosity, acid value, and iodine value, WCO has identical attributes to create biodiesel and meet the quality criteria (Cordero-Ravelo & Schallenberg-Rodriguez, 2018). Figure 1 illustrates the uses of feedstock for biodiesel.

When choosing a catalyst to employ in the manufacturing process, WCO has a higher concentration of free fatty acids (FFA) than vegetable oil (Gaur et al., 2021).

Table 1 Amount of WCO generated in Asian countries

Country	Quantity (million tonnes per year)	References
China	5	Cordero-Ravelo and Schallenberg-Rodriguez (2018)
Malaysia	0.54	Zhang et al. (2020a, 2020b)
Taiwan	0.05–0.03	Azmi (2019)
Indonesia	0.9	Tsai (2019)

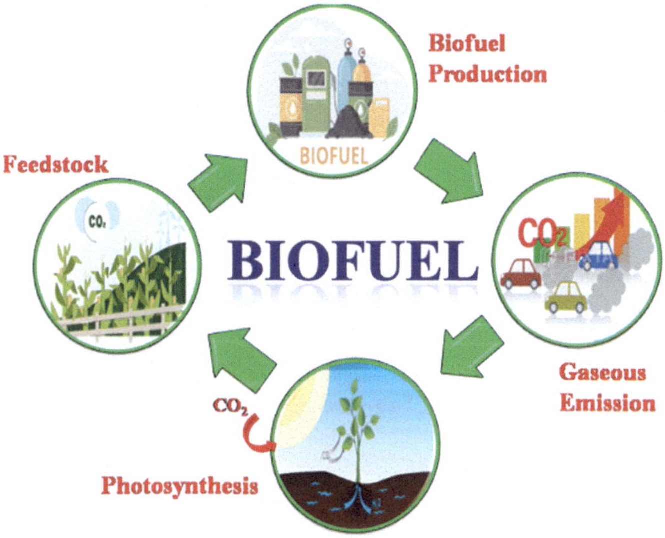

Fig. 1 Applications of feedstock for the biodiesel

Large-scale biodiesel production is constrained and soap manufacture is encouraged in environments with high FFA concentrations (Zahan & Kano, 2018). The parameters of kinematic viscosity, saponification, flash point, moisture content, and free fatty acids for WCO are very high, according to research (Suzihaque et al., 2022). The quantity and quality of biodiesel produced will influenced by these chemical properties. It is also difficult to feed fuel from the fuel tank into the engine if the oil has a high concentration or viscosity and will atomise (Khalisanni et al., 2008). The proportion of free fatty acids in the used cooking oil gathered by Khalisanni et al. (2008) and Foroutan et al. (2019), respectively, is shown in Table 2. "WCO is 2–3 times less expensive than fresh vegetable oil, which substantially lowers overall processing costs (Nur et al., 2015). The cost of raw materials has a considerable impact on the retail price of biodiesel because it accounts for up to 75–90% of the total cost of producing the fuel" (Nomanbhay et al., 2018). The usage of WCO as raw materials is unrelated to any crisis or problem, such as the argument over whether to utilise food or fuel, or the environment. Figure 2 focuses on how biofuels are used.

Table 2 Percentage of free fatty acid in WCO collected (Pålsson et al., 2017)

Type of free fatty acid	Value content (%)	
Oleic acid	43.67	41.04
Palmitic acid	38.35	31.88
Linoleic acid	11.39	17.98
Stearic acid	4.33	6.45
Myristic acid	1.03	0.77
Linolenic acid	0.37	0.43
Others	0.86	1.45
Total	100	100

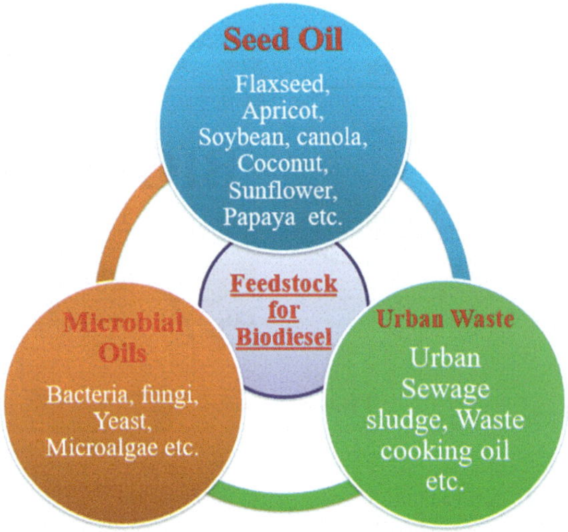

Fig. 2 Applications of biofuels

- WCO was successfully utilised as a cost-effective feedstock for the production of biodiesel, biojet fuel, and renewable diesel.
- To increase the biomethane production in an AD process, WCO was added to various slurries (Fig. 3).

3.2 Bio-Lubricants Production

Bio-lubricants act as anti-friction additives and have several advantages over based on petroleum lubricants, including lower volatility, increased viscosity, improved temperature tolerance, sustainability, and environmental friendliness. Waste cooking oil (WCO) has been used as a cheap basis material for making lubricants due to its excellent lubricating qualities, biodegradability, and low volatility. A bio-lubricant with improved lubrication performance and better oxidation stability is produced by

Fig. 3 WCO executions

chemically modifying WCO. "Hydrolysis, esterification, transesterification, hydrogenation, and epoxidation are notable chemical modification pathways for converting WCO into bio-lubricants. To improve the efficiency of the process, some of these steps may be combined and catalysed" (Zhang et al., 2020a, 2020b). One of the best ways to turn WCO into a bio-lubricant is through the transesterification process. It is not unexpected that numerous researchers have benefited from the simple method for creating different grades of bio-lubricants. However, it was shown that bio-lubricants made using the transesterification process had a low viscosity index and performed poorly at low temperatures. As a result, the transesterification method for making bio-lubricants has been modified. Zhang et al. (2020a, 2020b) and Sarno et al. (2019) used modified transesterification procedures using lipase, ionic fluids, and enhanced magnetite (Fe_3O_4-CA) as catalysts to make bio-lubricants. During the conversion process, several studies looked at the improved thermophysical and rheological properties of epoxidised WCO. Epoxidation-produced bio-lubricants exhibited greater rheological criteria and met ISO VG68 lubricant parameters (Paul et al., 2018, 2021; Vickers, 2017). In their own study findings, Jahromi et al. (2021) overcame the poor additive solubility, low oxidation stability, and low viscosity index indicated by bio-lubricants created by the transesterification technique using hydrolysis, Friedel–Crafts alkylation, dehydration, and hydro treatment (Table 3).

Table 3 Properties of waste cooking oil

Properties	Waste cooking oil value		
	Aworanti et al. (2019)	Al-Saadi et al. (2020)	Farooq et al. (2015)
Acid value (mg KOH/g)	28.5	1.86	35.4
Saponification value (mg KOH/g)	175.87	181.25	234.71
Viscosity (mm^2/s)	46.97	42.01	–
Density (kg/m^3)	908	–	916
Free fatty acid (%)	14.25	–	18
Flash point (°C)	223	234	–
Moisture content (%)	–	0.1	0.136

The study produced a product that met ISO VG68 criteria and improved pour point, kinematic, viscosity index and viscosity. Various researchers have investigated a variety of approaches and tactics to improve conversion efficiency, yield, and product qualities while minimising energy usage, response time, and other production-related aspects. The interactions between the process factors and response have been optimised and studied using a variety of mathematical and statistical tools, including artificial neural networks (ANN), response surface methodology (RSM), central composite design, genetic algorithms (GA), and others. Paul et al. (2021) used "Aspen HYSYS V9 to model the process, while Paul et al. (2018) employed ANN, RSM, and GA to optimise the synthesis of bio-lubricant from WCO by epoxidation". These technologies and methods have assisted in reducing material waste and production costs for bio-lubricants.

Key conclusions;

- WCO provides feedstock that is affordable, easily accessible, sustainable, and eco-friendly for the production of bio-lubricants.
- WCO can be chemically altered, catalysed, and transformed into bio-lubricants.

3.3 *Other Biotechnological Applications of WCO*

There are other ways to use WCO than the biotechnological ones that have already been highlighted, according to numerous researchers. To guarantee that the most benefits are obtained from this valuable waste, it is important to note that additional research is still being conducted in these areas. Figure 4 displays the uses for used cooking oil.

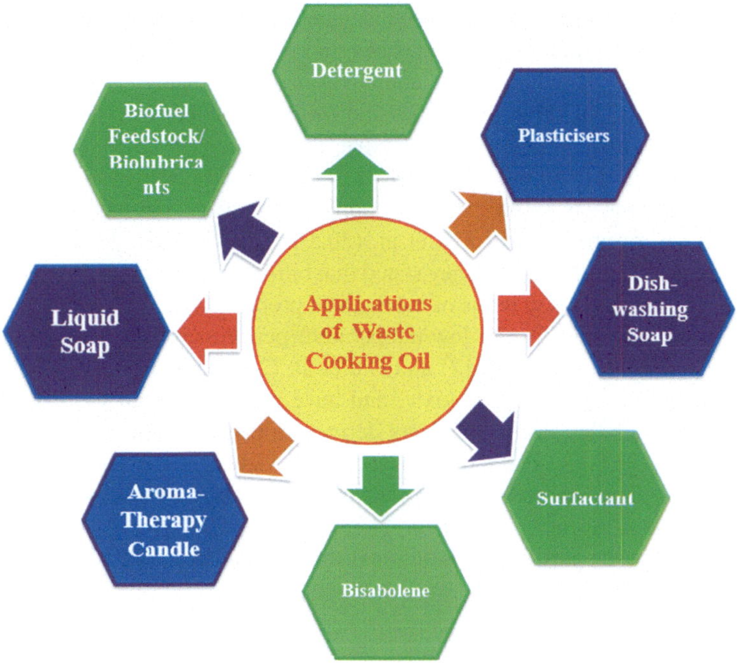

Fig. 4 Applications of WCO

3.3.1 Household Products

Detergents, dishwashing soaps, and aromatherapy candles have all been made with WCO. In a published experiment, KOH and WCO were fully combined to create dishwashing liquid soap for use in homes (Hartini et al., 2021). Two different studies Rini and Harahab (2018) and Astuti et al. (2021) combined WCO with activated charcoal to create aromatherapy candles in an effort to transform WCO into a more beneficial and higher-quality household product. In each of these instances, WCO was put to good use rather than being improperly disposed of, which would have produced offensive odours and polluted the environment. There is still a need for more creative methods to transform WCO into consumable household goods.

3.3.2 Plasticizers

WCO has also been effectively used as an additive and in the production of plastic toys, food packaging, and other products as well as plasticizers. The surge in plasticizer demand over the past ten years and the limitations of petroleum-based plasticizers have made it imperative to discover a feasible replacement. For young children and expecting mothers, dioctyl phthalate plasticizers dramatically raise the risk of poisoning when consumed. Because of this, numerous countries have outlawed their use in the production of toys and food packaging, which has shifted attention to bio-based plasticizers that are non-toxic and environmentally benign (Aboelazayem & Alayoubi, 2021). The emphasis has shifted to using inexpensive, renewable, and waste biomass to manufacture cost-effective, environmentally friendly plasticizers because using clean palm oil as a feedstock for bio-based plasticizers will increase food costs and put stress on the food supply. Epoxidized bio-based plasticizers from WCO were produced by Cai et al. (2020), Suzuki et al. (2018), Tan et al. (2019) and Zheng et al. (2018). They stated that bio-based plasticizers outperformed conventional plasticizers in terms of mechanical properties, thermal stability, adaptability, oxidation stability, and low-temperature performance. Furthermore, Feng et al. (2018), Huzaizi et al. (2017) and Liu et al. (2020) reported that plasticizers made from WCO are safer, non-toxic, and have superior overall mechanical and thermal characteristics than plasticizers derived from petroleum. There is general agreement that the WCO-based plasticizers road is viewed as the more environmentally friendly, secure, and economically advantageous one for the plastic industries. To develop more inventive and environmentally friendly methods for converting WCO into bio-based plasticizers, additional study is still required.

4 Surfactants

Surfactants are petroleum-based chemicals that are used to reduce the surface tension between liquid and liquid, liquid and gas, or liquid and solid phases. They have sulphonate groups. The benefits of employing renewable surfactants include reducing dependency on petroleum-based goods, the environmental effect of the feedstock, and the production process itself. Methyl ester sulphonate (MES) is the most well-known renewable surfactant (Aboelazayem & Alayoubi, 2021). WCO was pre-treated, transesterified into methyl esters, and then converted into MES using sodium bisulphite ($NaHSO_3$) as a sulphonating agent. Following this procedure, Permadani et al. (2018) were able to produce a product with a production 77.2% yield and very high surface activity. Using WCO as a feedstock, Khalaf et al. (2020) created cationic Gemini surfactants to be employed as a novel N80-steel corrosion inhibitor. The created surfactants showed a better cost-effective inhibitory efficiency of 97.86%. Furthermore, Permadani and Slamet (2019) used "WCO as a feedstock and titanium dioxide nanoparticles as a catalyst to generate a bio-detergent from methyl ester sulphonates surfactant. It was discovered that the MES and the bio-detergent both outperformed those made from petroleum-based feedstock". Similar experiments involved extracting MES from WCO and mixing it with powdered ZnO to create a liquid bio-detergent. The stability and stain degradation of the liquid detergent were 88.22% and 82.36%, respectively (Djohan et al., 2019). As a still-emerging study area, better and more inventive pathways must be developed for the process of valorising WCO into bio-surfactants. Environmentally friendly technologies are required to convert WCO into bio-based surfactants and investigate its broader range of commercial uses.

5 Conclusion and Future Perspectives

WCO is produced in vast quantities throughout the world in numerous countries but is regrettably improperly disposed of. Improper WCO disposal harms the environment by causing unpleasant stinks, degrading the habitats on land, and contaminating aquatic ecosystems. The current study revealed three viable avenues for the biotechnological use of WCO, as well as instances that succeeded in application. For the manufacturing of biofuel, bio-lubricants, bisabolene, liquid detergent, dish soap, and aromatherapy candles, WCO is a useful, affordable, and eco-friendly feedstock.

The study shows the advancements made in the use of WCO as a feedstock for the production of plasticizers, polyurethane foaming, surfactants, and bituminous rejuvenators. To fully grasp these prospects, however, concerted efforts and laws, including subsidies for WCO collection, sophisticated pre-treatment, and conversion technologies, are required. It is necessary to conduct more focused and important study on the collection, transportation, purification, and use of WCO. The potential for using WCO will be increased through the establishment and enhancement of subsidy programmes, "life cycle assessment studies, and monitoring of hazardous emissions during biofuel production and other WCO conversion activities". Governmental organisations, company owners, and other interested parties should be invited to help with awareness-raising and sensitisation initiatives for the conversion of waste cooking oil (WCO) into residential and commercial commodities. To lessen the effects of improper disposal of the continuously increasing WCO generation and increase the amount of resources accessible to support the expanding industrial sector, modern and creative pathways for the biotechnological conversion and exploitation of WCO should be explored. Without a doubt, WCO will make a significant contribution to the implementation of the circular economy.

References

Aboelazayem, O., & Alayoubi, R. (2021). Sustainable valorization of waste cooking oil into biofuels and green chemicals: Recent trends, opportunities and challenges. In *Sustainable solutions for environmental pollution: Waste management and value-added products* (Vol. 1, pp. 199–227).

Ahmad, A. L., Yasin, N. M., Derek, C. J. C., & Lim, J. K. (2011). Microalgae as a sustainable energy source for biodiesel production: A review. *Renewable and Sustainable Energy Reviews, 15*(1), 584–593.

Al-Saadi, A., Mathan, B., & He, Y. (2020). Biodiesel production via simultaneous transesterification and esterification reactions over SrO–ZnO/Al_2O_3 as a bifunctional catalyst using high acidic waste cooking oil. *Chemical Engineering Research and Design, 162*, 238–248.

Astuti, A. Y., Linarti, U., & Budiarti, G. I. (2021). Pengolahan Limbah Minyak Jelantah Menjadi Lilin Aromaterapi Di Bank Sampah Lintas Winongo, Kelurahan Bumijo, Kecamatan Jetis, Kota Yogyakarta. *Jurnal Pengabdian Kepada Masyarakat: Teknologi dan Aplikasi e-ISSN, 2723*, 8016.

Awogbemi, O., & Komolafe, C. A. (2011). Potential for sustainable renewable energy development in Nigeria. *The Pacific Journal of Science and Technology, 12*(1), 161–169.

Awogbemi, O., Kallon, D. V. V., Onuh, E. I., & Aigbodion, V. S. (2021a). An overview of the classification, production and utilization of biofuels for internal combustion engine applications. *Energies, 14*(18), 5687.

Awogbemi, O., Von Kallon, D. V., & Aigbodion, V. S. (2021b). Trends in the development and utilization of agricultural wastes as heterogeneous catalyst for biodiesel production. *Journal of the Energy Institute, 98*, 244–258.

Awogbemi, O., Von Kallon, D. V., Aigbodion, V. S., & Panda, S. (2021c). Advances in biotechnological applications of waste cooking oil. *Case Studies in Chemical and Environmental Engineering, 4*, 100158.

Aworanti, O. A., Ajani, A. O., & Agarry, S. E. (2019). Process parameter estimation of biodiesel production from waste frying oil (vegetable and palm oil) using homogeneous catalyst. *Journal of Food Processing & Technology, 10*, 1–10.

Azmi, Z. (2019). Malaysians use too much cooking oil. *Mole*. http://mole.my/malaysians-use-too-much-cooking-oil/

Cai, D. L., Yue, X., Hao, B., & Ma, P. C. (2020). A sustainable poly (vinyl chloride) plasticizer derivated from waste cooking oil. *Journal of Cleaner Production, 274*, 122781.

Cordero-Ravelo, V., & Schallenberg-Rodriguez, J. (2018). Biodiesel production as a solution to waste cooking oil (WCO) disposal. Will any type of WCO do for a transesterification process? A quality assessment. *Journal of Environmental Management, 228*, 117–129.

Dhanasekaran, R., Ganesan, S., Kumar, B. R., & Saravanan, S. (2019). Utilization of waste cooking oil in a light-duty DI diesel engine for cleaner emissions using bio-derived propanol. *Fuel, 235*, 832–837.

Djohan, G., Ibadurrohman, M., & Slamet, S. (2019). Synthesis of eco-friendly liquid detergent from waste cooking oil and ZnO nanoparticles. *AIP Conference Proceedings, 2085*(1).

Farooq, M., Ramli, A., & Naeem, A. (2015). Biodiesel production from low FFA waste cooking oil using heterogeneous catalyst derived from chicken bones. *Renewable Energy, 76*, 362–368.

Feng, G., Hu, L., Ma, Y., Jia, P., Hu, Y., Zhang, M., & Zhou, Y. (2018). An efficient bio-based plasticizer for poly (vinyl chloride) from waste cooking oil and citric acid: Synthesis and evaluation in PVC films. *Journal of Cleaner Production, 189*, 334–343.

Foo, W. H., Koay, S. S. N., Chia, S. R., Chia, W. Y., Tang, D. Y. Y., Nomanbhay, S., & Chew, K. W. (2022). Recent advances in the conversion of waste cooking oil into value-added products: A review. *Fuel, 324*, 124539.

Foroutan, R., Esmaeili, H., Mousavi, S. M., Hashemi, S. A., & Yeganeh, G. (2019). The physical properties of biodiesel-diesel fuel produced via transesterification process from different oil sources. *Physical Chemistry Research, 7*(2), 415–424.

Ganesan, K., Sukalingam, K., & Xu, B. (2019). Impact of consumption of repeatedly heated cooking oils on the incidence of various cancers—A critical review. *Critical Reviews in Food Science and Nutrition, 59*(3), 488–505.

Gaur, A., Mishra, S., Chowdhury, S., Baredar, P., & Verma, P. (2021). A review on factor affecting biodiesel production from waste cooking oil: An Indian perspective. *Materials Today: Proceedings, 46*, 5594–5600.

Gupta, A. R., & Rathod, V. K. (2018). Calcium diglyceroxide catalyzed biodiesel production from waste cooking oil in the presence of microwave: Optimization and kinetic studies. *Renewable Energy, 121*, 757–767.

Hajam, Y. A., Kumar, R., & Kumar, A. (2023). Environmental waste management strategies and vermi transformation for sustainable development. *Environmental Challenges*, 100747.

Hartini, S., Fiantika, Y., Widharto, Y., & Hisjam, M. (2021). Optimal treatment combination for dishwashing liquid soap based on waste cooking oil according to the requirement of Indonesian quality standards.

Haruhiro, F., Wataru, I., Katsuyuki, N., Joko, P., Hiroe, T., & Genshiro, K. (2015). A comparative study of waste cooking oil recycling programs in Bogor and Niigata cities and GHG emission reduction by recycling. In *2015 AASRI International Conference on Circuits and Systems (CAS 2015)* (pp. 169–172).

Hosseini, M., Nikbakht, A. M., & Tabatabaei, M. (2012). Biodiesel production in batch tank reactor equipped to helical ribbon-like agitator. *Modern Applied Science, 6*(3), 40.

Huzaizi, R. M., Tahir, S. M., & Mahbor, K. M. (2017). Electrical, thermal and structural properties of plasticized waste cooking oil-based polyurethane solid polymer electrolyte. *AIP Conference Proceedings, 1901*(1).

Jahromi, H., Adhikari, S., Roy, P., Shelley, M., Hassani, E., & Oh, T. S. (2021). Synthesis of novel biolubricants from waste cooking oil and cyclic oxygenates through an integrated catalytic process. *ACS Sustainable Chemistry & Engineering, 9*(40), 13424–13437.

Khalaf, M. M., Tantawy, A. H., Soliman, K. A., & Abd El-Lateef, H. M. (2020). Cationic Gemini-surfactants based on waste cooking oil as new 'green' inhibitors for N80-steel corrosion in sulphuric acid: A combined empirical and theoretical approaches. *Journal of Molecular Structure, 1203*, 127442.

Khalisanni, K., Khalizani, K., Rohani, M. S., & Khalid, P. O. (2008). Analysis of waste cooking oil as raw material for biofuel production. *Global Journal of Environmental Research, 2*(2), 81–83.

Kheang, L. S., May, C. Y., Foon, C. S., & Ngan, M. A. (2006). Recovery and conversion of palm olein-derived used frying oil to methyl esters for biodiesel. *Journal of Oil Palm Research, 18*, 247.

Kuniyil, M., Kumar, J. S., Adil, S. F., Assal, M. E., Shaik, M. R., Khan, M., & Siddiqui, M. R. H. (2021). Production of biodiesel from waste cooking oil using ZnCuO/N-doped graphene nanocomposite as an efficient heterogeneous catalyst. *Arabian Journal of Chemistry, 14*(3), 102982.

Liang, S., Liu, Z., Xu, M., & Zhang, T. (2013). Waste oil derived biofuels in China bring brightness for global GHG mitigation. *Bioresource Technology, 131*, 139–145.

Liu, D., Jiang, P., Nie, Z., Wang, H., Dai, Z., Deng, J., & Cao, Z. (2020). Synthesis of an efficient bio-based plasticizer derived from waste cooking oil and its performance testing in PVC. *Polymer Testing, 90*, 106625.

Lopes, M., Miranda, S. M., & Belo, I. (2020). Microbial valorization of waste cooking oils for valuable compounds production—A review. *Critical Reviews in Environmental Science and Technology, 50*(24), 2583–2616.

Lu, F., & Wu, X. (2014). China food safety hits the "gutter." *Food Control, 41*, 134–138.

Nomanbhay, S., Hussein, R., & Ong, M. Y. (2018). Sustainability of biodiesel production in Malaysia by production of bio-oil from crude glycerol using microwave pyrolysis: A review. *Green Chemistry Letters and Reviews, 11*(2), 135–157.

Nur, H., Ani, F. N., & Said, M. (2015). Review of the production of biodiesel from waste cooking oil using solid catalysts. *Journal of Mechanical Engineering and Sciences (JMES), 8*, 1302–1311.

Orsavova, J., Misurcova, L., Vavra Ambrozova, J., Vicha, R., & Mlcek, J. (2015). Fatty acids composition of vegetable oils and its contribution to dietary energy intake and dependence of cardiovascular mortality on dietary intake of fatty acids. *International Journal of Molecular Sciences, 16*(6), 12871–12890.

Pålsson, H., Pettersson, F., & Hiselius, L. W. (2017). Energy consumption in e-commerce versus conventional trade channels—Insights into packaging, the last mile, unsold products and product returns. *Journal of Cleaner Production, 164*, 765–778.

Paul, A. K., Borugadda, V. B., Bhalerao, M. S., & Goud, V. V. (2018). In situ epoxidation of waste soybean cooking oil for synthesis of biolubricant basestock: A process parameter optimization and comparison with RSM, ANN, and GA. *The Canadian Journal of Chemical Engineering, 96*(7), 1451–1461.

Paul, A. K., Borugadda, V. B., & Goud, V. V. (2021). In-situ epoxidation of waste cooking oil and its methyl esters for lubricant applications: Characterization and rheology. *Lubricants, 9*(3), 27.

Pelemo, J., Awogbemi, O., Inambao, F., & Onuh, E. I. (2020). In-situ hybridization of waste palm oil: A physicochemical, thermal, and spectroscopic analysis. *Jordan Journal of Mechanical & Industrial Engineering, 14*(4).

Pelemo, J., Awogbemi, O., Inambao, F., & Onuh, E. I. (2021). Hybridization of waste cooking oil: An innovative technique for improved feedstock. *International Journal of Mechanical and Production Engineering Research and Development, 11*(3), 441–454.

Permadani, R. L., Ibadurrohman, M., & Slamet. (2018, January). Utilization of waste cooking oil as raw material for synthesis of Methyl Ester Sulfonates (MES) surfactant. In *IOP Conference Series: Earth and Environmental Science* (Vol. 105, p. 012036). IOP Publishing.

Permadani, R. L., & Slamet. (2019). Development of nanofluid detergent based on methyl ester sulfonates surfactant from waste cooking oil and titanium dioxide nanoparticles. *IOP Conference Series: Materials Science and Engineering, 509*, 012120.

Rini, M., & Harahab, N. (2018). The influence of endorser in social media toward consumer decision making with AISAS model (attention, interest, search, action, and share). *ECSOFiM (Economic and Social of Fisheries and Marine Journal), 6*(1), 106–118.

Sarno, M., Iuliano, M., & Cirillo, C. (2019). Optimized procedure for the preparation of an enzymatic nanocatalyst to produce a bio-lubricant from waste cooking oil. *Chemical Engineering Journal, 377*, 120273.

Shahbandeh, M. (2013). Consumption of vegetable oils worldwide from 2013/14 to 2021/2022, by oil type.

Suzihaque, M. U. H., Alwi, H., Ibrahim, U. K., Abdullah, S., & Haron, N. (2022). Biodiesel production from waste cooking oil: A brief review. *Materials Today: Proceedings.*

Suzuki, A. H., Botelho, B. G., Oliveira, L. S., & Franca, A. S. (2018). Sustainable synthesis of epoxidized waste cooking oil and its application as a plasticizer for polyvinyl chloride films. *European Polymer Journal, 99*, 142–149.

Tan, J., Zhang, S., Lu, T., Li, R., Zhong, T., & Zhu, X. (2019). Design and synthesis of ethoxylated esters derived from waste frying oil as anti-ultraviolet and efficient primary plasticizers for poly(vinyl chloride). *Journal of Cleaner Production, 229*, 1274–1282.

Tsai, W. T. (2019). Mandatory recycling of waste cooking oil from residential and commercial sectors in Taiwan. *Resources, 8*(1), 38.

Vickers, N. J. (2017). Animal communication: When I'm calling you, will you answer too? *Current Biology, 27*(14), 713–715.

Yaakob, Z., Mohammad, M., Alherbawi, M., Alam, Z., & Sopian, K. (2013). Overview of the production of biodiesel from waste cooking oil. *Renewable and Sustainable Energy Reviews, 18*, 184–193.

Zahan, K. A., & Kano, M. (2018). Biodiesel production from palm oil, its by-products, and mill effluent: A review. *Energies, 11*(8), 2132.

Zhang, W., Ji, H., Song, Y., Ma, S., Xiong, W., Chen, C., & Zhang, X. (2020a). Green preparation of branched biolubricant by chemically modifying waste cooking oil with lipase and ionic liquid. *Journal of Cleaner Production, 274*, 122918.

Zhang, Y., Wei, J., Zhu, Y., & George-Ufot, G. (2020b). Untangling the relationship between corporate environmental performance and corporate financial performance: The double-edged moderating effects of environmental uncertainty. *Journal of Cleaner Production, 263*, 121584.

Zheng, T., Wu, Z., Xie, Q., Fang, J., Hu, Y., Lu, M., & Ji, J. (2018). Structural modification of waste cooking oil methyl esters as cleaner plasticizer to substitute toxic dioctyl phthalate. *Journal of Cleaner Production, 186*, 1021–1030.

Chapter 13
Enzymes as Emerging Biocatalysts for Biotransformation Processes

Rafael Vargas-Bernal

Abstract The increase in the generation of biological waste derived from agricultural, industrial, and household activity has an upward trend due to the lifestyle of the world population. This situation is guiding scientific research to exploit these residues, increasing sustainability, reducing environmental pollution, and thus achieving the zero-waste strategy. Enzymes are being proposed as biocatalysts to produce biotransformations that perform efficient conversions of biological waste into high-value-added products. This chapter discusses the role that enzymes are developing in the exploitation of these residues in the production of biofuels, biodegradable plastics, prebiotics, bioactive compounds, biosurfactants, and other products using biotransformation processes. In addition, future perspectives that will allow guaranteeing sustainable production through the gradual reduction of biological waste thanks to its exhaustive use are analyzed. After the review carried out, it was concluded that enzymes represent one of the viable technological alternatives for biological waste management to achieve the zero-waste strategy.

1 Introduction

The exponential increase in the world population has led to an enormous amount of food and energy production, which must be exploited most efficiently. Due to the large amount of biological waste produced by agricultural, forest, domestic, and industrial activity, all human beings in their daily activities must seek strategies to achieve the zero-waste concept. Traditional waste management strategies are landfills, incineration, composting, and animal feeding. The latter is the most interesting from the economic point of view; however, it is necessary to develop other strategies to fully exploit waste to obtain high-added-value products and lead to a zero-waste trend. Waste in its natural state can be considered as a raw material with zero or low

R. Vargas-Bernal (✉)
Departamento de Ingeniería Electrónica, DICIS, Universidad de Guanajuato, 36885 Salamanca, Guanajuato, Mexico
e-mail: r.vargas@ugto.mx

R. A. Bhat et al. (eds.), *Zero Waste Management Technologies*,
https://doi.org/10.1007/978-3-031-57275-3_13

economic contribution, so innovative paradigms must be developed to obtain high-value-added products (Bilal & Iqbal, 2019). The main factors that allow food waste to be used to produce high-value-added products are sustainability, carbon footprint, processing cost, efficiency, availability and renewability, effective cost ratio, resource competitiveness, and cradle-to-grave implications. Food waste processing can be done in solid, semi-solid, and liquid media. Liquid wastes make use of large amounts of water for processes such as sanitation, transportation, cooking, temperature regulation, or cleaning. The effluents from these processes contain suspended solids, oils, greases, nitrogen, organic matter, and inorganic materials. In the food industry, the effluents contain whey from the milk industry and its derivatives, residual water from the processing of vegetables and fruits, the soft drink industry, brewery, fruit pulp residues, etc. in diluted form. Solid waste includes bread and tortilla waste, vegetable and fruit waste, lignin, starch, cellulose, monosaccharides, etc. The sustainable reconversion of food waste into high-value-added products using enzymes is one possible option (Bilal et al., 2022). Lignocellulosic waste can be used as biomass to produce biofuels thanks to its composition of sugars and carbohydrates. The biofuel par excellence that can be obtained is bioethanol, and the steps for its synthesis involve a pretreatment process, enzymatic hydrolysis, and a fermentation process (Barbosa et al., 2020; Constantino et al., 2021; Han et al., 2020). Researchers around the world continually propose new and innovative approaches to develop biotransformations that optimize efficiency and profitability to produce high-value-added products. Food waste streams contain biomolecules such as proteins, organic acids, carbohydrates, and/or lipids, which through enzymes can be bioconverted into materials, industrial chemicals, and fuels (Bilal et al., 2022). Approximately, 40% of edible food is wasted and although the percentage and type of food differ around the world, by-products also vary in the supply chain. According to the Food and Agriculture Organization of the United Nations (FAO), each year worldwide 1.3 billion tons of food are wasted, which corresponds to a third of the food consumed by humans (Tsang et al., 2019). Food is wasted during collection, processing, distribution, marketing, and consumption. Traditional techniques for processing food waste, such as incineration, composting, landfilling, anaerobic digestion, animal feeding, and land spreading, have low economic and ecological impacts. At the municipal level, food waste is considered a significant fraction of solid waste. Thermochemical and thermal techniques such as pyrolysis, hydrothermal oxidation, incineration or gasification, or the production of biological by-products such as anaerobic digestion or fermentation can be used to convert food waste into energy.

The urgency of developing green processes to maintain environmental sustainability is necessary for enzyme-based biocatalysis to become one of the emerging advanced technologies of this century (Haque et al., 2022). The main types of enzymes that have been immobilized to work with food waste are *hydrolases*, especially *lipases*, *proteases*, *carbohydrates*, and non-hydrolytic enzymes (Bilal & Iqbal, 2019). Support matrices to immobilize enzymes on their surface are based on polysaccharides such as chitosan, alginate, agar, and agarose. Inorganic matrices such as titanium dioxide (TiO_2) and porous silica (SiO_2) have also been proposed due to their attractive mechanical and compatibility properties. Enzyme immobilization can be achieved through covalent bonding, entrapment, adsorption, and cross-linking making use of specific functional groups provided by support materials. The support material can be based on a nanomaterial, an inorganic material, or an organic material. The support material must be tolerant to organic solvents, have tensile strength, and have greater adhesion to the interface. Thanks to the sustainable and ecologically friendly nature, biocatalysis using enzymes makes it possible to achieve biodegradation of both toxic organic pollutants as well as bioenergetic wastes. The cost of enzymes ranges between 50 and 500 dollars per kilogram, so their reuse and efficiency are a priority to reduce the costs of high-value-added products generated through biocatalytic reactions. Lignocelluloses derived from brewer's grain waste, coconut fiber, spent grain, and rice straw can serve as renewable support materials to immobilize enzymes such as laccase, invertase, trypsin, and lipases. These enzymes can be produced efficiently from microbial species that grow in organic waste. Through this strategy, it is possible to achieve various cost-effective biotechnological applications using emerging and novel manufacturing techniques. The techniques to produce these enzymes are regularly based on submerged fermentation (SmF) or solid-state fermentation (SSF). From an ecological point of view, the second is preferred because this type of fermentation requires less water, which makes it favorable from various points of view (Haque et al., 2022). The main attraction of food waste conversion is the production of high-value-added chemicals used in various technological sectors such as enzymes, functional sugars, organic acids, dietary fibers, biopolymers, and biofuels.

The objective of this chapter is to highlight the relevance that enzymes have as biocatalysts to biotransform food waste into high-added-value products to achieve the zero-waste strategy. The rest of the chapter has been divided as follows: In Sect. 2, the basic concepts of enzymes as biocatalysts are examined. The uses of enzymes in the biotransformation of food residues are studied in Sect. 3. In Sect. 4, future research directions on enzymes as biocatalysts to be exploited by various commercial and industrial processes that will be developed in this century are discussed. Finally, the conclusions of this chapter are proportionate.

2 Basic Concepts of Biocatalysts

The chemical conversion of organic compounds using enzymes isolated or living within living cells is known as biocatalysis (Pyser et al., 2021). Biocatalysis has the advantages of simplicity of the procedure, tunability of the reactivity and selectivity of the enzyme, safety, and sustainability. Through biocatalysis, it is possible to use enzymes that allow a synthesis route that satisfies the reactivity and selectivity conditions required to achieve the transformation of raw materials into high-added-value products. To carry out an environmentally friendly organic synthesis, it is necessary to use a toolbox with a set of processes to develop the biochemical transformation called biocatalysis (de María, 2021). The processes can be carried out solvent-free, in water, or biogenic solvents (component, secretion, and metabolite of plants or animals) or neoterics solvents (ionic liquids, supercritical fluids, deep eutectic solvents), and their environmental impact is different on a case-by-case basis. Although solvent-free systems have the highest efficiency, they are not commonly implemented. The search for enzymes that perform efficiently, especially in continuous flow processes, is one of the research topics that could lead to the application of biocatalysts on an industrial scale. Biocatalysis is a biotransformation of feedstock into target compounds in the small laboratory scale range to the large industrial scale range (Wohlgemuth, 2021). This involves the application of a chemical reaction controlled by enzymes through a synthetic route that involves one or multiple steps. Currently, it is possible through computer tools, to reduce the time to choose the required enzyme, the most convenient enzyme production system, and the appropriate bioprocess tools, before developing the complete production process. During biotransformation, at least one coupling reaction is performed which involves designing both enzyme engineering and reaction engineering. This transformation is accomplished either by developing a step-oriented synthesis process or a biocatalytic total synthesis. Thanks to enzymes, all forms of life can develop activities that involve natural biotransformations when interacting with other living organisms and with the environment. Enzymes act as catalysts for biosynthesis as well as the biodegradation of natural species. Biotransformations involve activities that depend on the molecular weight of the initial feedstock and its origin, which can be either fossil-based or bio-based. When choosing the catalysts, it is necessary to look for high selectivity, as well as their specific activation or inhibition ratio required by the biotransformation. Thanks to immobilization, enzymes are attached or confined in solid materials (Benítez-Mateos et al., 2021). Among the advantages of immobilized enzymes are stability with temperature, organic solvents, and acidity, isolation of biocatalyst products, reuse or recycling in other reactions or processes, and increased reaction rates. Through the immobilization of the enzymes, their reuse is possible, reduces waste, allows their use in reactors in a continuous way, and provides cost-effectiveness of the process.

Food waste comes from its stages of production, processing, storage, and consumption (Wohlgemuth, 2021). The elimination of this waste is not only beneficial for the environment but also opens the possibility of turning it into products with high added value. An enormous variety of enzymes can be used to produce safe and convenient intermediates, components, or final products for large-scale industrial applications. Biocatalysis has been used to improve the quality of life through environmental care promoted in industries such as life sciences, pharmaceuticals, agrochemicals, and chemistry.

Reactions catalyzed by a single enzyme make use of crude or purified enzymes to perform regeneration by cofactors such as transfer reactions, reductions, or oxidations (Wohlgemuth, 2021). Oxygen can be used to generate oxidations that weaken organic chains while enzymes form selective catalytic chiral centers. The possible biocatalytic oxidation will depend on the specific kind of substrate material that will be transformed. Therefore, enzymes act as catalysts to produce selective oxidation on carbon-based feedstocks. In the case of chemical reductions, the catalytic centers are the carbonyl groups present in the raw material to be biotransformed, which are processed by transforming the nitriles into amines, dearomatizations, deoxygenates, reductive cycling, or reductive couplings. Transfer reactions allow selective and sustainable biocatalysis using a wide diversity of functional groups that allow biotransformation to be carried out using enzymes without requiring protection and deprotection steps.

The most widely used biocatalytic reactions are based on hydrolase enzymes that are economical and accessible on a large scale (Wohlgemuth, 2021). These enzymes catalyze the transformation through bond-cleaving reactions with water to digest the raw material, breaking the nutrients into smaller units, which is called hydrolysis. Among the qualities that this biotransformation presents are stereoselectivity, high chemical activity, suitability for different substrates, and stability. Thanks to its simple design and the absence of regeneration cofactors, this type of transformation is useful for developing chemical synthesis methodologies for large-scale applications. Due to these attractive characteristics, researchers around the world have been looking for alternatives related to the use of hydrolase to increase its performance for different types of raw materials. Given the enormous diversity of substrates, a wide diversity of reactions is generated, giving rise to many possible enzymes and selective biocatalytic reactions possible. Through these biocatalytic addition reactions, it is possible to catalyze the inclusion of water toward carbon–carbon double bonds and carbon–nitrogen triple bonds. Two or more biocatalytic reactions can be coupled to improve biotransformation processes to achieve much more complete control, even involving inhibition of the reaction, or recovery or purification of the product using thermodynamic equilibrium reactions. In unicellular and multicellular living beings, their basic functions involve biocatalysis by enzymes, so a complete domain of them and their properties can be exploited to carry out biotransformations of bio-based products.

Numerous species of microbes grow in organic food waste, so it can be used to produce in a non-toxic, sustainable, and environmentally friendly way (Haque et al., 2022). As biocatalysts, natural enzymes favor the sustainable synthesis of high-added-value products, are renewable, reduce energy consumption, and have low toxicity. Therefore, biocatalysts allow the development of processes that allow the production of bioenergy, either as liquid or gaseous biofuels, as well as promoters of bioremediation for the elimination of food waste. Different microbes can be used to obtain enzymes for biocatalytic processes for sectors such as energy, cosmetics, paper, textiles, pharmaceuticals, animal feed, beverages, and food (Bilal & Iqbal, 2020). Enzymes allow the decomposition of food waste and contribute to or alter the physicochemical and rheological properties of food. The main enzymes used in food processes are *esterases*, *peroxidases*, *glucose oxidases*, *laccases*, *arabinofuranosidases*, *cellulases*, *reductases*, *glucanases*, *aminases*, *lactases*, *proteases*, *pectinases*, *xylanases*, *lipases*, *amylases*, etc.

For their growth, microorganisms convert inorganic and organic waste into biochemicals and energy using enzymes and metabolic pathways (Intasian et al., 2021). Food waste can be transformed thanks to enzymes from bio-based products such as lignocellulose, fatty acids, triglycerides, glycerol, phenolic acid, and sugar and greenhouse gases such as carbon dioxide (CO_2) and methane (CH_4), into products with high added value. Greenhouse gases obtained from the anaerobic digestion of food waste can be valuable products through enzymatic reactions using limited oxygen conditions. The advantages of the enzymatic approach are its high product specificity and purity. Disadvantages of the enzymatic approach are the need for purification, cofactor and/or co-substrate regeneration, and costly enzyme preparation and operation process. Agricultural wastes containing sugars and phenolic acids can be transformed through enzymatic reactions into pharmaceutical ingredients, food additives, and biochemicals. Food waste through enzymatic reactions can be converted into lignocellulose, starch, and triglycerides. Triglycerides containing glycerol and fatty acids through enzymes can be converted into biochemicals and biofuels.

3 Uses of Enzymes in the Biotransformation of Food Waste

Implementing a biorefinery is a viable and technical option to obtain high-value-added products to reduce costs and waste volumes (Khatami et al., 2021; Tsang et al., 2019). This process can be used to produce biodegradable plastics such as polyhydroxyalkanoates (PHAs) from food waste in a cost-effective manner relative to expensive synthesis routes based on pure substrates such as glucose (Tsang et al., 2019). Figure 1 summarizes the main types of waste from the food industry. Waste processing not only contributes to producing zero-waste but also energy recovery, giving rise to gaseous, solids, or liquid biofuels and producing high-value-added materials or chemicals such as bioplastics, prebiotics, bioactive compounds, and biosurfactants. The physical processes used to take advantage of food waste are size reduction and/or pelleting, and extraction. The biochemical processes used to use food waste are fermentation and anaerobic digestion. The thermal and chemical processes used to exploit food waste are conversion, gasification, pyrolysis and/or liquefaction, and direct combustion. The production of polyhydroxyalkanoates is derived from various sources such as cooking oil waste or cheese whey. Food waste is made up of water with a percentage by weight in the range of 75 to 85% and the remaining solid percentage with a range of 60 to 70% organic matter on a dry basis. High lipid content is ideal for regularly obtaining biodiesel from waste such as palm oil, sunflower seeds, soybeans, and rapeseed oil. For high nitrogen contents, high levels of biochemical oxygen demand (BOD) are required by microorganisms to break down organic materials. Furthermore, high levels of chemical oxygen demand (COD) are required for the breakdown of organic material by oxidation to be able to process waste intestines, fats, and blood of animals. High content of fatty acids, fat, starch, and protein are obtained from mixed edible wastes such as cheese whey, bread waste, dried fruit shells, and nuts. Succinic acid, soluble sugar, and cellulose are produced from fruit pulp, peels, stubble, and straw.

Polyhydroxyalkanoates (PHAs) are intracellular biopolymers with attractive physicochemical, biodegradable, and biocompatible properties (Khatami et al.,

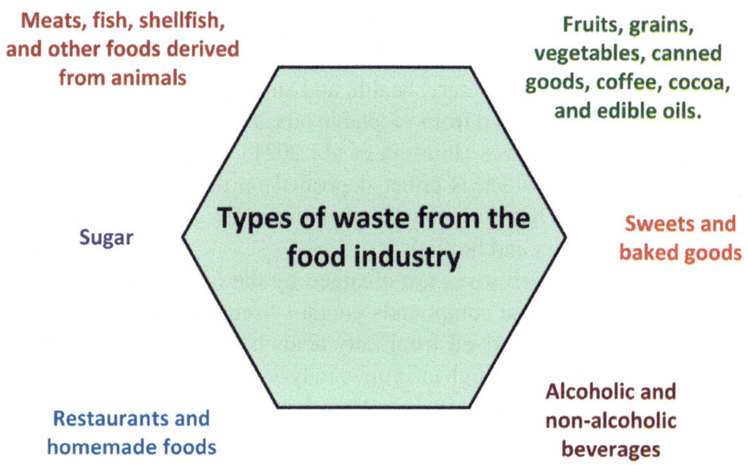

Fig. 1 Main types of waste from the food industry

2021). The bio-based approach seeks to reduce costs through synthesis based on microbial cultures. These polyester-based biopolymers have tunable physical and mechanical properties but above all, they are environmentally friendly. They have applications in biochemicals, biofuels, latex paints, and in the agricultural, biomedical, and packaging sectors. The enzyme used in the synthesis of the PHA establishes the specificity of the substrate and the length of the chain related to the number of carbon atoms that can be obtained. To achieve a high rate of polymer synthesis, the chemical expression of the polymer-enzyme complex must be high to optimize the metabolic pathway. The synthesis process if not controlled can lead to inconsistency in the polymer structure, unreliable monomer ratios, and differences in molecular weight from batch to batch. Polyhydroxyalkanoates can be synthesized from glucose or fatty acids by the bacterium *Pseudomonas putida* LS46 using a $PhaC1_{16}$ enzyme (Sharma et al., 2017). Both the bacteria and the enzyme have a direct influence on the length and properties of the obtained biopolymer.

Table 1 summarizes the derivatives of food waste and the high-added-value products that can be produced. Lignocellulose-based materials come from residues from forestry and agricultural activities (Intasian et al., 2021; Nawaz et al., 2020). Lignocellulose contains aromatic polymers and carbohydrates. These polymers have monomeric structures based on sugar-like five- and six-carbon chains. Lignocellulose is considered a mixed biopolymer composed of lignin (20–30%), hemicellulose (30–35%), and cellulose (40–45%). To biotransform lignocellulose and thus obtain value-added products, it is necessary to pretreat it and then hydrolyze it. The breakdown of cellulose and hemicellulose gives rise to sugars as glucose while lignin provides polyphenols. Among the sugars that can be obtained are xylose (to produce ethanol), arabinose (to produce fructose), and galactose (to produce sucrose). Lignin can be transformed into phenolic compounds that serve as precursors for the chemical, cosmetic, pharmaceutical, and food industries. In these industries, phenolic compounds offer antioxidant, dietary, health, and antimicrobial properties.

Triglycerides can be obtained from vegetable oils, animal fats, cooking oils, milk, etc. as well as their derivatives (Intasian et al., 2021). A large amount of wasted cooking oil is produced, which is either deposited on the soil or discharged into aquifers, which produces harmful effects on health and the ecosystem. These can be transformed into polymers and biofuels.

Fatty acids and their derivatives are obtained by the hydrolysis of triglycerides (Intasian et al., 2021). These compounds contain from 8 to 20 carbons. Because the large amount of waste derived from fatty acids has effects on land and water, they can be biotransformed through enzyme-catalyzed chemical reactions, making it possible to produce chemicals and fuels and use them in the food and pharmaceutical industries.

Glycerol is obtained by hydrolysis of triglycerides from food waste and used oils (Intasian et al., 2021). Biodiesel is derived from the transesterification of triglycerides and alcohols, also giving rise to glycerol as a by-product. Glycerol can be transformed to produce chemicals, pharmaceuticals, polymers, and biofuels.

Carbon dioxide is a final product of several industrial processes as well as aerobic digestion or yeast fermentation; it can produce bioethanol (Intasian et al., 2021). This gas by chemical catalysis can also be transformed into formic acid and methanol, compounds of 1 carbon atom. To reduce the energy required to trap CO_2, organisms such as plants, algae, and bacteria can be used to fix CO_2 in these materials through enzyme-controlled biotransformation. Products such as chemicals, polymers, and biofuels can be obtained by specific chemical reactions.

Methane (CH_4) occurs naturally in up to 40% of wetlands, with the remainder produced by human and animal activities (Intasian et al., 2021). Among human activities are household and agricultural activities such as anaerobic digestion of animal manure, agro-industrial sludge, sewage sludge, as well as composting of organic waste. It can be used to produce polymers and biofuels through catalytic reactions.

Proteolytic enzymes using hydrolysis can produce polymers from egg waste, seed oils, grains, soybeans, and other organic wastes (Bilal & Iqbal, 2019). Nanomaterials such as calcium oxide (CaO) and magnesium oxide (MgO) nanoparticles can be used as enzyme support materials to carry out heterogeneous catalysis to transform food waste into biodiesel through the transesterification reaction (Noraini et al., 2014). Carbon nanotubes can also be used as support materials to immobilize lipase to achieve the transesterification of ethyl butyrate, and thanks to this support it is possible to have catalytic activity up to 97% after several cycles of use. Enzymes, such as *β-amylase* and *glucoamylase*, allow the transformation of the starch polymer into glucose (Bilal & Iqbal, 2020). *Glucoamylases* are enzymes that can convert starchy flour residues into maltose and fermentable sugars that can be converted to glucose and then bioethanol. When bioethanol is fermented with the yeast *Saccharomyces cerevisiae,* it is possible to produce soy sauce and alcohol-free light beer. Microbe-based lipases can be used to degrade ester bonds found in food waste to triglycerides

Table 1 Main high-added-value products that are obtained from food waste by-products

		High-added-value products								
		Biofuels	Polymers	Pharmaceuticals	Chemicals	Food additives	Nutraceuticals	Fragrances	Material components	Preservatives
Food waste by-product	Lignocellulosic biomass (sugar, phenolic acid)			×		×	×	×	×	
	Methane (CH_4)	×	×							
	Carbon dioxide (CO_2)	×	×		×					×
	Glycerol	×	×	×	×					
	Fatty acids	×	×		×					
	Triglyceride	×	×							

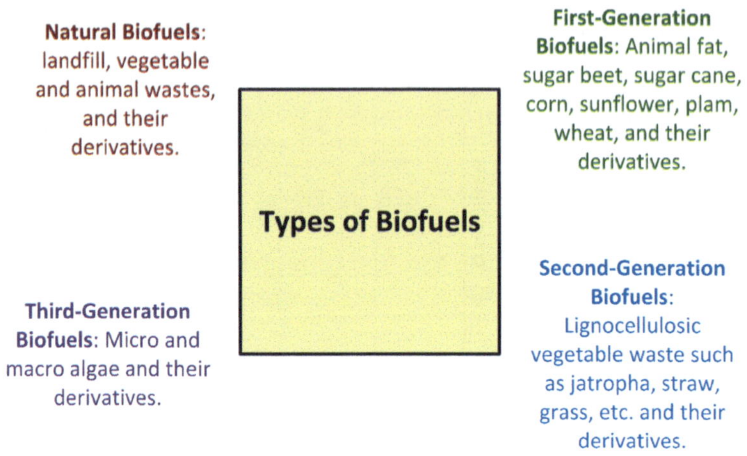

Fig. 2 Summary of the types of biofuels that can be obtained depending on the origin of the initial raw material

due to their availability and ease of synthesis. Foods containing carbohydrates can use carbohydrase to catalyze their transformation into sugar.

Biofuel is an ecological fuel of plant origin that can be produced from biological matter containing fats, oil, starch, and/or sugar. These biofuels are economical, with low greenhouse gas emissions, less toxic, and environmentally friendly (Noraini et al., 2014). Among the most common biofuels are biodiesel, bioethanol, and biogas. Biofuels are obtained from biological resources such as jojoba, palm, corn, soybean, algae, rice bran, coconut, food waste, etc. Biofuels can be classified into four groups: namely natural, first-generation, second-generation, and third-generation types as illustrated in Fig. 2. When enzymes are used to produce biodiesel, both the immobilization method and the efficiency of the enzyme are the main factors to consider within the transformation process.

To reduce the effect of greenhouse gases, bioethanol is proposed as a solution to replace fossil fuels in chemical and motor fuels (Han et al., 2020). Although bioethanol can be obtained from the production of sugar cane and corn, this is avoided so as not to threaten global food sources and increase crop prices. In this way, food waste or parts of vegetables not consumed as food are used to produce biofuel and other products with high added value.

It is expected that by the year 2050, the population of the planet will exceed 9 billion people and that biodiesel production will be in the range of 36 to 39 billion liters at a global level for the period from 2017 to 2027 (Kumar et al., 2020). The raw materials for producing biodiesel can be classified into three categories: oil from plants, oil from animal fats, and cooking oil. Thanks to the free oxygen present in excess in biological fuels, complete combustion is favored, and with this, it is possible to reduce toxic emissions. The transesterification reaction to produce biodiesel can make use of enzymatic, acid, or alkaline catalysts. Edible palm, soybean, and linseed oils use alkaline catalysts for biodiesel production. The operating temperature of an enzymatic catalyst is lower than that required by an alkaline catalyst. The alkaline catalyst par excellence is sodium methoxide.

Lipases resist denaturation and inhibition by alcohols such as methanol, which makes these enzymes interesting for biofuel production (Sasso et al., 2016). They present constant stability and catalytic activity, which optimizes their use in transesterification. The advantage of waste oils is that they have structures with short-chain alcohols, which increases the speed and efficiency of their biotransformation.

The use of the co-immobilization of enzymes has increased the efficiency of the biotransformation of palm oil to obtain biodiesel (Shahedi et al., 2021). Lipases such as *Thermomyces lanuginosa* and *Candida antarctica* were allowed to reach an efficiency of 94% using butanol and methanol. Lipases catalyze the hydrolysis of long-chain triglycerides and do not require cofactors. During the conversion, covalent acyl-enzyme intermediates are produced which, thanks to the alcohol, release the alkyl esters. The use of more than one enzyme generates fewer by-products and reduces the reaction time required to perform the biotransformation. The disadvantage of using multiple enzymes lies in high costs, acceptable activity and stability, and complications to find the right conditions to use both effects individually and collectively.

Organic wastes have a variety of differences in acidity, textures, etc. (Sufficiency et al., 2022). This makes it difficult for the same strategies to be used generically for all types of waste. The presence of lignin and cellulose in food waste facilitates its role as self-support materials for biotransformation procedures. Through microorganism fermentation, glucose and xylose from cellulose are transformed into ethanol, and lignin through anaerobic digestion and pyrolysis produces hydrogen and methane. Through acidogenic fermentation of food waste containing amino acids and acetate give rise to ethanol, lactate, butyrate, propionate, hydrogen, and sugars, it is possible to obtain bioplastics. Through composting, the synthesis of biofertilizers is possible.

Biodiesel production is commonly carried out by transesterification catalyzed by lipase-type enzymes thanks to the fact that little energy is consumed, they do not involve side reactions, and above all, they are environmentally friendly (Noraini et al., 2014). The main disadvantages of enzymatic reactions are their high cost, product contamination, and short operation time.

Lipases are water-soluble ester hydrolase enzymes with the classification *triacylglycerol hydrolases* E.C. 3.1.1.3 capable of hydrolyzing triacylglycerols to glycerol and fatty acids (Noraini et al., 2014). These are biocatalysts with outstanding biochemical and physiological properties that are found in nature or are produced by microorganisms, animals, and plants. The lipases of microbial origin based on fungi and/or bacteria are the most used for the transformation of organic materials. Among the advantages of lipases are the following: high reaction specificity, complete transformation, high efficiency, reduced energy consumption, low losses of by-products and/or waste, mild operating conditions, and low operating temperature. Lipases require aqueous conditions for the carboxylic ester linkages found in triacylglycerols to be converted to fatty acids and glycerol via a transesterification reaction. Figure 3 outlines the basic principle of the transesterification reaction based on the lipase enzyme. Among the conditions that can modify the reaction are enzyme–substrate

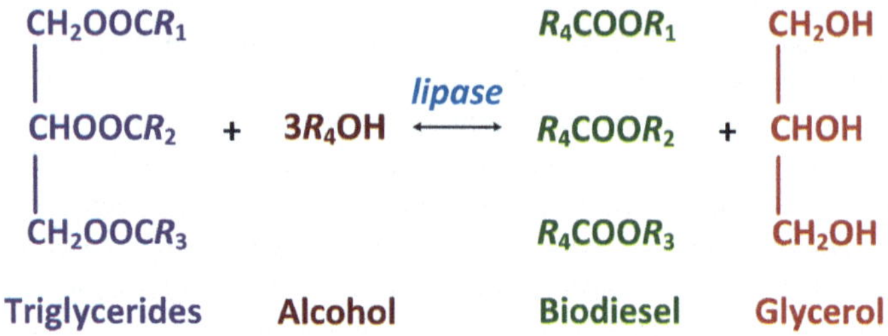

Fig. 3 Chemical reaction to obtain biodiesel by enzymatic transesterification

separation, enzyme activity, temperature, substrate composition, and acidity (pH). For each enzyme, there are optimal operating conditions, and these depend entirely on its origin. For efficient use of the enzyme, it must be immobilized on a substrate to maintain its stability and control the rate of the enzymatic reaction. Its reuse may involve up to 10 cycles, but if it is mixed with some other chemical component of the reaction, then even additional processes may be required to separate them or may even inhibit their catalytic capacity. Under low humidity conditions, this reaction can be reversed to become one of esterification, alcoholysis, and acidolysis.

A summary of the enzymes that have been used to produce biofuels has been compiled in Table 2. Microbe-based lipases are the main enzymes for transforming industrial and environmental wastes (Verma et al., 2021). As of 2021, more than 200 microbial-based enzymes are reported commercially, and 20 kinds of enzymes are exploited on an industrial scale. Lipases from microorganisms are classified as intracellular, extracellular, solubilized, and immobilized. For the most part, commercially available lipases are used immobilized.

Lignocellulose biomass can be transformed into bio-based products such as biofuels, chemicals, and materials using *deep eutectic solvents* (DESs) that are cost-effective, sustainable, and green (Bjelić et al., 2022). These solvents are polar liquids based on renewable, biodegradable, and environmentally friendly natural molar compounds with hydrogen bonding interactions. The solvents are hydrogen bond acceptor (HBA) and hydrogen bond donor (HBD)-based catalysts to increase selectivity, yield, and reaction rates of the desired product. *Natural deep eutectic solvents* (NADESs) can be used as catalysts and co-catalysts can be recycled and

Table 2 Main lipase enzymes to produce biofuels

Enzyme	References
Burkholderia cepacia lipase (BCL)	Sasso et al. (2016)
Pseudomonas cepacia lipase (PCL)	Kumar et al. (2020)
Thermomyces lanuginose lipase (TLL)	Shahedi et al. (2021)
Candida antarctica lipase B (CALB)	Shahedi et al. (2021)
Rhizopus oryae lipase R1 (ROL)	Helal et al. (2021)

recovered with negligible effect on their activity. These are enzyme-tolerant media and are suitable to produce biopolymers or oligomers via hydrolysis.

Lignocellulolytic enzymes have been proposed to transform agro-industrial waste into bioethanol considering a zero-waste policy of ecological and sustainable processes (Carrillo-Nieves et al., 2020). A Mexican white root fungus called *Trametes hirsut*a CS5 was used as a lignin degrader in three different substrates based on the orange peel (OP), brewer's spent grain (BSG), and wheat bran (WB). These raw materials are used as carbon sources in the chemical-free fermentation process. Ethanol production was highest for wheat bran, and it was capable of degrading to pentoses, and hexoses found in cellulose- and hemicellulose-based polymers. The use of this fungus will allow the process to reach a commercial level and will eliminate the need to use exogenous hydrolytic enzymes as well as pretreatment steps producing zero waste. Furthermore, it is possible to produce nutraceuticals and phenolic compounds using predigested substrates. In this way, not only resources are optimized, but all kinds of products are obtained with the same constituents under different quantities and process steps.

Hydrolysis is a chemical reaction where the water breaks one or more chemical bonds from a large molecule into a shorter chemical species (Noraini et al., 2014). Microalgae can be biotransformed by enzymatic hydrolysis. In the case of cellulose, the *cellulase* enzyme acts as a highly specific biocatalyst to reduce sugars including glucose. The hydrolysis process is controlled by different factors, among which are the surface area of the substrate, crystallinity, mass transfer, porosity, and content of the raw material to be transformed. Under this type of hydrolysis, it is possible to reach 100% hydrolyzed cellulose. The use of enzymes to degrade biomass operates as a biocatalysis process with a low hydrolysis rate. For materials containing lignocellulose, the hydrolysis process is expensive if *cellulase* is used as a biocatalyst. To speed up the transformation process, dilute acid hydrolysis uses temperatures in the range of 40 to 50 °C and acidity in the range of 4.5 to 5.0. The cost of this type of hydrolysis decreases the cost of the enzyme.

Microalgae considered the third generation of raw material for biofuel production is abundant and extracted from the sea at a tolerable cost (Constantino et al., 2021). To convert this type of biomass into ethanol, it is necessary to break the carbohydrates by hydrolysis into sugars which can be fermented to obtain bioethanol. The transformation into sugars is carried out through acid hydrolysis, biocatalysis using enzymes, or a combination of both processes. Bioethanol is obtained as the final metabolic product of sugar fermentation using yeasts, microorganisms, and enzymes. The hydrolysis process was carried out using a combination of enzymes: *Aspergillus niger* and *α-amylase* and was tested for three different algae: *Skeletonema sp.*, *Tetraselmis sp.*, and *Chlorella sorokiniana*, which reached a high bioethanol production, the last being the most efficient. The application of this strategy aims to disaggregate the protective cover of microalgae to expose the exposure of internal carbohydrates to the action of enzymes and thus increase the efficiency of biotransformation.

Wheat straw has been proposed to produce biogas, lignin, and bioethanol making partial or total use of the following sequence of steps: subcritical water pretreatment, enzyme-based hydrolysis, and a fermentation process, as shown in Fig. 4 (Chen

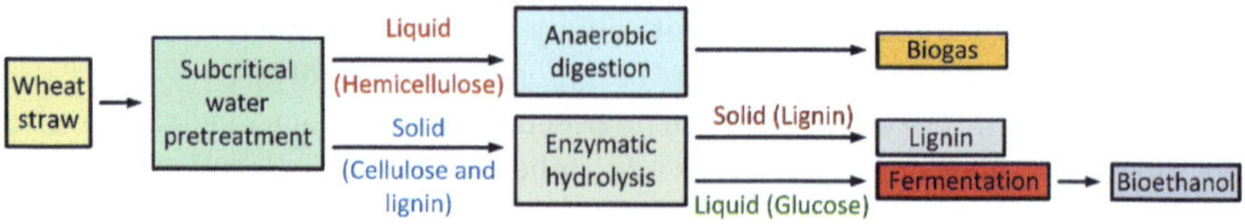

Fig. 4 Wheat straw biotransformation process using enzymes to obtain biogas, lignin, and bioethanol

et al., 2021). Through this conversion process, energy can be obtained, and the environmental impact of waste can be considerably reduced. Three high-added-value products can be synthesized using the same initial raw material. The efficiency of the hydrolysis and fermentation processes were evaluated for five commercial *cellulases* and reached the ranges of 77.85 to 89.59% and 93.34 to 96.18%, respectively. The advantages of this raw material considered agricultural waste are that it is abundant, cheap, and renewable.

A combination of enzymes based on *lytic polysaccharide monooxygenases* (LPMOs) (copper enzymes), *endoglucanases* (of the cellulase type), and *cellobiose dehydrogenases* (CDH) (extracellular hemoflavoenzyme) was used to transform lignocellulose from sugarcane straw by enzymatic hydrolysis to reduce to glucose that will be used to produce bioethanol (Barbosa et al., 2020). The origin of biocatalysts from different sources seeks to reduce the risk of enzyme contamination, fermentation times, and the possibility of inhibiting the action of enzymes. Both the determination of the percentage of each of the enzymes as well as the identification of the synergistic effect produced by each of the constituents is one of the challenges of the studies required to take full advantage of the biocatalysis of enzymes.

Cellulose nanocrystals obtained from sugar beet pulp, embedded with magnetite (Fe_3O_4) nanoparticles, through a hydrolysis process with a mixed enzyme of *mannanase* and *xylanase*, were modified by dopamine to produce a drug to control neurotransmitter functions (Ariaeenejad et al., 2021). The addition of magnetite makes the nanomaterial become a magnetic nanocarrier capable of immobilizing the enzyme and separating dopamine using a magnet and being able to be used 8 more times with 50% of the original activity of the enzyme. The test carried out used other substrates such as coffee waste and rice straw to reduce sugars during a period of 168 h, reaching a better performance for coffee waste. This type of study demonstrates that waste can be used to produce high-value-added products used for biomedical applications.

The use of multi-enzymatic cascade reactions for the conversion of amines to alcohols has been demonstrated as a zero-waste strategy (Contente & Paradisis, 2018). This system uses the immobilization of redox enzymes and *transaminases* (intracellular enzymes) to convert amines into alcohols with high added value, such as the antioxidant called hydroxytyrosol, through the online recovery of by-products and recirculation of aqueous media containing biocatalysts.

Hamburger waste, one of the most consumed fast foods in the world, can be biotransformed by enzymatic hydrolysis (using the commercial enzyme *α-amylase*) and fermentation to produce bioethanol (Han et al., 2020). Residue reduction reached a range of 76.8 to 80.4% using high oxygen demand; fermentation required 50 h, with the highest ethanol production being 0.271 g of ethanol per gram of hamburger.

The *Rhizopus oryzae* R1 lipase using chitosan as a support material, reused for 12 cycles, and maintaining the efficiency of 99.1%, was used to produce biodiesel using oleic acid esterification and olive oil transesterification (Helal et al., 2021). The researchers used monovalent metal ions such as sodium and potassium to promote the enzyme's activity. The properties that must be taken care of in the selection of the support material to increase the efficiency of the enzyme are surface functionality, porosity, and surface area. Other required properties for the support material are biocompatibility, biodegradability, antibacterial, non-toxicity, physiologically inert, and high affinity to proteins.

The wastewater from the sugarcane process contains toxic organic substances that must be removed before discharging into bodies of water (Kumar et al., 2020). Microbes can remove these contaminants through enzyme-based bioremediation, even to obtain high-value-added industrial products from these residues. Among the products that can be obtained after fermentation are xylitol, solvents, L-arabinose, and succinic acid. The main by-product derived from the sugar industry is xylose mother liquor (XML), a low-cost waste that can be very toxic to water and which in China represents a total of 50,000 tons per year. Lignocellulose-containing wastes are hydrolyzed using enzymes to convert them into monosaccharides; these can be fermented in lactic acid, butanol, xylitol, and/or ethanol using different bacteria (Zhao et al., 2020). Lignin represents the useless part of lignocellulose in the biotransformation process to generate biofuels, because its use is aimed at producing biopolymers.

Bioethanol can be obtained by hydrolyzing and fermenting household food waste that contains more than 50% carbohydrates (Ntaikou et al., 2021). A mixture of cellulolytic and amylolytic enzymes has been used to transform food waste at a rate of 141.06 ± 6.81 g of ethanol per kilogram of waste. *Wickerhamomyces anomalous* yeast was used for the fermentation of the liquid part obtained during hydrolysis. The household food waste used in this experiment consisted of biodegradable waste and meat bones.

Potato waste can be used as raw material to obtain bioethanol using hydrolysis and fermentation using the enzyme alpha-amylase as a catalyst (Suresh et al., 2020). The process was controlled by a genetic algorithm using an artificial neural network to determine the optimal parameters and using ultrasonication as a pretreatment for the raw material. The performance of the neural network turned out to be more effective than the Box-Behnken response surface (RSM) methodology in the search for parameters to develop biotransformation.

The cases of biotransformation processes described in this section allow us to conclude that the application of enzymes is still valid in many experiments and scientific investigations carried out by various specialists around the world. As a research topic, it is still current and will continue to develop because not all food waste or all enzymes, support materials, and processes have been fully studied and used with maximum efficiency to achieve the zero-waste strategy in all sectors where they are generated.

4 Future Research Directions on Enzymes as Biocatalysts

Because the efficiency of the enzymes must be increased, researchers around the world should carry out studies of the biocatalytic behavior when multiple enzymes are used, in such a way that it is possible to determine the percentage that must be added to the set of each one of the enzymes, as well as how to identify the synergistic effect produced by each of the constituents. Until now, some food waste has been exhaustively studied, but in this century, other types of waste must be used to develop both different products with high added value as well as other applications not yet conceived by current developments.

Future enzymes should increase their specific reaction times up to 10^6 times without losing their high selectivity (Hauer, 2020). Through enzymes, it is possible to achieve cost reduction, innovation, and process optimization. To be used on a large scale, enzymes must overcome three main challenges related to their performance, cost, and timeline, as well as their market intelligence. Regarding performance, it is necessary to expand the portfolio of both natural and synthetic substrates in which the use of enzymes is feasible with the help of computational and experimental tools, and to improve the stability of the biocatalyst through quality and purity control. Regarding the timeline and cost, the design and development time of the enzymes should be reduced using computational and synthesis tools that allow the development of libraries, which allow, in an economical and fast way, to carry the biocatalysis process at commercial levels, and the sale of enzymes at industrial production level. Market intelligence involves platform improvements of fewer reaction steps, less waste, and energy consumption, and in the value chain, it relates to selecting high-value products with a much smarter sequence of steps.

5 Conclusions

Humanity must achieve a circular economy using sustainable development based on the minimization of waste, recovery, and efficiency of resources, products, and processes, as well as recycling. Food waste must be converted into high-added-value products using biocatalytic systems with immobilized enzymes to promote environmental and economic sustainability, making the zero-waste strategy a reality. Despite the advances achieved so far, for enzyme-based biocatalysts to reach maximum efficiency and take the processes to commercial and industrial scale, it will be necessary for the challenges described in the chapter to be overcome through scientific research and technological development that will be taking place in this century around the world. Currently, in many processes, the cost of the support matrix to perform the immobilization is almost 50% of the cost of the enzymatic biocatalyst system, which must change if a transcendental change in the management of food waste is desired. The biocatalytic activity of enzymes must be retained and repeatable through a significant number of operation cycles to increase their productivity for a high cost–benefit ratio. Multi-enzyme catalytic systems have been shown to increase the efficiency of the transformation of raw materials into high-value-added products if the requirements for water activity, inhibitors, solvents, substrates, and cofactors are considered for each enzyme and process. The use of rationally designed enzymes has a profound impact on food sustainability in the value chain of the use of food waste, so multidisciplinary specialists will have to contribute in all steps to achieve a zero-waste strategy in all countries regardless of the nature of the waste produced in each of them.

References

Ariaeenejad, S., Kavousi, K., Maleki, M., Motamedi, E., Moosavi-Movahedi, A. A., & Salekdeh, G. H. (2021). Application of free and immobilized novel bifunctional biocatalysts in biotransformation of recalcitrant lignocellulosic biomass. *Chemosphere, 285*, 131412. https://doi.org/10.1016/j.chemosphere.2021.131412

Barbosa, F. C., Kendrick, E., Brenelli, L. B., Arruda, H. S., Pastore, G. M., Rabelo, S. C., Damasio, A., Franco, T. T., Leak, D., & Goldbeck, R. (2020). Optimization of cello-oligosaccharides production by enzymatic hydrolysis of hydrothermally pretreated sugarcane straw using cellulolytic and oxidative enzymes. *Biomass and Bioenergy, 141*, 105697. https://doi.org/10.1016/j.biombioe.2020.105697

Benítez-Mateos, A. I., Contente, M. L., Padrosa, D. R., & Paradisi, F. (2021). Flow biocatalysis 101: Design, development and applications. *Reaction Chemistry and Engineering, 6*(4), 599–611. https://doi.org/10.1039/d0re00483a

Bilal, M., & Iqbal, H. M. N. (2019). Sustainable bioconversion of food waste into high-value products by immobilized enzymes to meet bio-economy challenges and opportunities—A review. *Food Research International, 123*, 226–240. https://doi.org/10.1016/j.foodres.2019.04.066

Bilal, M., & Iqbal, H. M. N. (2020). State-of-the-art strategies and applied perspectives of enzyme biocatalysis in food sector—Current status and future trends. *Critical Reviews in Food Science and Nutrition, 60*(12), 2052–2066. https://doi.org/10.1080/10408398.2019.1627284

Bilal, M., Mehmood, T., Nadeem, F., Barbosa, A. M., Lucena, R., de Souza, G., Pompeu, B., Meer, B., Ferreira, L. F. R., & Iqbal, H. M. N. (2022). Enzyme-assisted transformation of lignin-based food bio-residues into high-value products with a zero-waste theme: A Review. *Waste and Biomass Valorization, 13*, 1807–1824. https://doi.org/10.1007/s12649-021-01618-4

Bjelić, A., Hočevar, B., Grilc, M., Novak, U., & Likovar, B. (2022). A review of sustainable lignocellulose biorefining applying (Natural) deep eutectic solvents (DESs) for separations, catalysis and enzymatic. *Reviews in Chemical Engineering, 38*(3), 243–277. https://doi.org/10.1515/revce-2019-0077

Carrillo-Nieves, D., Saldarriaga-Hernandez, S., Gutiérrez-Soto, G., Rostro-Alanis, M., Hernández-Luna, C., Alvarez, A. J., Iqbal, H. M. N., & Parra-Saldívar, R. (2020). Biotransformation of agro-industrial waste to produce lignocellulolytic enzymes and bioethanol with a zero waste. *Biomass Conversion and Biorefinery, 12*, 253–264. https://doi.org/10.1007/s13399-020-00738-6

Chen, J., Wang, X., Yang, Y., Song, Y., Zhang, F., Liu, B., Zhou, Y., Yi, Y., Shan, Y., & Lü, X. (2021). Integrating enzymatic hydrolysis into subcritical water pretreatment optimization for bioethanol production from wheat straw. *Science of the Total Environment, 770*, 145321. https://doi.org/10.1016/j.scitotenv.2021.145321

Constantino, A., Rodrigues, B., Leon, R., Barros, R., & Raposo, S. (2021). Alternative chemo-enzymatic hydrolysis strategy applied to different microalgae species for bioethanol production. *Algal Research, 56*, 102329. https://doi.org/10.1016/j.algal.2021.102329

Contente, M. L., & Paradisis, F. (2018). Self-sustaining closed-loop multienzyme-mediated conversion of amines into alcohols in continuous reactions. *Nature Catalysis, 1*, 452–459. https://doi.org/10.1038/s41929-018-0082-9

de María, P. D. (2021). Biocatalysis, sustainability, and industrial applications: Show me the metrics. *Current Opinion in Green and Sustainable Chemistry, 31*, 100514. https://doi.org/10.1016/j.cogsc.2021.100514

Han, W., Liu, Y., Xiaobin, X., Huang, J., He, H., Chen, L., Qiu, S., Tang, J., & Hou, P. (2020). Bioethanol production from waste hamburger by enzymatic hydrolysis and fermentation. *Journal of Cleaner Production, 264*, 121658. https://doi.org/10.1016/j.jclepro.2020.121658

Haque, S., Singh, R., Harakeh, S., Teklemariam, A. D., Alharthy, S. A., Tripathi, S. C., Singh, R. P., Hassan, A. A., Srivastava, N., & Gupta, V. K. (2022). Enzyme-based biocatalysis for the treatment of organic pollutants and bioenergy production. *Current Opinion in Green and Sustainable Chemistry, 38*, 100709. https://doi.org/10.1016/j.cogsc.2022.100709

Hauer, B. (2020). Embracing. Nature's catalysts: A viewpoint on the future of biocatalysis. *ACS Catalysis, 10*(15), 8418–8427. https://doi.org/10.1021/acscatal.0c01708

Helal, S. E., Abdelhady, H. M., Abou-Taleb, K. A., Hassan, M. G., & Amer, M. M. (2021). Lipase from *Rhizopus oryzae* R1: In-depth characterization, immobilization, and evaluation in biodiesel production. *Journal of Genetic Engineering and Biotechnology, 19*, 1. https://doi.org/10.1186/s43141-020-00094-y

Intasian, P., Prakinee, K., Phintha, A., Trisrivirat, D., Weeranoppanant, N., Wongnate, T., & Chaiyen, P. (2021). Enzymes, in vivo biocatalysis, and metabolic engineering for enabling a circular economy and sustainability. *Chemical Reviews, 121*(17), 10367–10451. https://doi.org/10.1021/acs.chemrev.1c00121

Khatami, K., Perez-Zabaleta, M., Owusu-Agyeman, I., & Cetecioglu, Z. (2021). Waste to bioplastics: How close are we to sustainable polyhydroxyalkanoates production? *Waste Management, 119*, 374–388. https://doi.org/10.1016/j.wasman.2020.10.008

Kumar, D., Das, T., Giri, B. S., & Verma, B. (2020). Preparation and characterization of novel hybrid bio-support material immobilized from *Pseudomonas cepacia* lipase and its application to enhance biodiesel production. *Renewable Energy, 147*, 11–24. https://doi.org/10.1016/j.renene.2019.08.110

Nawaz, M. Z., Bilal, M., Tariq, A., Iqbal, H. M. N., Alghamdi, H. A., & Cheng, H. (2020). Bio-purification of sugar industry wastewater and production of high-value industrial products with a zero-waste concept. *Critical Reviews in Food Science and Nutrition, 61*(21), 3537–3554. https://doi.org/10.1080/10408398.2020.1802696

Noraini, M. Y., Ong, H. C., Badrul, M. J., & Chong, W. T. (2014). A review on potential enzymatic reaction for biofuel production from algae. *Renewable and Sustainable Energy Reviews, 39*, 24–34. https://doi.org/10.1016/j.rser.2014.07.089

Ntaikou, I., Antonopoulou, G., & Lyberatos, G. (2021). Sustainable second-generation bioethanol production from enzymatically hydrolyzed domestic food waste using *Pichia anomala* as biocatalyst. *Sustainability, 13*, 259. https://doi.org/10.3390/su13010259

Pyser, J. B., Chakrabarty, S., Romero, E. O., & Narayan, A. R. H. (2021). State-of-the-art biocatalysis. *ACS Central Science, 7*(7), 1105–1116. https://doi.org/10.1021/acscentsci.1c00273

Sasso, F., Natalello, A., Castoldi, S., Lotti, M., & Santambrogio, C. (2016). *Burkholderia cepaci* Lipase is a promising biocatalysts for biofuel production. *Biotechnology Journal, 11*(7), 954–960. https://doi.org/10.1002/biot.201500305

Shahedi, M., Habibi, Z., Yousefi, M., Brask, J., & Mohammadi, M. (2021). Improvement of biodiesel production from palm oil by co-immobilization of *Thermomyces lanuginose* lipase and *Candida antarctica* lipase B: Optimization using response surface methodology. *International Journal of Biological Macromolecules, 170*, 490–502. https://doi.org/10.1016/j.ijbiomac.2020.12.181

Sharma, P. K., Munir, R. I., Blunt, W., Dartiailh, C., Chen, J., Charles, T. C., & Levin, D. B. (2017). Synthesis and physical properties of polyhydroxyalkanoate polymers with different monomer compositions by recombinant pseudomonas putida LS46 expressing a novel PHA synthase ($PhaC1_{16}$) enzyme. *Applied Sciences, 7*, 242. https://doi.org/10.3390/app7030242

Sufficiency, E., Qamar, S. A., Ferreira, L. F. R., Franco, M., Iqbal, H. M. N., & Bilal, M. (2022). Emerging biotechnological strategies for food waste management: A green leap towards achieving high-value products and environmental abatement. *Energy Nexus, 6*, 100077. https://doi.org/10.1016/j.nexus.2022.100077

Suresh, T., Sivarajasekar, N., Balasubramani, K., Ahamad, T., Alam, M., & Naushad, M. (2020). Process intensification and comparison of bioethanol production from food industry waste (Potatoes) by ultrasonic assisted acid hydrolysis and enzymatic hydrolysis: Statistical modelling and optimization. *Biomass and Bioenergy, 142*, 105752. https://doi.org/10.1016/j.biombioe.2020.105752

Tsang, Y. F., Kumar, V., Samadar, P., Yang, Y., Lee, J., Ok, Y. S., Song, H., Kim, K.-H., Kwon, E. E., & Jeon, Y. J. (2019). Production of bioplastic through food waste valoration. *Environment International, 127*, 625–644. https://doi.org/10.1016/j.envint.2019.03.076

Verma, S., Meghwanshi, G. K., & Kumar, R. (2021). Current perspectives for microbial lipases from extremophiles and metagenomics. *Biochimie, 182*, 23–36. https://doi.org/10.1016/j.biochi.2020.12.027

Wohlgemuth, R. (2021). Biocatalysis—Key enabling tools from biocatalytic one-step and multi-step reactions to biocatalytic total synthesis. *New Biotechnology, 60*, 113–123. https://doi.org/10.1016/j.nbt.2020.08.006

Zhao, C., Qiao, X., Shao, Q., Hassan, M., & Ma, Z. (2020). Evolution of the lignin chemical structure during the bioethanol production process and its inhibition to enzymatic hydrolysis. *Energy and Fuels, 34*(5), 5938–5947. https://doi.org/10.1021/acs.energyfuels.0c00293

Chapter 14
Biofunctionalized Nanomaterials a Zero Waste Approach for the Remediation of Pollutants

Vineet Kumar, Sunil Kumar, Attuluri Vamsi Kumar, and Younis Ahmad Hajam

Abstract Environmental deterioration is currently a major problem for both emerging and wealthy nations. Large-scale industrialization and intensive agricultural activity are the major concerns to cause contagion of environment. There are numerous conventional treatments for various environmental contaminants, but each has certain drawbacks. As a result, a different approach is needed, one that is efficient, less harmful, and produces better results. Because of advancement in nanotechnology, nanomaterials outperform more than traditional methods and are very helpful to in saving the environment. Nanomaterials in very small quantities lead to speedy decline in noxious waste material. An overview of several biofunctionalized nanomaterials with zero waste potential will be included in the current review in order to address potential future applications.

Keywords Environment · Industrialization · Nanomaterials · Waste · Pollutants

---

V. Kumar (✉)
Department of Zoology, Dr. Yashwant Singh Parmar Govt. P.G. College Nahan, District Sirmour, Nahan, Himachal Pradesh 173001, India
e-mail: vineetkatoch975@gmail.com

S. Kumar
ICMR-Regional Medical Research Centre, Gorakhpur, Uttar Pradesh 273013, India

A. V. Kumar
Department of Medical Lab Technology, Chandigarh University, Kharar, Punjab 140413, India

Y. A. Hajam
Department of Life Sciences and Allied Health Sciences, Sant Baba Bhag Singh University, Jalandhar, Punjab, India

R. A. Bhat et al. (eds.), *Zero Waste Management Technologies*,
https://doi.org/10.1007/978-3-031-57275-3_14

1 Introduction

The environment as it exists now is constantly decaying. Global environmental problems are escalating, and we must act as though there is an emergency on our world. We must adopt a fresh viewpoint and address disasters with fresh ideas and approaches, as well as with full awareness and seriousness (Marlon et al., 2019; Roy and Bharadvaja 2021; Roy et al., 2021). Nature takes millions of years to clean up contamination from the environment (Hajam et al., 2023; Roy et al., 2021). The majority of environmental pollution can be attributed to industry and automotive exhaust emissions (Adeleye et al., 2016a). Humanity is at risk from air pollutants such as NO (x), SO_2, and organic chemicals (Ambient Air, 2007). Carbon monoxide (CO) and certain other heavy metals when inhaled in large amounts can cause rapid poisoning amongst the illnesses brought on by the aforementioned substances leading to pulmonary problems (Araviiskaia et al., 2019). Nanotechnology appears to be a budding way out to these issues. Nanotechnology has the potential to greatly contribute to the creation of cleaner, greener technologies with considerable benefits equally for our surrounding and wellbeing (Adeleye et al., 2016b). The effectiveness of conventional environmental clean-up approaches is being improved, and nanotechnology techniques may be able to cope up contamination management and alleviation (Allen, 2002). Nanotechnology can be benefited to the environment in various ways like minimum energy production during industrialized processes, forming goods to be cast off after utilization and creating and utilizing ecologically beneficial materials (Mancebo & Wang, 2015). Although nanotechnology currently shows a lot of potential for tackling sustainability challenges (Marlon et al., 2019), the significance of various nanomaterials that can remove contaminants has been covered in this review chapter.

2 Pollutants

Particles that pollute the environment are referred to as pollutants. Life can be harmed by exposure to these toxins, and the effects on people and other living things occur both naturally and of human activities (Seewagen, 2020). Primary pollutants, and secondary pollutants, enter into the environment in various ways like air, water and soil (Adeleye et al., 2016a) (Fig. 1).

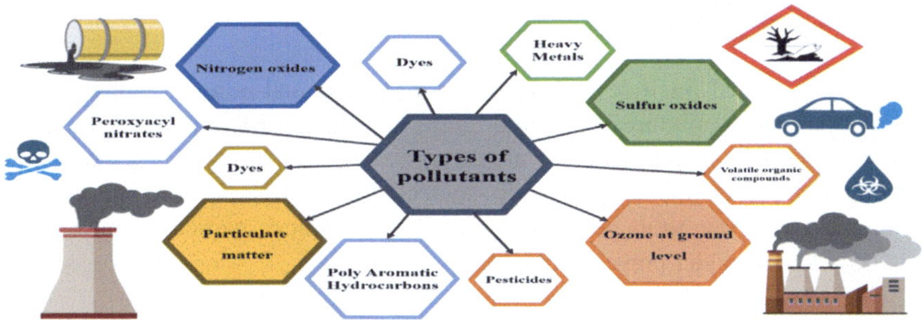

Fig. 1 Different types of pollutants

2.1 *Types of Pollutants*

2.1.1 Nitrogen Oxides or NO

When fuel is burned at high temperatures, a family of extremely reactive, deadly gases known as nitrogen oxides, or NOx, is created. NOx gases, which are often brown in colour, are released by both industrial and vehicular sources, including power plants, boilers, cement kilns, and turbines (Seewagen, 2020). Problematic chemical interactions between nitrogen oxides and volatile organic molecules occur in the atmosphere (Roy & Bharadvaja, 2021). On warm summer days, these processes cause smog to form NO and NO_2 are the two principal nitrogen oxides that are produced. Nitrogen is present in two important fossil fuels, coal and oil. NOx gases, which are often brown in colour, are released by both industrial and vehicular sources, including power plants, boilers, cement kilns, and turbines. Problematic chemical interactions between nitrogen oxides and volatile organic molecules occur in the atmosphere (Roy et al., 2021; Mancebo and Wang 2015). On warm summer days, these processes cause smog to form NO and NO_2 are the two principal nitrogen oxides that are produced (Adeleye et al., 2016a).

2.1.2 Sulphur Oxides or SO_x

A class of contaminants includes molecules of both sulphur and oxygen. The most prevalent type of sulphur in the lower atmosphere is sulphur dioxide or SO_2. Despite being colourless, sulphur dioxide has a unique taste and smell that can be noticed at high concentrations. Acid rain is a product of sulphur oxides, which easily dissolve in water and produce sulphurous acid or sulphuric acid in the atmosphere (Seewagen, 2020). The majority of sulphur oxides are created during the burning of sulphur-containing fuels. A significant source is also the roasting of metal sulphide ores. 35 to 65% of all sulphur dioxide emissions come specifically from natural sources (Lellis et al., 2008).

2.1.3 Particulate Matter

It sometimes called **particle pollution** or simply **PM** and is used to describe an airborne hazardous mixture of liquid droplets and solid particles. There are many sizes of particulate matter, and they are categorized as contaminants. Several types of particulate matter, including dust, dirt, soot, coal ash, and smoke, can be seen with the unaided eye. Due to its potential for extreme smallness, particulate matter can only be observed under powerful microscopes.

2.1.4 Ozone

Secondary pollutant found to be a highly reactive at the ground level. When primary pollutants like hydrocarbons and nitrogen oxides interact with sunlight, this pollutant is created. Ozone is a key element in photochemical smog and irritates people's lungs (Adeleye et al., 2016a). Photochemical smog's foul smell is caused by ground-level ozone, which can also impair lung function, particularly in young children.

2.1.5 Volatile Organic Compounds

Organic molecules, specifically hydrocarbons, are known as volatile organic compounds (VOCs), which are categorized as pollutants because they have harmful effects on the atmosphere. Due to their quick and simple air evaporation, they are classified as volatile (Lellis et al., 2008). VOCs arrive in a range of shapes and sizes, and they are released by both man-made structures like power plants and natural ones like plants and trees. Although roughly 85% of the VOCs in the air come from natural sources, those created by human activities are more reactive and thus more worrisome. The primary source of VOC emissions, particularly in urban areas with fewer plants, is human. These substances are essential to the creation of the dangerous pollutants ozone and photochemical smog (Briffa et al., 2020). Mercury is a harmful contaminant that can accumulate in food chains and cause severe biological harm in either its elemental form or when mixed with other elements (Manisalidis et al., 2020; Mancebo and Wang 2015; Roy et al., 2021).

2.1.6 Peroxyacyl Nitrates

They are also known as acyl peroxy nitrates or APN and are a byproduct of the atmospheric reaction of oxidized volatile organic molecules with nitrogen dioxide ($ceNO_2ceNO_2$) that results in photochemical smog (Bruns et al., 1974). Due to their formation in the atmosphere following the emission of primary pollutants, PANs are considered (Tchounwou et al., 2012) a secondary contaminant.

2.1.7 Dyes

They are a class of artificial organic chemicals that are employed in a number of sectors, including the textile industry and frequently cause industrial environmental contamination both during their production and later while colouring fabrics (Pereira & Alves, 2012). Synthetic colours in textile effluent reduce river light penetration, changing water vegetation's ability to photosynthesize and reduce thermal source for water life (Khan, 2020). The amount of dissolved oxygen is reduced when released dyes thinly coat the surface of receiving waters, which has a detrimental effect on aquatic life. Industrial effluents containing dyes increase biochemical oxygen demand (BOD). Dyes are essentially environmental persistent organic pollutants (POPs) (Lellis et al., 2019; Marlon et al., 2019).

2.1.8 Heavy Metals

Because of this pollution, a detrimental effect is a growing problem for land and air. (Briffa et al., 2020; Marlon et al., 2019). These harmful heavy metals are present in small amounts in nature on the surface of the Earth, but due to their tremendous increase in use, they are now rapidly accumulating in both the natural resources of land and water.

2.1.9 Pesticides

Since many years ago, pesticides have been used in agriculture to protect plants and animals from bug infestations and to maintain productivity. Pesticides are used to remove unwanted things due to their large-scale practice they proved to be hazardous to the environment (Hassaan & El Nemr, 2020). Pesticides impact on the environment by bioconcentration and biomagnifications (Hassaan & El Nemr, 2020). Pesticide receptors which are present in the adipose tissue of human body accumulate pesticides like DDT (dichlorodiphenyltrichloroethane) (Zhen et al., 2019). Pesticide pollutants are responsible for a number of issues in the organisms including malignancies, teratogens, DNA, and cell damage etc". (Jia et al., 2020).

2.1.10 Poly-Aromatic Hydrocarbons

They are constant and insidious environmental pollutants which are dangerous to public due to their poisonous, mutagenic, and carcinogenic prospective. The demand for petroleum products has increased and so have their emissions (Mishra et al., 2021). These have a high degree of constancy and resistance to deterioration (Manisalidis et al., 2020; Roy et al., 2021). The temperatures at which these aromatic compounds melt and boil are relatively high. Lower molecular weight poly-aromatic hydrocarbons are normally water soluble, but with each additional ring, they become more hydrophobic and aqueous insoluble (Masih et al., 2010; Mishra et al., 2021).

3 Health Consequences of Pollutants on People and the Environment

Over the past few decades, advanced technology has allowed for the identification and assessment of trace pollutants (Rather et al., 2023; Roy and Bharadvaja 2021). However, in the light of this, emerging environmental concerns must be taken into account (Ali et al., 2019). The path towards the discovery of multiple viable and long-lasting methods to address the issues posed by newly found and identified toxins, proves to be advantageous for future mankind (Behera et al., 2023).

3.1 Environmental Implications

3.1.1 Aquatic Life

Cultural eutrophication is one of the principal environmental contaminants that cause aquatic pollution. Nitrogen and phosphorus-rich chemicals have been released into aquatic bodies due to the population growth and increasing pace of food production (Dachs et al., 2000; Tilley et al., 2005). The growths of hazardous "algal blooms" are detrimental to native aquatic plant and animal species (Paso, 2022). Due to low oxygen percentages and high ratios of hazardous substances, eutrophication has unintentionally restricted the use of water for fisheries, industry, and drinking (Kelly et al., 2007). Biomagnification, or the build-up of moderately to extremely hydrophobic chemicals that are cycled through food webs, is another issue caused by persistent organic pollutants (Billé et al., 2013). A considerable decline in the population of locally important native species is crucial to the ecosystem (Adeleye et al., 2016a, 2016b, 2016c). Shellfish, zooplanktons, and coral reefs have all prematurely calcified as a result of it (Phale et al., 2019).

3.1.2 Terrestrial

Industries are the main causes of environmental pollution (Waldman & Shevah, 1993). The terrestrial ecology is under danger from the unreported dumping of contaminants such as plasticizers, xenobiotics, and aromatic hydrocarbons (Álvarez et al., 2017). Through eutrophication, an enrichment of nutrients in water demonstrates the rising danger of forest fires in radioactive pollution-contaminated areas

(Artiola et al., 2019). Deforestation rates have increased as a result of numerous pollutants, which have also caused loose soil and the displacement of sediments and nutrient-rich soil surface, which has increased the likelihood of flash floods (Jacobson, 2009).

3.1.3 Aerial

The sixth greatest cause of death worldwide, air pollution poses a danger to all of human being civilization. The largest environmental issues like global warming are accelerated by aerial pollutants (Singh & Agrawal, 2007). Ocean acidification, rising sea levels brought on by glaciers and snowmelt, intensified tropical storms, loss of productive farmland and animal habitats, and unreliable water supplies are all effects of global warming (Orru et al., 2017). Some pollutants, such as sulphur and nitrogenous oxides from burning, interact with the atmosphere and deposit acid as rain, fog, or snow. They endanger food stocks, undermine the integrity of the environment, and may have larger-scale socioeconomic repercussions. In developing nations like India, urban and periurban areas have significantly higher levels of air pollution (Corsi et al., 2018; Rather et al., 2023). They have consequently had an impact on the nearby farming areas and posed a danger to crop output tariffs. Elevated levels of particulate matter can alter emissions and lead to chemical and substantial changes in the environment (Ismail et al., 2019).

3.2 Health Implications

One of the most frequent health risks is "air pollutants such oxides, particulate matter, and ground-level ozone". Oxides like carbon monoxide (CO) reduce the amount of oxygen that reaches the body's organs, impair eyesight, and are lethal in high amounts. Sulphur oxides are naturally carcinogenic if exposed for an extended period of time. These pollutants have made a substantial contribution to health issues to human life. Along with adults, children have also suffered negative effects, including an increase in infant mortality cases, prenatal impacts, allergies, cardiovascular and behavioural issues, and endothelial dysfunction (Habtom et al., 2017).

3.2.1 Air Pollution Caused by Traffic

An imbalance in the concentration of free radicals, which results in oxidative stress and increases the risk of developing cancer, can be caused by a combination class of air pollutants. Organ inflammation can result from increased oxidative stress and impaired cardiovascular circulation, which can lead to issues with mental health. Ultrafine matter has the ability to penetrate the blood-brain barriers affecting vaso control and neural inflammation (Qadri & Faiq, 2020). Pesticides, fertilizers, dyes, detergents, and heavy metals all have the potential to negatively impact on the human health (Diagboya & Dikio, 2018). New contaminants are also emerging and entering into aquatic systems.

4 Pollution Control Using Both Conventional and Innovative Techniques

Traditional techniques for cleaning up dirty water include coagulation and flocculation. They are employed as a biological treatment's pretreatment, enhancing waste's biodegradability and eliminating macropollutant. The addition of substances like ferric chloride and polymers destabilizes the colloidal components, causing small particles to clump together into huge flocs. By creating flocs, a physiochemical method is used to remediate soluble chemicals and metal contaminants in industrial wastewater (Jayaswal et al., 2018; Poznyak et al., 2008; Siafaka et al., 2021) (Fig. 2).

Zeolite Adsorption: Zeolites are microporous, aluminosilicate minerals that are widely used as adsorbents due to their high surface area, uniform pore size, and ability to selectively adsorb molecules based on size, shape, and charge. Zeolites can be used to remove contaminants from water, such as heavy metals, ammonia, and nitrates.

Biosorption: It is a process that involves the use of biological materials such as bacteria, fungi, algae, and other living or dead biomass to remove pollutants or other contaminants from water, air or soil. The contaminants can be heavy metals, dyes,

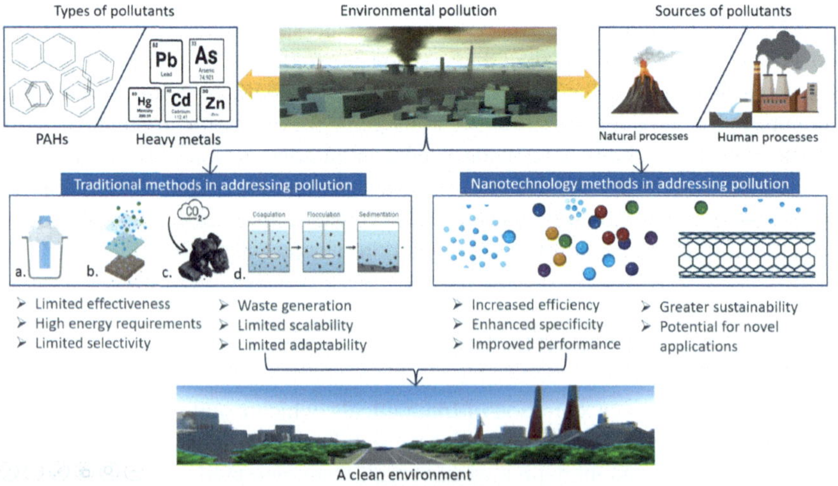

Fig. 2 Traditional and nanotechnological methods to remove hazardous pollutants to make a clean environment

pesticides, pharmaceuticals, and other organic or inorganic pollutants. The biological material used in biosorption contains functional groups such as carboxyl, amino, and hydroxyl groups that can bind to the contaminants present in the environment. These functional groups provide high surface area and high selectivity to the biological material, making them effective adsorbents for pollutants. The biosorption process has several advantages over other conventional methods of pollutant removal such as chemical precipitation, ion exchange, and reverse osmosis.

Lindholm-Lehto et al. (2015). The biological materials used for biosorption have a high degree of selectivity, meaning they can remove specific pollutants from the environment. biosorption applications include the removal of heavy metals from industrial wastewater, the removal of dyes from textile industry effluent, the removal of nutrients from agricultural wastewater, and the removal of pharmaceuticals from hospital effluent (Chedeville et al., 2009; de Oliveira et al., 2014; Dewil et al., 2017; James et al., 2014; Pi et al., 2018). It is feasible to achieve significant reductions in chemical oxygen demand. Traditional methods for removing pollutants have also included electrochemical procedures (Roy & Bharadvaja, 2021). Water is dosed by a metal sacrificial anode during electrocoagulation, has quick sedimentation rates, less sludge is produced, and requires fewer coagulant ions for treatment (Holt et al., 1999; Elazzouzi et al., 2017). Pollutant removal and complete mineralization are now possible with the help of electrochemical advanced oxidation processes (Martinez-Huitle & Ferro, 2006). A promising method for cleaning up low permeability matrices that have been contaminated with organic and inorganic contaminants is electrokinetic remediation. One common method for eliminating organic contaminants is electrochemical oxidation.

5 The Function of Nanomaterials in the Removal of Contaminants

Nanomaterials are materials with at least one dimension in the nanometer scale, typically between 1 and 100 nm. Due to their unique physicochemical properties, such as high surface area, large surface-to-volume ratio, and tunable surface reactivity, nanomaterials have been increasingly used in various environmental applications, including the removal of contaminants from water, air, and soil (Guerra et al., 2018; Janas & Koziol, 2016). It has been reported that different industries use nanoscale materials. The creation and use of more nanoscale have increased during the past few years. Numerous hazardous waste sites are harmed by chlorinated solvents or in oil spills; nanomaterials have been used to clean up polluted "soil and groundwater". The physicochemical, surface, and optical-electronic properties of manufactured nanoparticles provide solutions to problems that were previously challenging to solve using conventional techniques. It can come up with novel approaches to develop new procedures, swap out outdated equipment, and make new materials and chemicals that are highly effective and energy-efficient (Guerra et al., 2018). The function of nanomaterials in the removal of contaminants can be broadly categorized into three main mechanisms: adsorption, catalysis, and membrane filtration.

Adsorption: Nanomaterials such as activated carbon, metal–organic frameworks (MOFs), and graphene oxide have high surface areas and can effectively adsorb contaminants, such as heavy metals, organic pollutants, and dyes, from water and air. The adsorption mechanism can be enhanced by functionalizing the surface of nanomaterials with specific chemical groups or modifying their surface properties.

1. **Catalysis**: Nanomaterials such as metal nanoparticles, metal oxides, and carbon nanotubes can act as efficient catalysts for the degradation of contaminants in water and air. These nanomaterials can either adsorb the contaminants on their surface and then promote their degradation or directly oxidize the contaminants through redox reactions.
2. **Membrane Filtration**: Nanomaterials such as ceramic or polymeric membranes with nanopores can be used as efficient filters for the removal of contaminants from water and air. The nanometer-sized pores can effectively remove suspended particles, bacteria, and viruses, while allowing the passage of water or air.

6 Different Techniques for Creating Nanoparticles

In order to create novel properties that are distinct from those of bulk materials, science and innovation have made quick progress in the development of nanomaterials. It is possible to create nanoparticles through chemical, physical, and biological processes.

6.1 Chemical Methods

6.1.1 Nanoparticle Synthesis Using Polyols

Any natural substance that may be a polyol has a variety of hydroxyl organizations connected to it (Kovalska et al., 2020). A sol-gel approach provides oxide at an excessive temperature to create nanoparticles like titanium, vanadium, tin, lead, manganese, yttrium of different sizes (Quievryn et al., 2014).

6.1.2 Reduction in Solution (Colloidal Method)

One of the most crucial methods is the utilization of reducing agents like $Na_3C_6H_5O_7$. For the first time, Turkevich proposed this technique in 1951 for the production of Au monodisperse colloidal solutions to make it possible to produce spherical nanoparticles. Silver nanoparticles were later created using the same method, although their size distribution was wider, spanning from 60 to 200 nm. The manufacture of gold, iron, molybdenum, and silver nanoparticles is typically accomplished using this process. Chloroauric acid is dissolved in a solution and brought to a boil for gold. A reducing agent is trisodium citrate. Gold nanoparticles replaced the seed crystals after 10 min of mixing and swirling.

6.1.3 Electrochemical Synthesis

The principle benefits of this technology over a standard chemical reaction are the capacity to exactly manipulate the selected capacity and the rejection of the in all likelihood wasteful opportunity half-reaction (Cavaliere et al., 2021). There have been quite a few interests inside the electrochemical synthesis of Ag nanoparticles (Rodriguez-Sanchez et al., 2000). Ag nanoparticles with a length variety of 2 to 7 nm are produced by including electrons to oxidize the effective electrode of the cellular for soluble silver ions in acetonitrile containing tetra butyl ammonium. The electrochemical synthesis of Si nanoparticles with pink luminescence stabilized via styrene. Si nanoparticles glow when inspired through UV (Choi et al., 2014). Pt nanoparticles have a length of around 10 nm and are deposited at the electrode's surface (Islam & Islam, 2013).

6.2 Physical Methods

6.2.1 Plasma Method

This method is used to make nanoparticles. Plasma is produced by the use of radiofrequency (RF) heating coils, primary metal is positioned right into a pestle, that is then positioned into a room that has been evacuated (Singh et al., 2022). The steel is heated above its evaporation factor with excessive-voltage RF coils wrapped across the evacuated chamber. The technique makes use of helium, which enters the machine, to produce an excessive temperature near the coils, which leads to metal vapours condensing at the helium gasoline atoms in a cold collecting rod, in which nanoparticles are accrued and oxygen fuel passivates them (Cavaliere et al., 2021).

6.2.2 Microwave Irradiation

The usage of microwaves to warmth samples accelerates reactions and uses much less resources. Microwave generation also makes it viable to higher manage reaction methods. They are in the round shape, and the greater uniformity, the more powerful nanoparticles with more applicability are used to pollution clean-up (Wang et al., 2002). ZnS nanoparticles can be created with the aid of a microwave, and ZnS nanoparticles are also used to calculate the average length of the nanocrystallites (Fitzpatrick, 2017).

6.2.3 Radiation Dependent

Gamma radiation employs the fewest reagents and uses room temperature with the fewest artificial steps viable to produce monodisperse metal nanoparticles. The production of monosized and extensively dispersed metal clusters through radiolytic reduction has been tested to achieve success (93). The principle consequences of high-strength gamma rays interacting with a solution of steel ions are excitation and ionization of the solvent (Abidi & Remita, 2010).

6.3 Biological Method

Green synthesis is required to build reliable, lengthy-lasting, and environmentally pleasant production methods in order to save you the production of unwanted or dangerous by using-products. Natural assets and appropriate solvent systems are needed to reap this goal metal nanoparticles are produced sustainably to simply accept diverse biological components.

6.3.1 Plant

Plant leaf extracts have critical additives that may be employed inside the advent of nanoparticles. The plant extract is mixed with metal precursor solutions depending on the experimental conditions (Mittal et al., 2013). The formation of nanoparticles, as well as their output and balance, are known to be influenced with the aid of the parameters that manage how plant leaf extracts are constructed up. The main phytochemicals implicated in nanoparticle bioreduction are flavones, terpenoids, sugars, ketones, aldehydes, carboxylic acids, and amides (Prathna et al., 1857). In sweet basil extracts, the enol-to-keto-transformation controls the production of biogenic Ag nanoparticles. Metallic nanoparticles are also produced via plant extracts that include sugars like fructose and glucose (Panigrahi et al., 2004). The manufacture of steel nanoparticles uses plant extracts, which can be biomolecules product of proteins and carbohydrates, as a reducing agent (Mittal et al., 2013; Shankar et al., 2003; Singh et al., 2010).

6.3.2 Bacteria

Bacterial species had been utilized in commercial biotechnological tactics such environmental clean-up, genetic amendment, and biosorption (Gericke & Pinches, 2006).

Bacteria are huge prospects for the production of nanoparticles because they have got the capacity to reduce metallic ions. Synthesis gives a safe, riskless, and environmentally perfect method for producing steel nanoparticles due to its financial potential and viability (Zhang et al., 2020). Pure Au nanoparticles can be produced by using microorganism together with *Delftiaacidovorans*. Delftibactin is regarded as a non-ribosomal peptide (NRP), and due to its capacity to promote resistance to harmful Au ions, it is been linked to the synthesis of Au nanoparticles. The transition steel gold turned into not unsafe to microorganism as it formed inert gold nanoparticles connected to delftibactin (Pantidos & Horsfall, 2014).

6.3.3 Fungi

Growing ometallic/metal-oxide nanoparticles through the use of fungi are extra techniques for the manufacture of metal-based nanoparticles because they contain a selection of enzymes (Kalimuthu et al., 2008). Fungi can produce more nanoparticles than bacteria can on a steady foundation (Mittal et al., 2013). In contrast to different fungus, *Trichodermareesei* has the advantage of being a well-researched organism that may be manipulated to provide a big type of enzymes and might assist accelerate the synthesis of nanoparticles (Jacobson, 2009). Platinum nanoparticles can be biologically produced by the use of *Neurosporacrassa* (Mittal et al., 2013).

6.3.4 Algae

They are recognized to build up heavy metals, which can be exploited to supply metal nanoparticles biologically. Tetrahedral, decahedral, and icosahedral nanoparticles can be produced with the aid of the dried, unicellular alga *Chlorella vulgaris*, which is gathered on the floor. Amongst other matters, the proteins within the algal extract function as a stabilizing, lowering, and shape-controlling modifier (James et al., 2014). A marine alga known as *Sargassumwightii* has the capability to create extracellular Ag and Au nanoparticles (Pi et al., 2018). Considering that *Euglena gracilis* microalga cells are alive and had been grown both in a mixotrophic or autotrophic surroundings, the gold nanoparticles produced are of high yield (de Oliveira et al., 2014).

7 Types of Nanomaterials

7.1 *Basics of Origin*

Nanoparticles can be created artificially or naturally (synthetic). Natural nanoparticles are created by living things, biological processes, or human activities (Mittal et al., 2013). According to their names, synthetic or engineered nanoparticles are nanomaterials created by mechanical processes, intricate systems, or separate or combined chemical, biological, and physical processes (Prathna et al., 1857).

7.2 *Basics of 3D Structure*

Fall into one of three categories based on their structural differences. The electrons are dimensionless in zero D, unable to travel along any axes. In one D, they are able to move along a single axis, followed by two D and three D, where they are able to move in two and three axes, respectively (Iravani, 2011). The thin films that make up one-dimensional nanoparticles range in size from 1 to 100 nm. Nanotubes, such as carbon nanotubes, are typically included in two-dimensional nanoparticles. Dendrimers, quantum dots, and fullerenes (like carbon 60) make up three-dimension nanoparticles (Kovalska et al., 2020).

7.3 *Based on Material Type*

Nanomaterials can be divided into subcategories based on the type of material they are made of, including inorganic, organic, carbon-containing, or composite substances. The building blocks of inorganic nanomaterials are metals or their oxides, such as silicon-based compounds, titanium dioxide, gold, silver, or titanium dioxide. The source of organic nanomaterials, which can be micelles, liposomes, or polymeric particles, is either carbon- or inorganic-based resources. Multiphasic nanoparticles can combine with other comparable or separate nanoparticles in bulk and complex which is formed is a gold nanoparticle in composition with ceramic or a polymer. Carbon-based nanoparticles including fullerenes, nanotubes, and nanofibers are frequently seen in three-dimensional morphological patterns (Ambaye et al., 2020; Ansari et al., 2004; Amin et al., 2007; Bohu et al., 2019; Mügge et al., 2020; Rausch et al., 2005).

7.4 Characterization

The concept of surface plaque resonance (SPR), which is used in UV-seen spectroscopy, has been discussed in characterization methodologies (Pinel et al., 2008). The unique method of X-ray diffraction (XRD) is used for locating the crystalline section. The Fourier transform infrared technique, often known as FTIR spectroscopy, is useful for identifying a sample. In order to categorize nanoparticles according to length, shape, and surface charge, atomic force microscopy (AFM), scanning electron microscopy, and transmission electron microscopy were used. Dynamic mild scattering is used to determine the nanoparticles' sizes.

8 Different Applications of Nanoparticles

These are now widely used for anything from medical grafts to sensing devices. They have transformed the way that many techniques operate. The antibacterial properties of green manufactured particles, particularly silver nanoparticles have led to their usage in bandages, silicone replacement heart valves, catheters, bone prosthesis, and dental sets. The development of biosensors for faster disease diagnosis, monitoring the development of the issue, and monitoring the effectiveness of the prescribed medication has been facilitated by the essential property of nanoscale particles (Dwivedi & Gopal, 2010; Iravani, 2011; Mittal et al., 2013; Singh et al., 2022). The multifaceted antibacterial property of green nanoparticles has been shown in numerous researches. Typically, nanoparticles either work inside or outside of cells to change how quickly bacteria grow, for example, by inhibiting cytological transfer and signalling (Banasiuk et al., 2020; Pinel et al., 2008). Target therapies employ carriers including polymeric micelles, dendrimers, and liposomes in conjunction with anticarcinogenic nanomaterials (Huang et al., 2014) (Fig. 3).

9 Nanomaterials in Remediation of Soil

The degradation and infection of soil continue to pose a severe hazard to the environment, and remediation has grown to be a global problem. There is an urgent need to deal with the problem of degrading soil reservoirs as it has a full-size effect on global agricultural production and food protection (Bakshi & Abhilash, 2020). Agricultural productivity is appreciably impacted by way of both the need for accelerated production to feed extra humans and the mission of stopping in addition land deterioration (Bakshi & Abhilash, 2020). One of the most customarily used remedy methods is immobilization or adsorption (Arif et al., 2020; Gaur et al., 2014). Ferric oxide nanoparticles have a tremendous potential to absorb and immobilize heavy metals from various media samples, which include Cd and As (Peng et al., 2020).

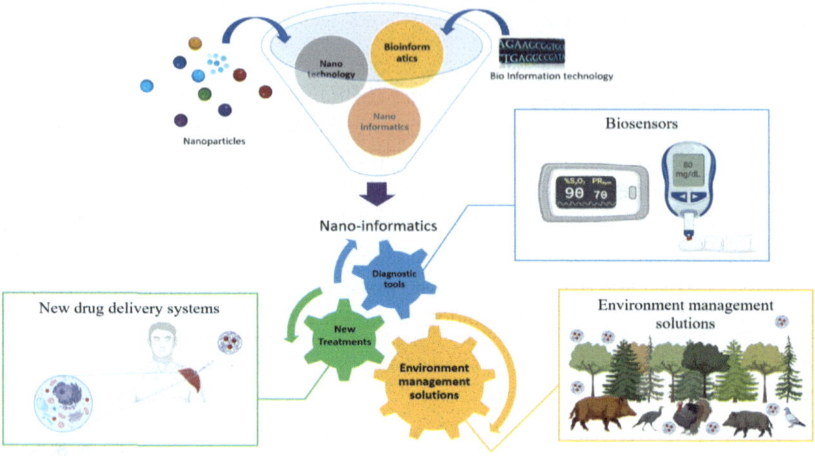

Fig. 3 Different applications of nanoparticles

The aquatic lifestyles are negatively impacted by means of oily sewage (Iravani, 2011). Total petroleum hydrocarbons (TPHs) may be efficaciously eliminated from water by the usage of iron nanoparticles, with advanced results of 88.34% (Yusefi et al., 2021). With the removal of heavy metals, excessive performance treatments based on nanotechnology can produce handled water with fewer contaminants and unsafe compounds.

The fundamental two-step approach for dye degradation the use of gold and silver nanoparticles is the build-up of electrons on the particle surface observed with the aid of the diffusion of the dye for a discount response (Sharma et al., 2015). Pollutant immobilization is used to clean up polluted soils (Sharma et al., 2015). Nanohydroxyapatite debris was a success in immobilizing steel concentrations in pore water by reducing the share of metals that had been exchangeable in contaminated sediments and soils. Reductive abiotic strategies affect the mobility and toxicity of redox-touchy factors as contaminants can be used on my own or at the side of immobilization techniques to remediate them. Additionally, nanomaterials are utilized in superior oxidation tactics (AOPs), which use a diffusion of oxidants to interrupt down natural pollution in infected soil (Ajmal et al., 2018; Bastús et al., 2014). Green nanoparticles have numerous potential for treating localized environments which can be infected with evidently harmful steel ions and for fighting environmental contaminants. Remediation and clean-up of waste websites can be gained from those objects. Ferrous nanoparticles are efficient fertilizers and may be used to disinfect heavy metal-infected water in addition to soil (Iravani, 2011).

10 Nanomaterials in Remediation of Water Pollution

Nanomaterials have shown great potential for the remediation of water pollution, especially in the removal of contaminants such as heavy metals, organic pollutants, and pathogens. The unique properties of nanomaterials, such as high surface area, high reactivity, and tunable surface chemistry, make them effective for water treatment applications. (Adeleye et al., 2016a). Some of the ways nanomaterials are used for water pollution remediation are:

1. **Adsorption**: Nanomaterials such as activated carbon, graphene oxide, and metal–organic frameworks (MOFs) can absorb a wide range of contaminants from water. These materials have high surface areas and can be functionalized to enhance their adsorption capacity and selectivity. For example, carbon nanotubes have been shown to adsorb organic pollutants, heavy metals, and pharmaceuticals from water.
2. **Catalysis**: Nanomaterials such as metal nanoparticles and metal oxides can be used as catalysts for the degradation of organic pollutants in water. These materials can either adsorb pollutants on their surfaces and then promote their degradation or directly oxidize pollutants through redox reactions. For example, titanium dioxide nanoparticles have been used to degrade organic pollutants such as dyes, pesticides, and pharmaceuticals.
3. **Membrane filtration**: Nanomaterials such as ceramic or polymeric membranes with nanopores can be used as efficient filters for the removal of contaminants from water. These materials can remove suspended particles, bacteria, viruses, and even dissolved pollutants such as heavy metals and organic compounds. For example, graphene oxide membranes have been shown to remove heavy metals from water.
4. **Disinfection**: Nanomaterials such as silver nanoparticles and copper nanoparticles can be used as disinfectants to kill bacteria and viruses in water. These materials have a high antimicrobial activity and can be used to disinfect water for drinking or agricultural use. For example, silver nanoparticles have been used to disinfect water contaminated with E. coli bacteria.

Overall, nanomaterials offer several advantages for water pollution remediation, including high efficiency, selectivity, and low energy consumption. However, it is important to ensure the safe and responsible use of nanomaterials to minimize potential environmental risks. (Adeleye et al., 2016b). In addition to zinc and titanium oxides, a variety of remediation procedures, including lysis, filtration, adsorption, and oxidation, use carbon nanotubes, submicron particles, ceramic, nanowire, and polymeric membranes (Iravani, 2011). Using hydroxyl radicals (OH), textile effluents are removed. When exposed to light, semiconductors have the potential to break down a variety of organic (dye) impurities.

11 Nanomaterials in Remediation of Air Pollution

Nanotechnology can be a useful resource in air purification. The ineffectiveness of air waste elimination to the floor made it tough to cast off the filtered chemicals (Jain & Singh, 2012). Filters are useful when handling air pollutants. Three mechanisms, floor contact with the filter structure, inertial force generated with the aid of particle and clear-out production, and electric rate, are utilized by the membrane of the filter out to deal with pollutants (Zhu et al., 2017). The main problem with filters is they result in a strain decrease, which requires a massive amount of power to remedy. Membranes fabricated from carbon nanotubes are more effective at setting apart CO_2 from different gases (Skoulidas et al., 2006). In comparison to standard gasoline separation techniques, carbon nanotubes can capture gases 100 instances faster (Chaturvedi et al., 2012).

In a variety of industries, nanomembrane technology is used to separate and purify gases and polluting vapours as well as to stop their release into the environment (Zhang et al., 2015). A state-of-the-art tool that can recognize and respond to a physical change that can be felt on a nanometer scale is a nanosensor. Small 3D circular molecules called nanoparticles are utilized as sensors to detect the presence of toxins in an environment (Peng et al., 2014). With the application of technological progress to the development of quicker and more precise sensors, biosensor can catch hazardous gas molecules using single or many membrane nanotubes. A novel sort of nanosensor is smart dust. Ultra-compact nanocomputers and nanoscale silicon-based particles enable wireless data collection and transmission to the main server. Solar-powered nanosensors are easily able to hang in the air for extended periods of time (Peng et al., 2014).

12 Conclusion

Nanoparticles have altered litigation methods for biotechnology and environmental cases. The increase of humanity in nature has been followed with the aid of an increase in industrialization and resource use. There is a urgent need to get rid of those contaminants with the least quantity of damage to the planet. Various nanoparticles are powerful for clearing out pollution from environment, like sewage spills and landmines. Additionally, they have minimal harmful effects during production rates, allowing for further study to make them more compatible with future decreases in production costs.

References

Abidi, W., & Remita, H. (2010). Gold based nanoparticles generated by radiolytic and photolytic methods, 170–88.

Adeleye, A. S., Conway, J. R., Garner, K., Huang, Y., Su, Y., & Keller, A. A. (2016b). Engineered nanomaterials for water treatment and remediation: Costs, benefits, and applicability **286**, 640–62.

Adeleye, A. S., Conway, J. R., Garner, K., Huang, Y., Su, Y., & Keller, A. A. (2016c). Engineered nanomaterials for water treatment and remediation: Costs, benefits, and applicability. *Chemical Engineering Journal*, **286**(January), 640–62. Elsevier. https://doi.org/10.1016/J.CEJ.2015.10.105

Adeleye, A. S., Conway, J. R., Garner, K., Huang, Y., Su, Y., & Keller, A. A. (2016a). Engineered nanomaterials for water treatment and remediation: Costs, benefits, and applicability. *Chemical Engineering Journal, 286*, 640–662.

Ajmal, Z., Muhmood, A., Usman, M., Kizito, S., Lu, J., Dong, R., & Wu, S. (2018). Phosphate removal from aqueous solution using iron oxides: Adsorption, desorption and regeneration characteristics. *Journal of Colloid and Interface Science*, **528**, 145–55. Elsevier.

Ali, M. U., Liu, G., Yousaf, B., Ullah, H., Abbas, Q., & Munir, M. A. M. (2019). A systematic review on global pollution status of particulate matter-associated potential toxic elements and health perspectives in urban environment. *Environmental Geochemistry and Health, 41*(3), 1131–62. Springer Netherlands. https://doi.org/10.1007/S10653-018-0203-Z

Allen, D. T. (2002). Air pollution. *Kirk-othmer encyclopedia of chemical technology*, January. John, Inc. https://doi.org/10.1002/0471238961.01091823151206.A01.PUB2

Álvarez, X., Valero, E., Santos, R. M. B., Varandas, S. G. P., Fernandes, L. F. S., & Pacheco, F. A. L. (2017). Anthropogenic nutrients and eutrophication in multiple land use watersheds: Best management practices and policies for the protection of water resources. *Land Use Policy, 69*, 1–11. Elsevier.

Ambaye, T. G., Vaccari, M., Castro, F. D., Prasad, S., & Rtimi, S. (2020). Emerging technologies for the recovery of rare earth elements (REEs) from the end-of-life electronic wastes: A review on progress, challenges, and perspectives. *Environmental Science and Pollution Research, 27*(29), 36052–74. Springer. https://doi.org/10.1007/S11356-020-09630-2

Ambient Air. (2007). *Van Nostrand's scientific encyclopedia*, May. Wiley, Inc. https://doi.org/10.1002/0471743984.VSE9142

Amin, S. A., Küpper, F. C., Green, D. H., Harris, W. R., & Carrano, C. J. (2007). Boron binding by a siderophore isolated from marine bacteria associated with the toxic dinoflagellate gymnodinium catenatum. *Journal of the American Chemical Society, 129*(3), 478–479. https://doi.org/10.1021/ja067369u

Ansari, M. Z., Yadav, G., Gokhale, R. S., & Mohanty, D. (2004). NRPS-PKS: A knowledge-based resource for analysis of NRPS-PKS megasynthases. *Nucleic Acids Research, 32*(WEB SERVER ISS), W405–13. https://doi.org/10.1093/nar/gkh359

Araviiskaia, E., Berardesca, E., Bieber, T., Gontijo, G., Sanchez Viera, M., Marrot, L., Chuberre, B., & Dreno, B. (2019). The impact of airborne pollution on skin. *Journal of the European Academy of Dermatology and Venereology, 33*(8), 1496–1505. Blackwell Publishing Ltd. https://doi.org/10.1111/JDV.15583

Arif, Z., Sethy, N. K., Kumari, L., Mishra, P. K., & Upadhyay, S. N. (2020). Recent advances in functionalized polymer-based composite photocatalysts for wastewater treatment. *Nano-Materials as Photocatalysts for Degradation of Environmental Pollutants*, 39–64. Elsevier.

Artiola, J. F., Walworth, J. L., Musil, S. A., & Crimmins, M. A. (2019). Soil and land pollution. In *Environmental and pollution science* (pp. 219–35). Elsevier.

Bakshi, M., & Abhilash, P. C. (2020). Nanotechnology for soil remediation: Revitalizing the tarnished resource. In *Nano-materials as photocatalysts for degradation of environmental pollutants* (pp. 345–70). Elsevier.

Banasiuk, R., Krychowiak, M., Swigon, D., Tomaszewicz, W., Michalak, A., Chylewska, A., Ziabka, M., Lapinski, M., Koscielska, B., & Narajczyk, M. (2020). Carnivorous plants used for green synthesis of silver nanoparticles with broad-spectrum antimicrobial activity. *Arabian Journal of Chemistry, 13*(1), 1415–28. Elsevier.

Bastús, N. G, Merkoçi, F., Piella, J., & Puntes, V. (2014). Synthesis of highly monodisperse citrate-stabilized silver nanoparticles of up to 200 Nm: Kinetic control and catalytic properties. *Chemistry of Materials, 26*(9). 2836–46. ACS Publications.

Behera, D., Krishnakumar, S., & Anoop, A. (2023). Occurrence, distribution and sources of petroleum contamination in reef-associated sediments of the Gulf of Mannar, India. *Marine Pollution Bulletin, 196*(November). Elsevier Ltd. https://doi.org/10.1016/j.marpolbul.2023.115576

Billé, R., Kelly, R., Biastoch, A., Harrould-Kolieb, E., Herr, D., Joos, F., Kroeker, K., Laffoley, D., Oschlies, A., & Gattuso, J.-P. (2013). Taking action against ocean acidification: A review of management and policy options. *Environmental Management, 52*(4), 761–79. Springer.

Bohu, T., Anand, R., Noble, R., Lintern, M., Kaksonen, A. H., Mei, Y., Cheng, K. Y., et al. (2019). Evidence for fungi and gold redox interaction under earth surface conditions. *Nature Communications, 10*(1). Nature Publishing Group. https://doi.org/10.1038/S41467-019-10006-5

Briffa, J., Sinagra, E., Blundell, R. (2020) Heavy metal pollution in the environment and their toxicological effects on humans. *Heliyon, 6*(9), e04691.

Bruns, A., Chen, H. S., & Turnbull, D., Dolnicar, S., Chapple, A., Adorno, T., Horkheimer, M., Negus, K., et al. (1974). The role of personality in organizational life: Issues and evidence. *Sage Publications*. https://doi.org/10.1163/_q3_SIM_00374.

Cavaliere, P. D., Perrone, A., & Silvello, A. (2021). Water electrolysis for the production of hydrogen to be employed in the ironmaking and steelmaking industry. *Metals, 11*(11), 1816. MDPI.

Chaturvedi, S., Dave, P. N., & Shah, N. K. (2012). Applications of nano-catalyst in New Era. *Journal of Saudi Chemical Society, 16*(3), 307–25. Elsevier.

Chedeville, O., Debacq, M., & Porte, C. (2009). Removal of phenolic compounds present in olive mill wastewaters by ozonation. *Desalination, 249*(2), 865–69. Elsevier.

Choi, J., Kim, K., Han, H., Hwang, M. P., & Lee, K. H. (2014). Electrochemical synthesis of red fluorescent silicon nanoparticles. *Bulletin of the Korean Chemical Society, 35*(1), 35–38. Korean Chemical Society.

Corsi, I., Winther-Nielsen, M., Sethi, R., Punta, C., Della Torre, C., Libralato, G., Lofrano, G., et al. (2018). Ecofriendly nanotechnologies and nanomaterials for environmental applications: Key issue and consensus recommendations for sustainable and ecosafe nanoremediation. *Ecotoxicology and Environmental Safety, 154*(June), 237–44. Academic Press. https://doi.org/10.1016/J.ECOENV.2018.02.037

Dachs, J., Eisenreich, S. J., & Hoff, R. M. (2000). Influence of eutrophication on air—water exchange, vertical fluxes, and phytoplankton concentrations of persistent organic pollutants. *Environmental Science and Technology, 34*(6), 1095–1102. ACS Publications.

de Oliveira, A. G., Ribeiro, J. P., de Oliveira, J. T., De Keukeleire, D., Duarte, M. S., & do Nascimento, R. F. (2014). Degradation of the pesticide chlorpyrifos in aqueous solutions with UV/H_2O_2: Optimization and effect of interfering anions. *Journal of Advanced Oxidation Technologies, 17*(1), 133–38. De Gruyter.

Dewil, R., Mantzavinos, D., Poulios, I., & Rodrigo, M. A. (2017). New perspectives for advanced oxidation processes. *Journal of Environmental Management, 195*, 93–99. Elsevier.

Diagboya, P. N. E., & Dikio, E. D. (2018). Silica-based mesoporous materials; Emerging designer adsorbents for aqueous pollutants removal and water treatment. *Microporous and Mesoporous Materials, 266*, 252–67. Elsevier.

Dwivedi, A. D., & Gopal, K. (2010). Colloids and surfaces a: Physicochemical and engineering aspects biosynthesis of silver and gold nanoparticles using chenopodium album leaf extract. *Colloids and Surfaces A: Physicochemical and Engineering Aspects, 369*(1–3), 27–33. Elsevier B.V. https://doi.org/10.1016/j.colsurfa.2010.07.020

Elazzouzi, M. K., & Elyoubi, M. S. (2017). Electrocoagulation flocculation as a low-cost process for pollutants removal from urban wastewater. *Chemical Engineering Research and Design, 117*, 614–26. Elsevier.

Fitzpatrick, R. W. (2017). *A criminal case study involnsference of acid sulfate soilving transference of acid sulfate chapter 8 a criminal case study involving tra.*

Gaur, N., Flora, G., Yadav, M., & Tiwari, A. (2014). A review with recent advancements on bioremediation-based abolition of heavy metals. *Environmental Science: Processes and Impacts, 16*(2), 180–93. Royal Society of Chemistry.

Gericke, M., & Pinches, A. (2006). Microbial production of gold nanoparticles. *Gold Bulletin, 39*(1), 22–28. Springer.

Guerra, F. D., Attia, M. F., Whitehead, D. C., & Alexis, F. (2018). Nanotechnology for environmental remediation: Materials and applications. *Molecules, 23*(7), 1760. MDPI.

Habtom, H., Demanèche, S., Dawson, L., Azulay, C., Matan, O., Robe, P., Gafny, R., Simonet, P., Jurkevitch, E., & Pasternak, Z. (2017). Soil characterisation by bacterial community analysis for forensic applications: A quantitative comparison of environmental technologies. *Forensic Science International: Genetics, 26*, 21–29. Elsevier Ireland Ltd. https://doi.org/10.1016/j.fsigen.2016.10.005

Hajam, Y. A., Kumar, R., & Kumar, A. (2023). Environmental waste management strategies and Vermi transformation for sustainable development. *Environmental Challenges*, 100747.

Hassaan, M. A., & El Nemr, A. (2020). Pesticides pollution: Classifications, human health impact, extraction and treatment techniques. *The Egyptian Journal of Aquatic Research, 46*(3), 207–20. Elsevier.

Holt, P., Barton, G., & Mitchell, C. (1999). *Electrocoagulation as a wastewater treatment. The Third Annual Australian Environmental Engineering Research Event, 1000*, 41–46. Victoria: Third Annual Australian Environ-mental Engineering Research Event.

Huang, H., Xu, Y., Tang, C.-J., Chen, J.-R., Wang, A.-J., & Feng, J.-J. (2014). Facile and green synthesis of photoluminescent carbon nanoparticles for cellular imaging. *New Journal of Chemistry, 38*(2), 784–89. Royal Society of Chemistry.

Iravani, S. (2011). Green chemistry green synthesis of metal nanoparticles using plants, 2638–50. https://doi.org/10.1039/c1gc15386b

Islam, A., & Islam, M. S. (2013). Electro-deposition method for platinum nano-particles synthesis, *1*(2), 8–18.

Ismail, I. N., Jalaludin, J., Bakar, S. A., Hisamuddin, N. H., & Suhaimi, N. F. (2019). Association of traffic-related air pollution (TRAP) with DNA damage and respiratory health symptoms among primary school children in Selangor. *Asian Journal of Atmospheric Environment (AJAE), 13*(2).

Jacobson, M. Z. (2009). Review of solutions to global warming, air pollution, and energy security. *Energy and Environmental Science, 2*(2), 148–73. Royal Society of Chemistry.

Jain, C. K., & Singh, R. D. (2012). Technological options for the removal of arsenic with special reference to South East Asia. *Journal of Environmental Management, 107*, 1–18. Elsevier.

James, C. P., Germain, E., & Judd, S. (2014). Micropollutant removal by advanced oxidation of microfiltered secondary effluent for water reuse. *Separation and Purification Technology, 127*, 77–83. Elsevier.

Janas, D., & Koziol, K. K. (2016). Carbon nanotube fibers and films: Synthesis, applications and perspectives of the direct-spinning method. *Nanoscale, 8*(47), 19475–90. Royal Society of Chemistry.

Jayaswal, K., Sahu, V., & Gurjar, B. R. (2018). Water pollution, human health and remediation. In *Water remediation*, (pp. 11–27). Springer.

Jia, Z.-Q., Zhang, Y.-C., Huang, Q.-T., Jones, A. K., Han, Z.-J., & Zhao, C.-Q. (2020). Acute toxicity, bioconcentration, elimination, action mode and detoxification metabolism of broflanilide in Zebrafish, Danio Rerio. *Journal of Hazardous Materials, 394*, 122521. Elsevier.

Kalimuthu, K., Babu, R. S., Venkataraman, D., Bilal, M., & Gurunathan, S. (2008). Biosynthesis of silver nanocrystals by bacillus licheniformis. *Colloids and Surfaces B: Biointerfaces, 65*(1), 150–53. Elsevier.

Kelly, B. C., Ikonomou, M. G., Blair, J. D., Morin, A. E., & Gobas, F. A. P. C. (2007). Food web–specific biomagnification of persistent organic pollutants. *Science, 317*(5835), 236–39. https://doi.org/10.1126/SCIENCE.1138275

Khan, S. H. (2020). Green nanotechnology for the environment and sustainable development. In *Green materials for wastewater treatment*, (pp. 13–46). Springer.

Kovalska, E., Luxa, J., Hartman, T., Antonatos, N., Shaban, P., Oparin, E., Zhukova, M., & Sofer, Z. (2020). Non-aqueous solution-processed phosphorene by controlled low-potential electrochemical exfoliation and thin film preparation. *Nanoscale, 12*(4), 2638–47. Royal Society of Chemistry.

Lellis, B., Favaro-Polonio, C. Z., Pamphile, J. A., & Polonio, J. C. (2008). Effects of textile dyes on health and the environment and bioremediation potential of living organisms. *Update in Anaesthesia.*

Lellis, B., Fávaro-Polonio, C. Z., Pamphile, J. A., & Polonio, J. C. (2019). Effects of textile dyes on health and the environment and bioremediation potential of living organisms. *Biotechnology Research and Innovation, 3*(2), 275–90. Elsevier.

Lindholm-Lehto, P. C., Knuutinen, J. S., Ahkola, H. S. J., & Herve, S. H. (2015). Refractory organic pollutants and toxicity in pulp and paper mill wastewaters. *Environmental Science and Pollution Research, 22*(9), 6473–99. Springer.

Mancebo, S. E., & Wang, S. Q. (2015). Recognizing the impact of ambient air pollution on skin health. *Journal of the European Academy of Dermatology and Venereology, 29*(12), 2326–32. Wiley. https://doi.org/10.1111/JDV.13250

Manisalidis, I., Stavropoulou, E., Stavropoulos, A., & Bezirtzoglou, E. (2020). Environmental and health impacts of air pollution: A review. *Frontiers in Public Health, 8*(February). Frontiers Media S.A. https://doi.org/10.3389/FPUBH.2020.00014

Marlon, J. R., Bloodhart, B., Ballew, M. T., Rolfe-Redding, J., Roser-Renouf, C., Leiserowitz, A., & Maibach, E. (2019). How hope and doubt affect climate change mobilization. *Frontiers in Communication, 4.* Frontiers Media S.A. https://doi.org/10.3389/FCOMM.2019.00020

Martinez-Huitle, C. A., & Ferro, S. (2006). Electrochemical oxidation of organic pollutants for the wastewater treatment: Direct and indirect processes. *Chemical Society Reviews, 35*(12), 1324–40. Royal Society of Chemistry.

Masih, J., Masih, A., Kulshrestha, A., Singhvi, R., & Taneja, A. (2010). Characteristics of polycyclic aromatic hydrocarbons in indoor and outdoor atmosphere in the north central part of India. *Journal of Hazardous Materials, 177*(1–3), 190–98. Elsevier.

Mishra, S., Lin, Z., Pang, S., Zhang, W., Bhatt, P., & Chen, S. (2021). Recent advanced technologies for the characterization of xenobiotic-degrading microorganisms and microbial communities. *Frontiers in Bioengineering and Biotechnology, 9*, 632059. Frontiers Media SA.

Mittal, A. K., Chisti, Y., & Banerjee, U. C. (2013). Synthesis of metallic nanoparticles using plant extracts. *Biotechnology Advances.* Elsevier B.V. https://doi.org/10.1016/j.biotechadv.2013.01.003

Mügge, C., Heine, T., Baraibar, A. G., van Berkel, W. J. H., Paul, C. E., & Tischler, D. (2020). Flavin-dependent N-hydroxylating enzymes: Distribution and application. *Applied Microbiology and Biotechnology, 104*(15), 6481–99. Springer. https://doi.org/10.1007/S00253-020-10705-W

Orru, H., Ebi, K. L., & Forsberg, B. (2017). The interplay of climate change and air pollution on health. *Current Environmental Health Reports, 4*(4), 504–13. Springer.

Panigrahi, S., Kundu, S., Ghosh, S. K., Nath, S., & Pal, T. (2004). General method of synthesis for metal nanoparticles (i), 411–14.

Pantidos, N., & Horsfall, L. E. (2014). Biological synthesis of metallic nanoparticles by bacteria, fungi and plants. *Journal of Nanomedicine and Nanotechnology, 5*(5), 1. OMICS Publishing Group.

Paso, K. G. (2022). Constructing thermodynamic models of toxic metal biosorption. *Microbial biodegradation and bioremediation: Techniques and case studies for environmental pollution, January* (pp. 109–43). Elsevier. https://doi.org/10.1016/B978-0-323-85455-9.00020-5

Peng, F., Su, Y., Zhong, Y., Fan, C., Lee, S.-T., & He, Y. (2014). Silicon nanomaterials platform for bioimaging, biosensing, and cancer therapy. *Accounts of Chemical Research, 47*(2), 612–23. ACS Publications.

Peng, Y., Liu, Y., Dai, J., Cao, L., & Liu, X. (2020). A sustainable strategy for remediation of oily sewage: Clean and safe. *Separation and Purification Technology, 240*, 116592. Elsevier.

Pereira, L., & Alves, M. (2012). Dyes—Environmental impact and remediation. In *Environmental protection strategies for sustainable development*, 111–62. Springer.

Phale, P. S., Sharma, A., & Gautam, K. (2019). Microbial degradation of xenobiotics like aromatic pollutants from the terrestrial environments. In *Pharmaceuticals and personal care products: Waste management and treatment technology* (pp. 259–78). Elsevier.

Pi, Y., Li, X., Xia, Q., Wu, J., Li, Y., Xiao, J., & Li, Z. (2018). Adsorptive and photocatalytic removal of persistent organic pollutants (POPs) in water by metal-organic frameworks (MOFs). *Chemical Engineering Journal, 337*, 351–71. Elsevier.

Pinel, N., Davidson, S. K., & Stahl, D. A. (2008). VerminephrobacterEiseniae Gen. Nov., Sp. Nov., a nephridial symbiont of the earthworm eisenia foetida (Savigny). *International Journal of Systematic and Evolutionary Microbiology, 58*(9), 2147–2157. https://doi.org/10.1099/ijs.0.65174-0

Poznyak, T., Bautista, G. L., Chaírez, I., Córdova, R. I., & Ríos, L. E. (2008). Decomposition of toxic pollutants in landfill leachate by ozone after coagulation treatment. *Journal of Hazardous Materials, 152*(3), 1108–14. Elsevier.

Prathna, T. C., Mathew, L., & Chandrasekaran, N. (1857). Biomimetic synthesis of nanoparticles: science, technology and applicability (pp. 1–21).

Qadri, R., & Faiq, M. A. (2020). Freshwater pollution: Effects on aquatic life and human health. In *Fresh water pollution dynamics and remediation*, (pp. 15–26). Springer.

Quievryn, C., Bernard, S., & Miele, P. (2014). Polyol-based synthesis of praseodymium oxide nanoparticles. *Nanomaterials and nanotechnology* (vol. 4, p. 7). SAGE Publications Sage UK: London, England.

Rather, M. A., Bhuyan, S., Chowdhury, R., Sarma, R., Roy, S., & Neog, P. R. (2023). Nanoremediation strategies to address environmental problems. *Science of the Total Environment, 886*(August), 163998. Elsevier. https://doi.org/10.1016/J.SCITOTENV.2023.163998

Rausch, C., Weber, T., Kohlbacher, O., Wohlleben, W., & Huson, D. H. (2005). Specificity prediction of adenylation domains in nonribosomal peptide synthetases (NRPS) using transductive support vector machines (TSVMs). *Nucleic Acids Research, 33*(18), 5799–5808. https://doi.org/10.1093/nar/gki885

Rodriguez-Sanchez, L., Carmen Blanco, M., & López-Quintela, M. A. (2000). Electrochemical synthesis of silver nanoparticles. *The Journal of Physical Chemistry B, 104*(41), 9683–88. ACS Publications.

Roy, A., Sharma, A., Yadav, S., Jule, L. T., & Krishnaraj, R. (2021). Nanomaterials for remediation of environmental pollutants. *Bioinorganic Chemistry and Applications*. Hindawi Limited. https://doi.org/10.1155/2021/1764647

Roy, A., & Bharadvaja, N. (2021). Efficient removal of heavy metals from artificial wastewater using biochar. *Environmental Nanotechnology, Monitoring and Management, 16*(December). Elsevier B.V. https://doi.org/10.1016/J.ENMM.2021.100602

Seewagen, C. L. (2020). The threat of global mercury pollution to bird migration: Potential mechanisms and current evidence. *Ecotoxicology, 29*(8), 1254–67. Springer US. https://doi.org/10.1007/s10646-018-1971-z

Shankar, S. S., Ahmad, A., Pasricha, R., & Sastry, M. (2003). Bioreduction of chloroaurate ions by geranium leaves and its endophytic fungus yields gold nanoparticles of different shapes. *Journal of Materials Chemistry, 13*(7), 1822–26. Royal Society of Chemistry.

Sharma, K., Singh, G., Kumar, M., & Bhalla, V. (2015). Silver nanoparticles: Facile synthesis and their catalytic application for the degradation of dyes. *RSC Advances, 5*(33), 25781–88. Royal Society of Chemistry.

Siafaka, P. I., Okur, N. Ü., Karantas, I. D., Okur, M. E., & Gündoğdu, E. A. (2021). Current update on nanoplatforms as therapeutic and diagnostic tools: A review for the materials used as nanotheranostics and imaging modalities. *Asian Journal of Pharmaceutical Sciences, 16*(1), 24–46. Shenyang Pharmaceutical University. https://doi.org/10.1016/J.AJPS.2020.03.003

Singh, A., & Agrawal, M. (2007). Acid rain and its ecological consequences. *Journal of Environmental Biology, 29*(1), 15. Pragati Press-India.

Singh, A. K., Talat, M., Singh, D. P., & Srivastava, O. N. (2010). Biosynthesis of gold and silver nanoparticles by natural precursor clove and their functionalization with amine group. *Journal of Nanoparticle Research, 12*(5), 1667–75. Springer.

Singh, K., Bang, P., Pattebahadur, R., Singh, A., Fitzpatrick, R. W., Raven, M. D., Forrester, S. T., et al. (2022). A systematic approach to soil forensics: Criminal case studies involving transference from crime scene to forensic evidence. *Criminal and Environmental Soil Forensics, 8*(2), 228–32. Shenyang Pharmaceutical University. https://doi.org/10.1007/978-1-4020-9204-6_8

Skoulidas, A. I., Sholl, D. S., & Johnson, J. K. (2006). Adsorption and diffusion of carbon dioxide and nitrogen through single-walled carbon nanotube membranes. *The Journal of Chemical Physics, 124*(5), 054708. American Institute of Physics.

Tchounwou, P. B., Yedjou, C. G., Patlolla, A. K., & Sutton, D. J. (2012). Heavy metal toxicity and the environment. *Molecular, clinical and environmental toxicology* (pp. 133–64). Springer.

Tilley, S. J., Orlova, E. V., Gilbert, R. J. C., Andrew, P. W., & Saibil, H. R. (2005). Structural basis of pore formation by the bacterial toxin pneumolysin. *Cell, 121*(2), 247–56. Elsevier.

Waldman, M., & Shevah, Y. (1993). Biodegradation and leaching of pollutants: Monitoring aspects (technical report). *Pure and Applied Chemistry, 65*(7), 1595–1603. De Gruyter.

Wang, H., Xu, J., Zhu, J., & Chen, H. (2002). Preparation of CuO nanoparticles by microwave irradiation *244*(May), 88–94.

Yusefi, M., Shameli, K., Yee, O. S., Teow, S. Y., Hedayatnasab, Z., Jahangirian, H., Webster, T. J., & Kuča, K. (2021). Green synthesis of Fe3o4 nanoparticles stabilized by a garcinia mangostana fruit peel extract for hyperthermia and anticancer activities. *International Journal of Nanomedicine, 16*, 2515–32. Dove Medical Press Ltd. https://doi.org/10.2147/IJN.S284134

Zhang, L., Zhao, B., Jiang, C., Yang, J., & Zheng, G. (2015). Preparation and transport performances of high-density, aligned carbon nanotube membranes. *Nanoscale Research Letters, 10*(1), 1–8. Springer.

Zhen, X., Liu, L., Wang, X., Zhong, G., & Tang, J. (2019). Fates and ecological effects of current-use pesticides (CUPs) in a typical river-estuarine system of Laizhou Bay, North China. *Environmental Pollution, 252*, 573–79. Elsevier.

Zhu, M., Han, J., Wang, F., Shao, W., Xiong, R., Zhang, Q., Pan, H., Yang, Y., Samal, S. K., & Zhang, F. (2017). Electrospun nanofibers membranes for effective air filtration. *Macromolecular Materials and Engineering, 302*(1), 1600353. Wiley Online Library.

Chapter 15
Genetically Modified Microbes and Their Contribution in the Management of Solid Waste

Yaseen Maqbool Bhat and Gowhar Hamid Dar

Abstract In modern times the production of solid waste is increasing significantly due to different human activities and handling that much of waste is a challenging issue. A lot of waste management options like incineration, landfilling, etc., are already available to deal with such waste but at the same time these have more or less interference with the environment. Where the environment is at risk due to the waste accumulation followed by improper handling of the waste, the field demands an alternative way which could be used as sustainable waste management practice. The issue is supported by molecular biology which has made a way-out for the treatment of solid waste with the help of genetically modified microbes. Genetically modified microbes are the organisms produced through genetic engineering with the potential to degrade many kinds of contaminants and are now being used for waste decomposition through enzymatic action providing a better way for waste reduction without causing any harm to the environment.

Keywords Solid waste · Genetically modified microbes · Decomposition · Biofuels · Biochemical pathway · Enzymatic activity

1 Introduction

Solid waste is characterised as the unusable solid materials produced by human activity in urban, suburban, and rural settings. In the developing world, there is an increasing issue of controlling trash and identifying alternatives to dispose the solid waste especially food waste. The detrimental effects on the environment and human health are lessened or eliminated via solid waste management (Siddiqua et al., 2022). A viable strategy for managing solid waste is to use greenways (Kumari et al., 2022). One of the finest eco-friendly methods for managing solid waste is recycling (Jebaranjitham et al., 2022). By altering the environmental conditions to promote

Y. M. Bhat (✉) · G. H. Dar
Higher Education Department, Sri Pratap College, Jammu and Kashmir, India
e-mail: yaseenba29@gmail.com

R. A. Bhat et al. (eds.), *Zero Waste Management Technologies*,
https://doi.org/10.1007/978-3-031-57275-3_15

the growth of microorganisms and also destroy the target pollutants, bioremediation is a unique green method used to clean polluted media including soil, subterranean material, and solid waste (Mendoza-Burguete et al., 2023). By changing the physical characteristics of dumping sites, greening can be a significant step in the direction of preventing the build-up of solid waste in the region (Abubakar et al., 2022). In a regulated environment, it is the process through which organic wastes breakdown biologically (Lee & Khor, 2023). Another method for allowing microorganisms to break down organic materials both aerobically and anaerobically in controlled environments is composting (Aguilar-Paredes et al., 2023). Nowadays, the development of novel organisms with favourable characteristics that are suitable to the bioremediation of contaminants is the primary objective of genetic engineering procedures (Ayilara & Babalola, 2023). Through genetic manipulation, new strains of microorganisms with desired characteristics are constructed (Astola et al., 2023). The most effective, affordable, and ecologically sustainable method for treating different wastes is to use microorganisms in combination with other biotechnologies (Kumar et al., 2021a, 2021b; Patel et al., 2022). The application of Genetically Engineered Microbes is gaining popularity in managing the problems created due unprecedented production of solid waste (Mohanty et al., 2022).

2 Solid Waste Management

Waste management primarily refers to the handling, storage, collection, and disposal of waste. Since developing countries lack adequate garbage collection and disposal systems, the waste management (WM) scenario differs markedly among developed and developing nations (Debrah et al., 2021). Due to growing concern over environmental deterioration and sustainability, waste management has therefore become extremely important and is given emphasis (Abubakar et al., 2022). Unplanned, haphazard, and unscientific methods of disposing of waste on the urban outskirts and villages lead to overloading landfills which are not only impossible to restore to a suitable condition but also have devastating environmental repercussions in the form of soil and groundwater pollution and a contribution to global climate change (Ferronato & Torretta, 2019). Reducing waste's negative effects and ripple impacts on both the environment and human health is waste management's primary objective (Abubakar et al., 2022). Urban development, the industrialisation, and the booming population demand on natural resources continue to make it a growing challenge that has a negative impact on the environment worldwide (Voumik & Sultana, 2022). The fundamental ideas of waste management have been modified in accordance with the 4R paradigm (refuse, reduce, reuse, and recycle) (Jerin et al., 2022).

Landfills, Incineration, Composting, and Gasification are a few of the routinely adopted waste management procedures. The management of solid waste is a field in which developed nations excel and have set benchmarks. The current scenario mandates an effective or productive waste management system that is also sustainable on an economic and ecological level. There are numerous widely used waste management and treatment strategies that are practised such as the following:

3 Sanitary Landfill

In a sanitary landfill, solid wastes are collected in a trench and covered with dirt every day to allow natural decomposition. This technique is more effective and is used all around the world. The places where garbage is kept separate from the environment are called practically hygienic landfills.

4 Incineration

It is a method in which solid wastes a treated under intense heat. It is believed to be a waste-to-energy procedure.

5 Composting

It is a natural process in which organic solid wastes are broken down in presence of oxygen using microbes.

6 Gasification

It is a procedure in which biomass or solid waste is converted into gases like nitrogen, hydrogen, carbon dioxide, etc.

7 Avoidance and Reduction

Avoidance entails reusing used items, inventing reusable items, and, rather than purchasing new products, mending damaged ones. A decentralised method based primarily on each country's economic situation, solid waste management (SWM) is used to manage waste. In any waste management strategy, regardless of the location or nation, monitoring waste output is the first step. Recently, technologies including geographic information systems (GISs), radio-frequency identification (RFID), ultrasonic sensors, and global system for mobile and general radio packet service (GSM/

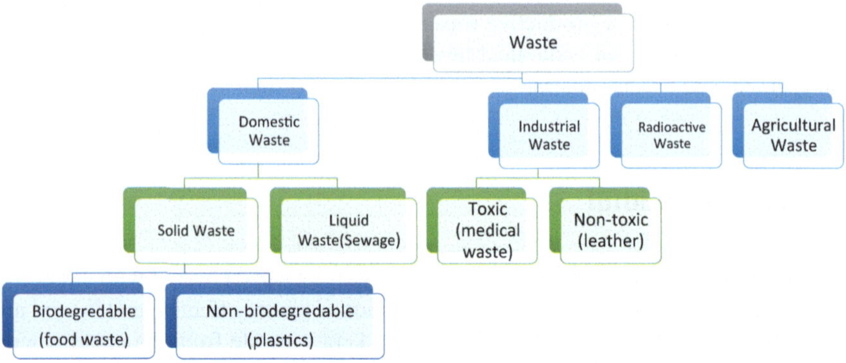

Fig. 1 Comprehensive diagrammatic representation of broad categories of different wastes

GPRS) have been created to improve trash pickup from bins and trucks. Due to their exorbitant cost, most of these cutting-edge procedures are not useful in developing nations. In reality, in roughly 63% of countries, rag-picking by unskilled individuals is the only method employed for waste collection or classification (Figs. 1 and 2).

8 Genetically Modified Microbes

Generally speaking, there are two sorts of microbes: wild-type (WT) species and engineered, or genetically modified microorganisms (GMMs) (Deckers et al., 2020). Although wild-type microorganisms use a small number of carbon sources and produce yields that are far below than those required for commercial viability, they nonetheless have the metabolic pathways present in their original form (Rodríguez et al., 2021). Because of these limitations, a lot of time and effort is put into bioprospecting and screening microbes to find ones with the ability to create unique compounds. The wild-type microorganism's capability for metabolism currently restricts optimization for high yields and carbon source selectivity to boost output (Peña-Castro et al., 2023).

Through the genetic engineering of microorganisms, new biochemical pathways may be added, the organism's overall metabolic performance can be improved, and the desired end products can be chosen (Hanlon & Sewalt, 2021). Increased yield and selectivity for certain products may be achieved for organisms with appropriate manipulation of the genes involved in the biosynthetic pathway and/or other controlled genes and pathways (Antar et al., 2021). In order to overexpress the desired trait, Genetically Engineered Microbes (GEMs) are created by modifying microorganisms with a more powerful protein. The best part of using biotechnological methods is that isolated genes from any organism—bacteria, viruses, fungi, plants, or animals—will work when inserted into selected plant species. Oil spills, camphor, hexane, naphthalene, toluene, octane, xylene, alobenzoates, and trichloroethylene

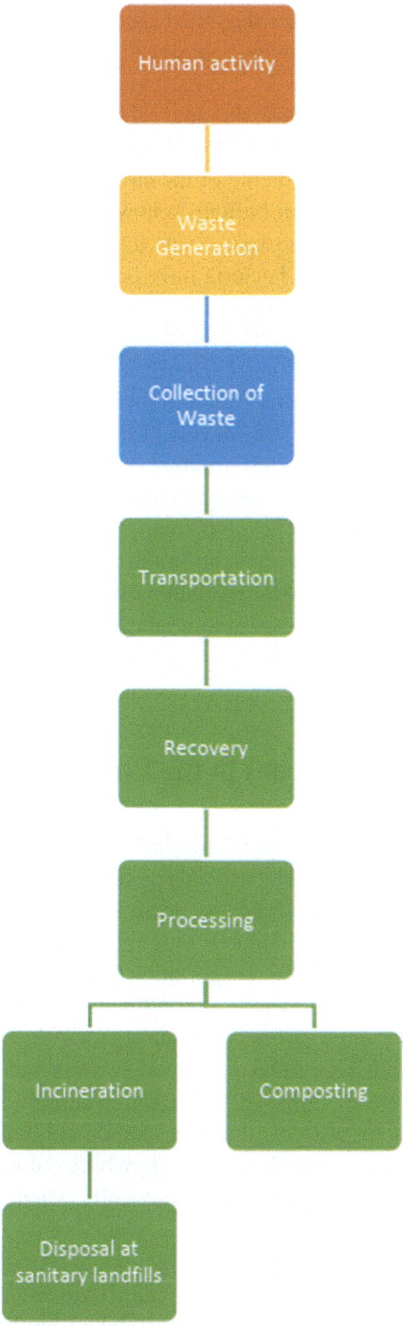

Fig. 2 Elaborative sketch of waste management plan

have all been degraded using Genetically Engineered Microbes (GEMs) such as bacteria, fungus, and algae. These artificially constructed microbes are more potent than their wild counterparts, have more potential for degradation, and can easily respond to a wide range of contaminants as substrates. There is a long and meaningful road ahead for the application of genetic engineering to create these creatures for the well-being of the environment and ultimately for public health.

Molecular biology is now commonly believed to have the potential to be used in creating microorganisms that can perform remedial jobs (Fig. 3). The Genetically Engineered bacteria have been successfully used to degrade a variety of contaminants under controlled settings and have a greater degradative capability. Different kinds of bacteria, including *Ralstonia eutropha*, *Pseudomonas putida*, *Mycobacterium marinum*, *Escherichiacoli*, *Sphingomonas desiccabilis*, and *Bacillus idriensis*, have genes inserted into their genomes that enable them to perform efficient bioremediation. The enzymes LiP, MnP, and lac have the ability to breakdown lignin as reported in fungi strains such as white-rot basidiomycetes, *Phanerochaete chrysosporium* and *Pleurotus ostreatus* (Zhang et al., 2011). In contrast to conventional chemical approaches, enzymatic activity during fungal pre-treatment entirely degrades lignin while preventing the synthesis of phenolic chemicals. Microorganisms may contribute to the oxidation of phenol by discharged fungal enzymes and the consumption of glucose by supplying reducing equivalents to fulfil their energy needs (Zhang et al., 2011).

9 Microbes in the Production of Biofuels from Solid Waste

The most popular biofuel at the moment, bioethanol, is made from either maize or sugar cane and is regarded as a first-generation biofuel (Fdez-Güelfo et al., 2011). However, the decreased energy content and hygroscopic characteristics of bioethanol-blended fuels cause them to have undesirable qualities, which restricts their usage in transportation (Fdez-Güelfo et al., 2011). In addition, it is corrosive and miscible in water. Second-generation biofuels were created as a result of the shortcomings of first-generation renewable fuels. Synthetic paraffinic kerosene (SPK), which are improved liquid aviation and diesel fuels, are created from biomass using Fischer–Tropsch conversion technique (Harvey et al., 2014). The scarcity of feedstock prevents the widespread deployment of second-generation biofuels (Howard et al., 2013). Due to the drawbacks of first- and second-generation biofuels, biofuels made from plentiful, generically rich non-food feedstocks such as organic waste and wastewater have been developed (Howard et al., 2013). Agronomic cellulosic waste and forest waste in particular are receiving more attention from European countries in order to create sustainable biofuels (Jovanovic Tews et al., 2010). Additional feedstock for the production of better biofuels can be obtained from the excessive food waste in Asian nations (Li et al., 2010). Because it contains a significant amount of lignocellulosic fibre, municipal solid waste (MSW) is a promising sustainable biofuel feedstock. When opposed to the manufacture of traditional fuels, biofuels

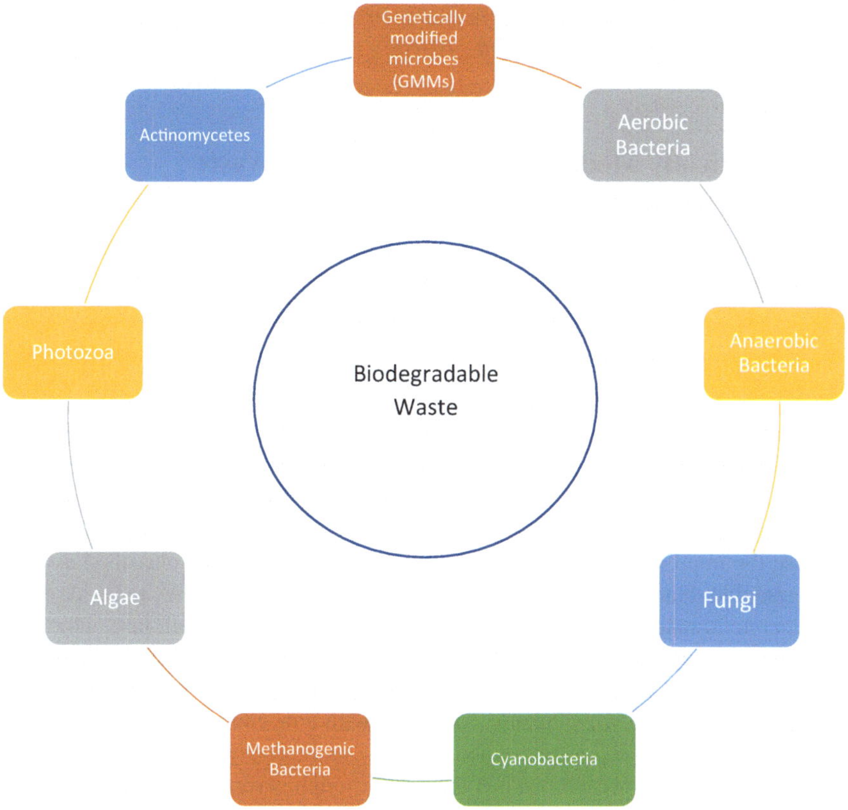

Fig. 3 Difference kinds of microbes in waste degradation

made from the organic fraction of municipal solid waste (OFMSW) may have less of an adverse effect on the environment. Sadly, poor production rates in the present advanced biofuel industry make commercial scaling-up attempts difficult.

Biofuels made by microorganisms can improve productivity while broadening the range of bioenergy molecules generated. Initial leftovers are often produced by seeding local forest products and food crops, such as maize stalks and straw. When harvests are fully developed and processed in their finished state, secondary residues are removed. Agriculture and food waste comes from sources including wood chips, palm kernel cakes, and sugar cane bagasse. After people or animals consume a product made from biomass, tertiary leftovers can be found. In practice, these waste streams occur in the form of wastewater containing soluble organic components or municipal solid waste. Utilising modified microbes will improve the quality of the biofuels produced from these sources of biomass and trash. Recombinant enzymes can be obtained from a variety of microorganisms and can even be created heterologous in a chosen host organism (Liao et al., 2016). The production of several advanced biofuel precursors can be increased by using genetic recombination techniques to introduce specific metabolic pathways into microorganisms like *Escherichia coli* and *Saccharomyces cerevisiae* (Liao et al., 2016).

10 Genetically Modified Microbes in Solid Waste Decomposition

Polyethylene Terephthalate (PET) is one of the most widely utilised plastics for packaging (such as the manufacture of bottles). PET is a polar, linear polymer made of ethylene glycol (EG) and repeating units of aromatic terephthalic acid (TPA). Bis(2-hydroxyethyl) terephthalate is the name of the PET monomer. Due to its superior mechanical and thermal qualities, PET is primarily employed in beverage bottles, foil, textile fibres, and food containers. The breakdown of plastic by Genetically Engineered Microbes has emerged as a possible alternative for the plastic circular economy to address the global issue of unchecked pollution caused by polyethylene terephthalate (PET). Numerous investigations have been carried out recently to increase the microbial ability for PET breakdown. Despite being classified as non-biodegradable, many studies have used microorganisms or enzymes to break down PET. *Ideonella sakaiensis* 201-F6 and the enzymes PETase and MHETase, which are the focus of many scientists due to very promising elements of future management of PET, are a prime example. According to Danso et al., scientists have uncovered 27 enzymes that break down synthetic polymers, with conventional serine hydrolases such cutinases (EC 3.1.1.74), lipases (EC 3.1.1.3), and carboxylesterases (EC 3.1.1.1) being involved in the breakdown of PET. Even though several enzymes are known, there are still many problems with their actual application to break down PET, such as limited thermal stability or transfer on a mass scale. To get over the restrictions, much more research is required on mutational changes of the enzyme's active region. High hydrophobicity, crystallinity, strong chemical bonding, and high molecular weight of plastics are the barriers to biodegradation. The biodegradation processes carried out by wild-type strains, directly isolated from various habitats, particularly from plastic-contaminated sites, were the main focus of the research. Most often, the enzymatic activity of cutinases (EC 3.1.1.74) or PETase (EC 3.1.1.101), with the assistance of MHETase, allows for the biodegradation of PET (EC 2.1.1.102). *Thermobifida fusca* (TfCut2), *Humicola insolens* (HiC), and *Ideonella sakaiensis* (PETase and MHETase), among others, all include cutinases with PET biodegradable activity. Cutinases are used extensively in studies of the breakdown of a wide variety of plastic polymers because they can hydrolyse both ester linkages found in aliphatic and aromatic polyesters. PETase, however, can hydrolyse ester linkages found solely in aromatic polyesters. PET-degrading enzymes are members of the esterase subclass and have the catalytic triad found in hydrolases (Ser-His-Asp). Numerous studies have recently been carried out to enhance and better understand the mechanisms of action of these enzymes, particularly PETase or MHETase. MHETase hydrolyses MHET into terephthalic acid (TPA) and ethylene glycol(EG), whereas PETase hydrolyses PET into oligomers of mono-2-hydroxyethyl terephthalate (MHET). Therefore, recombination and overexpression of those enzymes may be essential for more effective PET breakdown as well as monomer recycling (TPA and EG) and for bioconversion to high-value chemicals.

The scientific community is now considering this global issue as a result of the increasing volume of plastic garbage. According to published findings, naturally isolated bacteria can only degrade a small amount of plastic, and it might take decades for microbes to evolve the ability to utilise plastic as a carbon source. The published results on genetically altered microbes or chemically designed enzymes indicate that this area offers a promising strategy for managing plastic waste. Today, we can avoid cofactor supplementation with the simultaneous multifold improvement in their activity, decrease in reaction time, and increase in their thermal stability by chemically engineering enzymes, such as by changing the active site or by introducing new bonds. The idea of nano-immobilising enzymes to increase their endurance to temperature and pH is a really intriguing method that can be used to PET-degrading technology in future. Recently, the first attempts to employ immobilised enzymes were successful. The potential for recycling PET breakdown by products like EG and TPA to create new PET with qualities comparable to virgin PET is another viewpoint that should be discussed. As a result, there will be less need for fossil substrates to make plastic, and at the same time, this could be a field in which using immobilised enzymes can have further advantages. The long-known bioremediation, which appears to be a distant goal when utilising GMMs, may eventually be replaced by safer and simpler to manage modified proteins due to restrictions and worries over the usage of GMOs globally. One potential future method to lower the level of contamination may involve the use of improved mutants of PETase cutinase, and to our knowledge, this application is one of the most recent advancements in this sector.

Table showing different microbes used in biodegradation

Microorganism(s)	Waste material	References
Bacillus, *Staphylococcus*	Endosulfan	Mohamed et al.
Acinetobacter sp., *Microbacterium*	Aromatic hydrocarbons	Simarro et al.
Gloeophyllumtrabeum, *Trametes versicolor*	Hydrocarbons	Karigar and Rao
Pseudomonas putida	Benzene and xylene	Safiyanu et al.
Pseudomonas sp., *Micrococcusluteus*, *Bacillus subtilis*, *Streptococcus lactis*	Degradable plastic	Priyanka and Archana
Corynebacterium sp., *Pseudomonas* sp., *Arthrobacter globiformis*, *Bacillus* sp.	Polyurethane	Blake and Howard; El-Sayed et al.; Howard et al. and Kay et al.
Bacillus sp., *Micrococcus* sp., *Vibrio* sp., *Arthrobacter* sp., *Pseudomonas* sp.	High-density polyethylene (HDPE)	Balasubramanian et al.; Fontanella et al. and Kumar et al.

(continued)

(continued)

Microorganism(s)	Waste material	References
RhodococcusruberC208, *Brevibacillusborstelensis707*, *RhodococcusruberC208*, *Staphylococcus epidermidis*, *Bacillus cereus* C1	Low-density polyethylene(LDPE)	Chandra and Rustgi; Chatterjee et al.; Hadad et al.; Sharma and Sharma and Sivan et al.
Aspergillus flavus, *Mucor rouxii*	Disposable plastic films	El-Shafei et al.
Phanerochaetechrysosporium, *Lentinustigrinus*, *Aspergillus niger*, *A. sydowii*	PVC films	Ali
Streptomyces sp., *Sporichthya* sp., *Actinoplanes* sp.	Polyethylene	Sathya et al.

11 Challenges for Genetic Engineering in Solid Waste Management

The potential for environmental applications of genetically modified microorganisms is enormous (Rafeeq et al., 2023). The unforeseen negative effects of some, though, might be problematic (Qian et al., 2022). As self-replicating organisms, germs make it challenging to stop an unfavourable outcome by simply halting more organism discharges. One typical issue is that once released into the environment, recombinant genes can spread from their host bacterium to other bacteria (Bodor et al., 2020; Imron et al., 2020; Varjani et al., 2020). There has been research on horizontal gene transfer (via transformation, transduction, and conjugation) in many environmental settings (Searle & Malins, 2016). According to the findings, inter-specific horizontal gene transfer has been detected for numerous bacterial species, numerous genes, and in numerous different environmental settings. Estimates of low, or non-detectable, DNA transfer in laboratory testing may be misleading due to the potentially enormous number of bacteria expelled as well as their protracted rate of reproduction (Varjani et al., 2020). On the other hand, a number of strategies have been created to lessen horizontal gene transfer to other bacteria (Searle & Malins, 2016; Secretariat of the Convention on Biological Diversity, 2000; Sims et al., 2010).

However, it has been claimed that genetically altered bacteria will be weak competitors and unable to live in the wild due to energy inefficiency, disruption of genomic coadaptation, or domestication. Numerous studies have confirmed the notion that genetically altered microorganisms are less competitively fit than their ancestors. Nevertheless, there are a few significant outliers where genetic modifications surprisingly enhance competitive fitness. A modified species or its planned genes may have a higher chance of survival if certain genes' unfavourable effects are removed by subsequent evolution. A genetically modified microbe's or its designed genes' survival probability assessment is a challenging ecological and evolutionary task. Therefore, an efficient regulatory framework would only require such analyses in the event that there are strong indications that there could be serious adverse environmental effects. When creating microbial biofuel production processes that incorporate genetically modified microbes, the Cartagena Protocol on Biosafety, the Convention on Biological Diversity (Tripathi et al., 2019) and the International Plant Protection Convention (IPPC) all of which address the environmental aspects of Genetically Modified Organisms (GMOs), should be adhered.

12 Conclusion

Around the world, solid waste has grown to be a serious problem. Cities routinely have trash piles by roads, rivers, and several other open spaces, which is having a negative impact on the environment and public health. The top three nations that create municipal solid trash are the USA, China, and India. The main areas that require a focus of emphasis in order to achieve sustainability are waste management, preservation of natural resources and biodiversity, and treatment of toxins or pollutants. Recycling, incineration, waste-to-energy conversion, composting, or landfilling are all methods of managing municipal solid waste. In many towns around the world, landfilling is recommended for the disposal of solid waste. Waste undergoes physical, chemical, and biological transformations in landfills, which function as ecological reactors. In comparison with previous methods, using engineered microorganisms can offer a considerably safer and more affordable solution. Genetically Engineered Microbes (GEMs) are created by introducing a more robust protein into bacteria through genetic engineering and biotechnology in order to overexpress the desired trait. In order to digest oil spills, camphor, hexane, naphthalene, toluene, octane, xylene, halobenzoates, trichloroethylene, etc., Genetically Engineered Microbes (GEMs) such as bacteria, fungi, and algae have been used. These engineered microorganisms have better degradative capacities than natural strains and may quickly adapt to a variety of contaminants as substrates. The adoption of genetic engineering to create these organisms for the protection of the environment and, ultimately, public health, has a long and worthy road ahead of it. Finally, it can be said that the application of microorganisms and microbiological methods has opened up fresh perspectives in the domain of sustainable development, notably in the fields of environment and other significant environmental challenges.

13 Future Prospective

The published results on genetically altered microbes or chemically designed enzymes indicate that this area offers a promising strategy for managing plastic waste. Today, we can avoid cofactor supplementation with the simultaneous multifold improvement in their activity, decrease in reaction time, and increase in their thermal stability by chemically engineering enzymes, such as by changing the active site or by introducing new bonds. The idea of nanoimmobilising enzymes to increase their endurance to temperature and pH is a really intriguing method that can be used to PET-degrading technology in the future. A great deal of pressure has been placed on waste management globally by rapid population expansion, urbanisation, and industry. Therefore, care and sustainable development should be the guiding principles of the waste management. For a sustainable environment and development, waste management is gaining popularity all over the world. Prior to being disposed of in landfills, trash must first be separated from its source and put through a variety of recovery and recycling processes, which is the key to effective waste management. Since biotechnology and microorganisms can be employed successfully in waste management, more study and information will be generated in relation to their usage and function in waste management. Garbage production could be reduced by composting waste and turning it into electricity. The issues are thought to be significantly lessened if the waste's biodegradation component is somehow isolated. A long-term methodology and strategy should be developed and revised in the future to accommodate evolving lifestyles. Effective waste management is directly impacted by community involvement and knowledge advancement in the area of sustainability. From very small scale to large size, the community must establish the concept of recycling and reuse. The development and use of efficient Genetically Engineered Microbes will prove to be a game changer in coming time as the economical and ecologically sound products are produced through the application of Genetically Engineered Microbes on solid waste materials.

References

Abubakar, I. R., Maniruzzaman, K. M., Dano, U. L., AlShihri, F. S., AlShammari, M. S., Ahmed, S. M. S., Al-Gehlani, W. A. G., & Alrawaf, T. I. (2022). Environmental sustainability impacts of solid waste management practices in the global south. *International Journal of Environmental Research and Public Health, 19*(19), 12717. https://doi.org/10.3390/ijerph191912717. Published 2022 October 5.

Aguilar-Paredes, A., Valdés, G., Araneda, N., Valdebenito, E., Hansen, F., & Nuti, M. (2023). Microbial community in the composting process and its positive impact on the soil biota in sustainable agriculture. *Agronomy, 13*(2), 542. https://doi.org/10.3390/agronomy13020542

Antar, M., Lyu, D., Nazari, M., Shah, A., Zhou, X., & Smith, D. L. (2021). Biomass for a sustainable bioeconomy: An overview of world biomass production and utilization. *Renewable and Sustainable Energy Reviews, 139*, 110691.

Astola, A., Durán-Guerrero, E., Díaz, A. B., Lasanta, C., & Castro, R. (2023). Impact of the genetic improvement of fermenting yeasts on the organoleptic properties of beer. *European Food Research and Technology, 249*, 1677–1687. https://doi.org/10.1007/s00217-023-04251-8

Ayilara, M. S., & Babalola, O. O. (2023). Bioremediation of environmental wastes: the role of microorganisms. *Plant-Soil Interactions, 5*. https://doi.org/10.3389/fagro.2023.1183691

Bodor, A., Bounedjoum, N., Vincze, G. E., Erdeiné Kis, Á., Laczi, K., Bende, G., Szilágyi, Á., Kovács, T., Perei, K., & Rákhely, G. (2020). Challenges of unculturable bacteria: Environmental perspectives. *Reviews in Environmental Science and Bio/technology, 19*, 1–22.

Debrah, J. K., Vidal, D. G., & Dinis, M. A. P. (2021). Raising awareness on solid waste management through formal education for sustainability: A developing countries evidence review. *Recycling, 6*(1), 6. https://doi.org/10.3390/recycling6010006

Deckers, M., Deforce, D., Fraiture, M. A., & Roosens, N. H. C. (2020). Genetically modified microorganisms for industrial food enzyme production: An overview. *Foods (basel, Switzerland), 9*(3), 326. https://doi.org/10.3390/foods9030326

Fdez-Güelfo, L., Álvarez-Gallego, C., Sales, D., & Romero, L. (2011). The use of thermochemical and biological pretreatments to enhance organic matter hydrolysis and solubilization from organic fraction of municipal solid waste (OFMSW). *Chemical Engineering Journal, 168*, 249–254.

Ferronato, N., & Torretta, V. (2019). Waste mismanagement in developing countries: A review of global issues. *International Journal of Environmental Research and Public Health, 16*(6), 1060. https://doi.org/10.3390/ijerph16061060

Hanlon, P., & Sewalt, V. (2021). GEMs: Genetically engineered microorganisms and the regulatory oversight of their uses in modern food production. *Critical Reviews in Food Science and Nutrition, 61*(6), 959–970. https://doi.org/10.1080/10408398.2020.1749026

Harvey, B., Meylemans, H., Gough, R., Quintana, R., Garrison, M., & Bruno, T. (2014). High-density biosynthetic fuels: The intersection of heterogeneous catalysis and metabolic engineering. *Physical Chemistry Chemical Physics: PCCP, 16*, 9448–9457.

Howard, T. P., Middelhaufe, S., Moore, K., Edner, C., Kolak, D. M., Taylor, G. N., Parker, D. A., Lee, R., Smirnoff, N., Aves, S. J., et al. (2013). Synthesis of customized petroleum-replica fuel molecules by targeted modification of free fatty acid pools in *E. coli. Proceedings of the National Academy of Sciences of the United States of America, 110*, 7636–7641.

Imron, M. F., Kurniawan, S. B., Ismail, N. I., & Abdullah, S. R. S. (2020). Future challenges in diesel biodegradation by bacteria isolates: A review. *Journal of Cleaner Production, 251*, 119716.

Jebaranjitham, J. N., Christyraj, J. D. S., Prasannan, A., Rajagopalan, K., Chelladurai, K. S., Gnanaraja, J. K. J. S. (2022). Current scenario of solid waste management techniques and challenges in Covid-19—A review. *Heliyon, 8*(7), e09855. https://doi.org/10.1016/j.heliyon.2022.e09855. Published 2022 July 2.

Jerin, D. T., Sara, H. H., Radia, M. A., Hema, P. S., Hasan, S., Urme, S. A., Audia, C., Hasan, M. T., & Quayyum, Z. (2022). An overview of progress towards implementation of solid waste management policies in Dhaka, Bangladesh. *Heliyon, 8*(2), e08918. https://doi.org/10.1016/j.heliyon.2022.e08918

Jovanovic Tews, I., Jones, S. B., Santosa, D. M., Ziyu, D., Ramasamy, K., & Zhu, Y. (2010). A survey of opportunities for microbial conversion of biomass to hydrocarbon compatible fuels. *Pacific Northwest National Laboratory, 19704*, 01–33.

Kumar, A., Sharma, A., Chaudhary, P., & Gangola, S. (2021a). Chlorpyrifos degradation using binary fungal strains isolated from industrial waste soil. *Biologia, 76*(10), 3071–3080. https://doi.org/10.1007/s11756-021-00816-8

Kumar, G., Lal, S., Maurya, S. K., Bhattacherjee, A., Chaudhary, P., Gangola, S., et al. (2021b). Exploration *of Klebsiella pneumoniae* M6 for paclobutrazol degradation, plant growth attributes, and biocontrol action under subtropical ecosystem. *PLoS ONE, 16*(12), e0261338. https://doi.org/10.1371/journal.pone.0261338

Kumari, A., Aich, A. R., Kumari, S., Mohanty, S. (2022). Greenways for solid waste management. In C. Baskar, S. Ramakrishna, S. Baskar, R. Sharma, A. Chinnappan, & R. Sehrawat, (Eds.), *Handbook of solid waste management*. Springer. https://doi.org/10.1007/978-981-16-4230-2_8

Lee, B. H., Khor, S. M. (2023). Biodegradation versus composting. In: G. A. M. Ali, & A. S. H. Makhlouf (Eds.), *Handbook of biodegradable materials*. Springer. https://doi.org/10.1007/978-3-031-09710-2_69

Li, Q., Cai, H., Hao, B., Zhang, C., Yu, Z., Zhou, S., & Chenjuan, L. (2010). Enhancing clostridial acetone-butanol-ethanol (ABE) production and improving fuel properties of ABE-enriched biodiesel by extractive fermentation with biodiesel. *Applied Biochemistry and Biotechnology, 162*, 2381–2386.

Liao, J., Mi, L., Pontrelli, S., & Luo, S. (2016). Fuelling the future: Microbial engineering for the production of sustainable biofuels. *Nature Reviews Microbiology, 14*, 288–304.

Mendoza-Burguete, Y., de la Luz, P.-R., Ledesma-García, J., Campos-Guillén, J., Ramos-López, M. A., Guzmán, C., & Rodríguez-Morales, J. A. (2023). Global situation of bioremediation of leachate-contaminated soils by treatment with microorganisms: A systematic review. *Microorganisms, 11*(4), 857. https://doi.org/10.3390/microorganisms11040857

Mohanty, S., Saha, S., Santra, G. H., & Kumari, A. (2022). Future perspective of solid waste management strategy in India. In: C. Baskar, S. Ramakrishna, S. Baskar, R. Sharma, A. Chinnappan, & R. Sehrawat, (Eds.), Handbook of solid waste management. Springer, Singapore. https://doi.org/10.1007/978-981-16-4230-2_10

Patel, A. K., Singhania, R. R., Albarico, F. P. J. B., Pandey, A., Chen, C. W., & Dong, C. D. (2022). Organic wastes bioremediation and its changing prospects. *Science of the Total Environment, 824*, 153889. https://doi.org/10.1016/j.scitotenv.2022.153889

Peña-Castro, J. M., Muñoz-Páez, K. M., Robledo-Narvaez, P. N., & Vázquez-Núñez, E. (2023). Engineering the metabolic landscape of microorganisms for lignocellulosic conversion. *Microorganisms, 11*(9), 2197. https://doi.org/10.3390/microorganisms11092197

Qian, X., Lü, Q., He, X., Wang, Y., Li, H., Xiao, Q., Zheng, X., & Lin, R. (2022). *Pseudomonas* sp. TCd-1 significantly alters the rhizosphere bacterial community of rice in Cd contaminated paddy field. *Chemosphere, 290*, 133257.

Rafeeq, H., Afsheen, N., Rafique, S., Arshad, A., Intisar, M., Hussain, A., Bilal, M., Iqbal, H. M. N. (2023). Genetically engineered microorganisms for environmental remediation. *Chemosphere, 310*. https://doi.org/10.1016/j.chemosphere.2022.136751

Rodríguez, J. E. G., Brojanigo, S., Basaglia, M., Favaro, L., & Casella, S. (2021). Efficient production of polyhydroxybutyrate from slaughterhouse waste using a recombinant strain of *Cupriavidus necator* DSM 545. *Science of the Total Environment, 794*, 148754.

Searle, S., & Malins, C. (2016). Waste and residue availability for advanced biofuel production in EU member states. *Biomass and Bioenergy, 89*, 2–10.

Secretariat of the Convention on Biological Diversity. (2000). Cartagena protocol on biosafety to the convention on biological diversity: Text and annexes. *Montreal, 2000*, 1–19. ISBN 92-807-1924-6.

Siddiqua, A., Hahladakis, J. N., & Al-Attiya, W. A. K. A. (2022). An overview of the environmental pollution and health effects associated with waste landfilling and open dumping. *Environmental Science and Pollution Research, 29*, 58514–58536. https://doi.org/10.1007/s11356-022-21578-z

Sims, R., Mabee, W., Saddler, J., & Taylor, M. (2010). An overview of second-generation biofuel technologies. *Bioresource Technology, 101*, 1570–1580.

Tripathi, N., Hills, C. D., Singh, R. S., & Atkinson, C. J. (2019). Biomass waste utilisation in low-carbon products: Harnessing a major potential resource. *Climate and Atmospheric Science, 2*, 35.

Varjani, S., Rakholiya, P., Ng, H. Y., You, S., & Teixeira, J. A. (2020). Microbial degradation of dyes: An overview. *Bioresource Technology, 314*, 123728.

Voumik, L. C., & Sultana, T. (2022). Impact of urbanization, industrialization, electrification and renewable energy on the environment in BRICS: Fresh evidence from novel CS-ARDL model. *Heliyon, 8*(11), e11457. https://doi.org/10.1016/j.heliyon.2022.e11457

Zhang, F., Rodriguez, S., & Keasling, J. (2011). Metabolic engineering of microbial pathways for advanced biofuels production. *Current Opinion in Biotechnology, 22*, 775–783.

Chapter 16
Biostimulation of Microbes for Enhanced Oil Removal from Petroleum Hydrocarbon-Contaminated Soils: A Zero Waste Remediation Approach

J. A. Ruley and B. M. Tomor

Abstract The discovery of oil and its consequent exploitation have changed several dynamics in development of several countries, developed and developing alike. The critical importance of oil economy and purported developments have been greatly welcomed with optimism to the extent that major media houses at regional and international levels have devoted consideration space to cover the events. Despite this glorification, the oil economy also comes with pitfall of oil curse characterized by power struggles and collateral damage on the environment. The poking of the environment with several depressions and creation of spoil heaps has caused notable destruction of the environment's scenic beauty. Besides, the dynamics of the ecosystem that are also grossly affected include mortality of plants, livestock and humans. In this chapter, we present opportunities that have been unfolded through scientific innovations which are capable of mitigating the scotching effects of crude oil exploitation on the environment. We provide a thorough exposition of strides made in ecorestoration of crude oil-contaminated soils as opposed to the use of physical, chemical and electrical approaches of contaminated soil clean-up that have commonly been in use before. We zero on biostimulation as a novel strategy that is undertaken in the biological clean-up of polluted soil environments and highlight practicalities of deployment of animal dung as a zero remediation approach. We underscore the utility of cow dung as a catalyst for faster bioremediation of crude oil-contaminated soils. We also illustrate the chapter with notable studies in the recent past that have proven that biostimulation with animal dung is a sure way of cleaning up crude oil-contaminated soils. Observably, we qualify biostimulation with cow dung as a priority approach that ought to be given due attention by environmentalists as a credible zero remediation strategy.

J. A. Ruley (✉)
Department of Agricultural Sciences, University of Juba, P.O. Box 82, Juba, South Sudan
e-mail: janenajeb@yahoo.com

B. M. Tomor
Department of Wildlife, University of Juba, P.O. Box 82, Juba, South Sudan

R. A. Bhat et al. (eds.), *Zero Waste Management Technologies*,
https://doi.org/10.1007/978-3-031-57275-3_16

Keywords Soil · Hydrocarbons · Ecorestoration · Crude oil · Biostimulation · Bioremediation

1 Introduction

Soil is a base for man's survival, plants and biodiversity (International Soil Reference and Information Centre [ISRIC], 2023; Javed et al., 2022; Pinto et al., 2023). It is a critical natural resource that supports several environmental and agricultural functions such as filtering pollutants, supporting plant growth and food production. Other functions of soil include regulating hydrologic and bio-geochemical cycles, a natural conduit for recycling organic matter, mitigates global climatic change and is a reservoir of biodiversity (Elmi et al., 2023; Rezanezhad et al., 2023). These multifarious functions make soil a critical component of mother earth. However, the full realization of these benefits rests on preservation and conservation of soil as a key element of environmental systems. That stated, soil pollution protection is fundamental and an indispensable goal that leadership of countries should underline in their plans both in the short term and long run. This need is universal for both developed and developing countries. Despite this universality, the urgency is high in crude oil producing environments (Ruley et al., 2020a, 2020b, 2022). Olawale et al. (2020) note that hydrocarbon-contaminated soils pose a big threat to human life, animal life, soil macrobiota (moles, rabbits and other soil burrowing animals) and microbiota (bacteria, archaea, protista, fungi and viruses) and aquatic animals. The gross damage to local ecosystems is largely manifested in rampant spread of diseases, death and mutation of the progeny (Olawale et al., 2020).

Petroleum hydrocarbons (PHC) are majorly categorized into four structures: saturates, aromatics, asphaltenes and resins (Muneeswari et al., 2022; Rajasekaran et al., 2023). These enter the soil ecosystem through accidental or operational discharges, leaking underground storage tank (LUST) or accidental release of a fuel product from underground storage tank (UST). The soil ecosystem is the ultimate sink of hydrocarbon compounds such as BTEX, Aliphatic and PAHs (Ehiomogue et al., 2021; Manian, 2023). Petroleum hydrocarbon (PHC) pollutants originate from fractions of wastewater and oily sludge produced during extraction, distribution, dumping or storage (Pichtel, 2016). According to Environmental Dynamics International (EDI) (2021), about 3 to 5 barrels of wastewater are generated per barrel of crude oil. In crude oil producing environments, the contamination of soil also arises from secondary crude oil activities. These include oil spills during transportation, installation of refining plants and operational activities (Selva Filho et al., 2023). The contamination of soil occurs when either used or virgin oil enters soil ecosystem (Abdulyekeen et al., 2022). The oil-ridden soil ecosystem becomes toxic and becomes detrimental to plant, human and animal lives (Abdulyekeen et al., 2022). The discharge of complex and diverse petroleum organic and inorganic compounds introduces PHC into soil environment leading to degradation (Ruley, 2022). Over time, PHC contamination of soil leads to carcinogenic problems to humans (Gao et al., 2022;

Verma et al., 2023). Petroleum degraded soil inhibits plant growth by disrupting soil–water-plant relationships (Ruley, 2022) and eventually leads to a decline in agricultural productivity. PHC contamination also affects livestock and wildlife by altering growth of natural pastures (Chilvers & McClelland, 2023; Otten et al., 2023). Consequently, remediation measures to reclaim PHC-contaminated soils in order to restore ecosystem services for sustainable development have become a global matter of urgent concern.

PHC polluted soils contain persistent organic pollutants, halogenated organic chemicals and heavy metals and or toxic metalloids. In absolute and relative terms, these categories of pollutants affect human life and biodiversity (Otten et al., 2023). According to Xue et al. (2023) and Ali et al. (2023), PHC contamination largely alters soil redox potential ratio leading to increase pH. As a result, soil biota is suffocated while the toxic conditions created by the high pH content also affects the soil biological processes of biota as well as their survival (Atai et al., 2023). Consequently, PHC-contaminated soils have to be remediated in order to restore ecosystem services for sustainable development. Contaminated soils can be rehabilitated (Koolivand et al., 2020; Kuyukina et al., 2020; Okpashi et al., 2020). Several remediation technologies exist (Vidonish et al., 2016). Binary categorization is used in this chapter for conventional and biological remediation approaches.

2 Conventional Remediation Approaches for Petroleum Hydrocarbon-Contaminated Soil

Several technologies exist for remediation of PHC-contaminated soils (Akram & Deka, 2021; Hoang et al., 2021; Kiarmasi et al., 2020). One of the ways through which contaminated soils can be rehabilitated is by deployment of conventional engineering techniques, which are physical, chemical, electrical and thermal in nature (Ambaye et al., 2023; Xue et al., 2023). Physical remediation involves removal of soil-contaminated sites (soil excavation), deep burial (land filling) and encapsulation (Blenis et al., 2023). Chemical remediation entails direct injection of oxidants or surfactants into contaminated soil to either wipe them out or convert them into less toxic compounds (Cao et al., 2023; Yuan et al., 2023). For electrical remediation, electrical principles are applied to decontaminate particular sites though only limited to granular type of soil contaminated with heavy metals. The thermal methods are largely used in environments where the contaminants are highly volatile and include in situ vitrification, incineration and electrical pyrolysis (Haokip et al., 2023; Homayounpour, 2023).

3 Environmental Hazards Posed by Conventional Remediation Approaches

Although conventional methods described in the foregoing section have been widely used, they are very expensive (Pothula et al., 2023; Rafeeq et al., 2023; Ruley, 2022). Besides, they are difficult to implement in environments with large expanses of contaminated soil. Additionally, they require use of heavy equipment and substantial amounts of energy which expose soil to secondary degradation and contamination. The secondary contaminants have a permanent effect on soil properties. Also, they destroy biodiversity ultimately rendering soil unsuitable for plant growth (Feepee et al., 2023; Izah & Aigberua, 2023). Consequently, there is urgency for adopting cost-effective and less destructive biological remediation.

4 Biological Remediation of PHC-Contaminated Soil

Biological methods use living organisms to biodegrade contaminants (Koolivand et al., 2020; Kuyukina et al., 2020; Okpashi et al., 2020). The biological remediation techniques such as bioremediation (Hauptvogl et al., 2020). Bioremediation techniques were pioneered in 1930s for restoring hydrocarbon-contaminated soils on a large scale (Borah et al., 2023; Scully & Orlygsson, 2023). Bioremediation refers to the use of biologically mediated processes to detoxify, degrade or transform pollutants to an innocuous state (King et al., 2023; Ruley et al., 2022). Bioremediation, a short form for biological remediation, is a branch of biotechnology that uses biological processes to overcome environmental problems. The mechanism involves use of biological organisms such as plants and microbes (Pande et al., 2020). It is preferred over traditional remediation because it is very safe and cost-effective (Sharma, 2020).

In contraposition to conventional physicochemical remediation techniques, bioremediation is less invasive and more respectful of the soil environment and its biota (Lacalle et al., 2020). Given its environmental friendliness, Lacalle et al. (2020) called bioremediation techniques gentle remediation options (GROs). In recent years, bioremediation is the most commonly used method for restoration of contaminated soil. Several studies have identified various phytoremediation species (de Souza et al., 2023; Ruley, 2022; Thalassinos et al., 2023). Similarly, involvement of a multitude of bacteria and fungi in bioremediation process has been documented (Masotti et al., 2023; Thacharodi et al., 2023; Vaksmaa et al., 2023). The combined action of soil microorganisms and plants presents an effective bioremediation strategy (Masotti et al., 2023; Ruley et al., 2020a, 2020b, 2020c). Microorganisms break down organic pollutants and use them as carbon and energy sources (Ruley, 2022). With exception of heavy metals, microorganisms' biodegrade, transform or chelate various toxic compounds (Ruley et al., 2023). Although microorganisms fail to fully biodegrade heavy metals, they reduce on their water solubility consequently transforming them into less toxic products (Thai et al., 2023).

Much as bioremediation is a strategic eco-restoration approach in which the agents (plants and soil microorganisms) mineralize and detoxify the PAHs to innocuous compounds such as water and carbon dioxide (El-Gendy & Nassar, 2021), this process of biodegradation is a little slow. Scientific discoveries and proven potential of deployment of microorganisms for environmental clean-up spanning over 50 years (Hidalgo et al., 2022) have shown that to harness their potentiality, efforts must be made to provide microorganism with favorable conditions (Hidalgo et al., 2022). This accounts for why biostimulation is encouraged (El-Gendy & Nassar, 2021) because usually, PHC-contaminated soils are deficient in macro- and micro-nutrients (Ruley, 2022). Resultantly, the soil cannot support healthy plant growth and survival of diverse microorganisms. It is therefore critical to bio-stimulate-contaminated sites in order to improve on soil quality and microbial populations before introducing phytoremediation species (Ruley, 2022). Biostimulation will increase indigenous microbial activity by providing limiting nutrients (Landaeta-Hernández et al., 2023; Nivetha et al., 2023).

5 Biostimulation of PHC Degrading Microorganisms with Animal Dung

Biostimulation is the modification of physical and chemical parameters of soil (Nivetha et al., 2023; Ruley et al., 2020a). It involves addition of nutritional amendments (N and P), oxygenation, temperature and pH control and addition of surfactants in contaminated sites to stimulate growth of indigenous hydrocarbon biodegraders while also boosting biodegradation rates (Lima et al., 2019). Bioremediation is enhanced by addition of nutrients (e.g., biogas slurry, manure, spent mushroom compost, rice straw and corncob) or electron acceptors (phosphorus, nitrogen, oxygen, carbon). Animal dung is ubiquitous and richer in soil nutrients compared to other biostimulants. Biostimulation involves the use of animal dung to activate, amplify and square the potentials of plants to biodegrade petroleum hydrocarbons from soil ecosystems (Landaeta-Hernández et al., 2023; Ruley et al., 2020a).

Animal dung is excreta from animals. They are waste made up of undigested or partially digested residue of consumed food material (Vendatu, 2023). It is composed of a mixture of both feces and urine in the ratio of 3:1. The dung is collected from poultry, sheep, goat, pig and dairy farms from cows. For quite a long time, these dung have been conceived as waste posing a big problem to farmers regarding disposing them of. However, extant studies have confirmed that animal dung is rich in protein, energy, minerals and vitamins that known for conditioning the soil (Vendatu, 2023). Given several effects of animal dung on soil environment, environmentalists regard them as 'gold mine'. In this chapter, most reference is made of cow dung. Some of its major ingredients include lignin, cellulose and hemicellulose (Ding et al., 2023; Yang et al., 2023). Besides these fibers, animal dung is rich in many minerals. Notably, it has high contents of N, P (Fan et al., 2023; Liu & Zhang, 2023; Rout et al., 2023) and trace amounts of S, Fe, Mg, Cu, Co and Mn (Appiagyei Osei-Owusu et al., 2023; Beef Cattle Research Council of Canada [BCRC], 2023). Cow dung also contains environmentally beneficial microorganisms (Noviardi et al., 2023). Almost cost-free, some animal dung such as cow dung is largely and widely available in the environment (Kusmiyati et al., 2023). Lucky enough, all oil producing countries are endowed with cattle and therefore are blessed with abundance or prevalence of cow dung as illustrated in Table 1.

In the listed countries above, save for Muslim countries such as Iraq, Iran, Saudi Arabia and some parts of Nigeria, Pig dung is also used for biostimulation. Universally, there exists millions of heads of sheep, goats and chicken in both countries. These generate great amounts of animal dung. However, the most dominant animal dung is cow dung. Overall, Zhu et al. (2020) generalize the cow dung yield potential to an average of 25 kg per day. Lactating cows produce higher amounts of manure than milk to the extent that the quantity of cow dung produced is about three times more than the milk produced. The key reason for lactating cows producing more cow dung is largely because they are grazed in areas with better pastures, feeding and lounging barns as well as milking parlors where they are fed on concentrates and silage for higher milk production. Besides feeding on forages, carbohydrates, mineral fortified feeds and vitamins accounts for high production capacity of cow dung (Erickson & Kalscheur, 2020). One other unique attribute is that lactating cows especially those under managed grazing are fed on Total Mixed Ration (TMR), eat approximately 7% more nitrogen (N) and P and have a high propensity of excreting

Table 1 Cattle population in world crude oil producing countries

Country	Global rank in crude oil	Estimated heads (in millions)	Source(s)
China	7th	102.16	Textor (2023)
USA	1st	89.3	Cook (2020); USDA (2023)
Sudan	35th	107	Xinhua(2021)
Nigeria	13th	21.26	Sasu (2023)
Russia	3rd	17.9	Statista Research Department (2023)
Canada	8th	3.5	Shahbandeh (2023)
South Sudan	35th	49	Taha et al. (2023)
Iran	6th	10	Gharekani et al. (2020)
Angola[a]	14th	4.8	BBC (2020)
Iraq	5th	1.5	World Organization for Animal Health (2023)
Norway	15th	0.3	Ridder (2023)
Saudi Arabia	2nd	0.7	Puri-Marza (2023)

[a] Cattle population set to increase in the next six years given the increase by 75,000 heads that Chad government is paying to repay the loan of USD 100 Million (BBC, March 7 2020)

more nutrient rich manure than ordinary cows by 10% and excrete 10% more nutrient in manure (Ministry of Agriculture, Food & Rural Affairs Ontario, 2020).

6 Rationale of Animal Dung Biostimulation as a Zero Remediation Strategy

Management of livestock waste is challenging in different environments. Studies (such as Hanafiah et al., 2021; Lombardi et al., 2022; Wu et al., 2020) have established that if not well handled, cow dung and urine can bring about an increase in greenhouse gas emissions. The dung is toxic to nature if left in ubiquitous state (Li et al., 2023).To serve as eco-friendly biostimulants, livestock dung should be treated by either anaerobic digestion or composting. Therefore, efforts to realize zero management of animal dung as biostimulants require rejig of currently used practices. Awasthi and colleagues prescribe the need for rigorous scientific awareness campaigns about animal dung management and treatment and efforts to co-digest the substrates for better Biostimulation efficacy. Cox (2019) proffers, basing on the estimates of recent researches that by 2030, stakes are high that animal excreta to the tune of 5 billion tonnes will be produced each year with the vast majority constituted of livestock manure. This fecal output, Cox notes is likely to contain 100 million tonnes of P, 30 million tonnes of K and 18 tonnes of Ca. As long as all these valuable mineral components are tapped and recycled back to the global agriculture system, livelihoods of millions of people have a bright and sustainable future as this will guarantee employment, agrobased industrial growth, assured food security and environmental conservation (Cox, 2019). A further observation made by Cox (2019) revolves around the challenges facing some cattle rich countries like Netherlands who's over 80% of farms are producing excessive cow dung than is ideally required legally for use as fertilizer. In another incident, Cox (2019) reports about illicit disposal of cattle manure discharged into rivers with environmental scientists expressing concerns about the health threats of this practice. Thus, prudent management of animal dung provides hope for redeeming chunks of land lost to PHC contamination and is therefore a prominent zero remediation approach. A clear testimony of this contribution is provided by Emoyan (2020), who noted that biostimulation of PHC-contaminated soils with animal dung plays multiple roles in environmental clean-up (Fig. 1).

The various ways in which biodegradation is enhanced are also discussed in the following subsections.

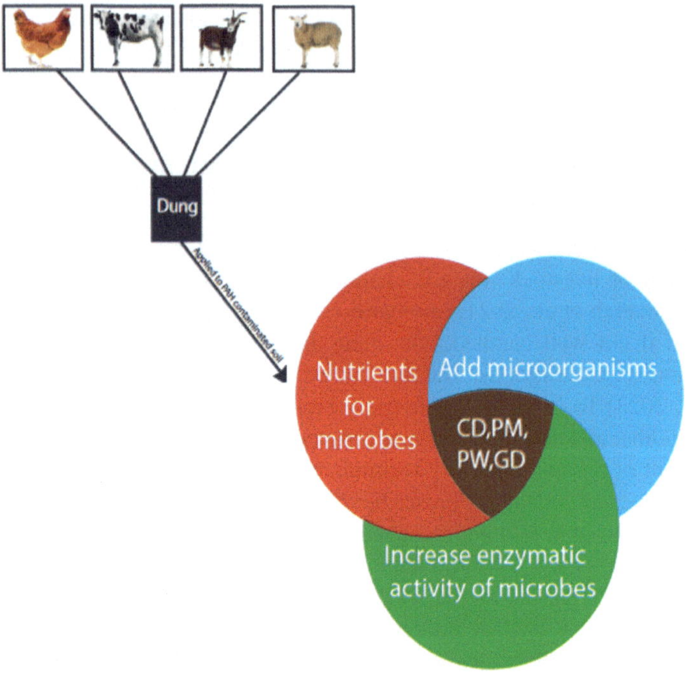

CD=Cow Dung, PM= Poultry Manure, PW=Sheep Manure, GD=Goat Dung

Fig. 1 Composite roles of animal dung in soil environment

6.1 Animal Dung as a Conditioner of PHC-Contaminated Soil

The fibrous nature of animal dung, microbial abundance and mineral richness qualifies it as more of substitute for inorganic chemical fertilizers to conditioner for soil (Dhiman et al., 2021; Fasake & Dashora, 2021). In their study, Wokem and Madufuro (2020) found out that cattle manure posted a reducing effect on PHC polluted soils by causing a reduction in soil PH levels (7.10–6.20). One major finding from their study was a 61% reduction of TPH from diesel-polluted soil. Besides introducing minerals and other nutrients in the soil, animal dung improves soil physicochemical properties such as porosity, aeration, and water retention.

6.2 Animal Dung as Stimulant of Luxuriant Growth of Phytoremediation Species

Animal fecal excreta is rich in easily-decomposable C and N compounds hence providing a rich source of readily available nutrients to biota and increased potential of soil to support and sustain plant life (Zhu et al., 2020). In environments susceptible to crude oil spillage as well as incidences where oil spills have been reported such as in Niger Delta (Olawale et al., 2020) and Sudd region (Ruley et al., 2020a), plant growth is severely affected especially due to below ground (root) stress. In excess, the stress results in generation of reactive oxygen species (ROS) O−, OH−, H_2O_2 which are most noticeable in chloroplasts and mitochondria (Dumanović et al., 2021; Mansoor et al., 2022). Plants in natural undisturbed land normally have superoxide (SOD), catalase (CAT) and peroxidases (POX) enzymes which according to Rajput et al. (2021) provide an antioxidant defense system enabling plants to scavenge ROS. However, in the presence of PHC contaminants, the vigor of plants to scavenge is inhibited with plants presenting less or no potential to wipe out the ROS.

6.3 Animal Dung as a Contributor and Stimulant of Soil Biota

In soil environment and mainly the rhizosphere, live hydrocarbonoclastic microorganisms (Adeleye et al., 2022). While these organisms have a biodegradation capability and can breakdown hydrocarbons, they need to be bioenhanced (Adeleye et al., 2022; Boopathy et al., 2021). Addition of biostimulants such as animal dung squares abundance of microorganisms by providing limiting nutrients which amplifies the capacity of the microbes to degrade PHC into innocuous compounds that are less hazardous in the soil environment. Wokem and Madufuro (2020) reported that cattle manure addition to diesel-polluted soils creates favorable conditions for the thriving of microbial consortium comprising of both hydrocarbon utilizing bacteria (HUB) and hydrocarbon utilizing fungi (HUF). Dominant HUB were *Bacillus*, *Pseudomonas*, *Proteus*, *Alcaligenes*, *Klebsiella*, *Enterobacter*, *Citrobacter*, *Corynebacterium* and *Escherichia coli* while most dominant HUF were *Fusarium*, *Aspergillus*, *Microsporium*, *Penicilium* and *Mucor*. Animal dung supplies limiting nutrients which activates potential of indigenous microbes to degrade PHC (Abdulyekeen et al., 2022). As long as the animal dung is applied in appropriate quantities, there is a corresponding rise in the activities of the inhabiting microbes (Ruley et al., 2020a). In a related study, Behera and Ray (2021) observed that dung contains a suite of microbes rich in metabolites. These are excellent at mineralization of hydrocarbons and thus turn them into innocuous and non-lethal byproducts to plants and other soil biota. Obi et al. (2022) provided a confirmation that the range of microbes contained in cow dung potentially degrade benzene to harmless compounds. In some instances, the microorganisms enzymatically attack the PHCcompounds helping to eliminate their complex mixtures from the soil. These benefits accrue when an optimal amount of cow dung is added. Ruley et al. (2020a) determined the optimal rate at $2tha^{-1}$. Surpassing the optimal rate would make cow dung application a chimera as its excess would divert the scavenging role of soil microbes.

6.4 Animal Manure as a Heavy Metal Clean-Up Reagent

Heavy metals were equally highly concentrated in areas near drilled abandoned oil wells compared to areas far away. This provides strong evidence that heavy metals are derived from crude oil extraction activities. Although heavy metals are known to be naturally occurring compounds, studies (such as Raimi et al., 2022; Ruley, 2022; Tian et al., 2022) have confirmed that anthropogenic activities such as crude oil exploitation introduce large quantities of heavy metals in the environment which threatens the ability of environment to support and sustain human, animal and plant health. Heavy metals such as arsenic, lead, cadmium, nickel, mercury, chromium, cobalt, zinc and selenium are associated with crude oil contamination (Fadguba et al., 2023) and are highly toxic even in minor quantities.

The supplementation of soil with cow dung improves nutrient availability in the soil leading to a reduction in the concentrations of heavy metals including Zn, Cd, Ni, Pb, Ar, Cr and Hg that are introduced in soil by PHC contaminants (Susianti et al., 2022). Besides, the population and diversity of both bacteria and fungi isolates in the polluted environment are boosted leading to enhanced remediation. Susianti et al. (2022) established that cow dung contains a suite of microorganisms play an immense role in the degradation of heavy metals such as chromium, strontium and arsenic. In a related study, Banerjee et al. (2023) established that use of dry cow dung powder removes chromium and adsorbs. Animal manure such as cow dung helps in the sequestering of heavy metals (zinc, copper, lead, nickel, cadmium, etc.) existing in an aqueous form especially where produced or wastewater is involved (Akyürek et al., 2022; Iwuozor et al., 2022).

Cow dung immobilizes heavy metals from contaminated soil (Jacob et al., 2022). By composition, cow dung ash as a bio-organic waste is highly rich in silica, aluminum oxide, iron oxide, calcium oxide, calcium oxide and magnesium oxide (Behera & Ray, 2021). The multifariousness of minerals in animal dung such as cattle manure ash makes it an eco-friendly and low cost adsorbent. The basis of its use is further recommended by Tallou et al. (2020) who proffered that its use as eco-friendly treatment for PHC-contaminated soils is beneficial to people wishing to live in harmony with nature as its use does not exert nature degrading side-effects to the environment. Deserving mention is the fact that in the heavy metal-contaminated soils, application of cow dung owing to its high silica content leads to considerable removal of metal ions. Silica has a high metal ions affinity (Iwuozor et al., 2022; Saraswat et al., 2020).

7 Biostimulation Cocktail as a Credible Zero Remediation Innovation

As noted from the preceding sections, PHC contamination impedes plant metabolic processes such as germination, transpiration and photosynthesis, consequently slowing their rate of growth and development (Obasi et al., 2021). Equally, the scotching effects in the soil greatly impact the metabolism of resident soil microorganisms such as bacteria and fungi. For excellent results, studies have established that bacteria and fungi cocktail is necessary. This innovation provides a ray of hope that zero remediation approaches will become more feasible especially with deployment of a consortia of waste to produce bacteria and fungi cocktail. This creates a synergy that speeds up the chelation potential of the resident soil microorganisms (Obasi et al., 2021). Where dung from both animals is available, it can be used as a consortium. Osinowo et al. (2020) refer to the consortium of animal dung as bacteria cocktail. Bacterial cocktail presents with greater degradation potential compared to application of a single amendment (Osinowo et al., 2020). Several studies have deployed bacterial cocktail and reported excellent results. For example, John et al. (2021) used goat and poultry dung in one screenhouse experiment. At the end of the cycle, they observed a manifold increase in abundancies and diversity of PAH utilizing bacteria, fungi and actinomycetes. Bacteria genera (*Bacillus, Pseudomonas, Acinetobacter, Alcaligenes* and *Serratia*) fungi (*Penicillium, Aspergillus* and *Cladosporium*) and actinomycetes (*Rhodococcus, Nocardia* and *Streptomyces*) were more abundant in the soil ecosystem after addition of the biostimulants. The remediation with soil-goat dung and soil-poultry dropping amendments compared favorably with soil-NPK fertilizer technique because microbial activities were enhanced to produce eco-friendly waste (John et al., 2021). Consortia of goat and rabbit dung were experimented by Ogoanah et al. (2020) and results showed a 76.77% remediation efficiency since TPH content before the experiment at 9984.0 mg/kg was removed up to 3170.6 mg/kg after. Aghalibe et al. (2020) used a consortium of cow dung, poultry dung and NPK. The synergy from the multiple amendments led to a significant biodegradation of PHC. It is therefore plausible to deduce that animal dung biostimulation supports the multiplication and degradation potential of hydrocarbon utilizing bacteria (HUB) (Awari et al., 2020; Raspor et al., 2023).

More studies have established that a cocktail for microorganisms can equally be produced using animal dung and other agricultural and non-agricultural waste. The deployment of this cocktail in the soil environment greatly improves biodegradation. For example, Boopathy et al. (2021) used poultry dung and sunflower seed husk in PHC-contaminated soil. Results showed a manifold rise of indigenous bacterial population in the soil. In relation, Adeleye et al. (2022) used a consortia of compost, processed cocoa husk and cow dung, and these removed over 70% of PHC contaminants from soil. Muhammad et al. (2022) used a consortia of chicken dung, cow dung and rice husk under laboratory conditions and achieved an average removal of PHC contaminants by about 78%. Further related evidence is provided by Aziz et al. (2020) who studied the enhancement potential of biochar prepared from fruit and vegetable waste and sewage sludge on bioremediation of diesel-polluted soil. When cattle manure was included in the mix as a biostimulant, the PHC removal efficiency was established at 82%.

8 Conclusion and Recommendation

In environments with soil contaminated with PHC, the innovative zero remediation approaches involving use of animal dung provide high hopes that the wastelands created by oil pollution will be redeemed. The ubiquity of animal dung in these countries and recent innovations such as cocktail biostimulation imply that soil conservation will be achieved. Hence, biostimulation will be a redeemer that will minimize the deleterious impacts posted by oil exploitation to the soil environment. Observably, we recommend more studies about cheaper and more effective ways of applying biostimulation cocktail for effective ecorestoration of PHC-contaminated soils to be conducted.

References

Abdulyekeen, K. A., Giwa, S. O., & Ibrahim, A. A. (2022). Bioremediation of used motor oil-contaminated soil using animal dung as stimulants. *Animal manure: Agricultural and biotechnological applications* (pp. 185–215). Springer International Publishing.

Adeleye, A. O., Yerima, M. B., Nkereuwem, M. E., Onokebhagbe, V. O., Shiaka, G. P., Amoo, A. O., Asaju, I. C., Yalwaji, B., & Ishaq, S. M. (2022). Effect of bacterial co-culture and organic amendments on the bioremediation of hydrocarbons in a soil contaminated with spent engine oil. *Acta Scientiarum. Biological Sciences, 44*(1), e62289.

Aghalibe, C. U., Igwe, J. C., Obike, A. I., & Onwuka, K. E. (2020). Polycyclic aromatic hydrocarbon (PAH) degradation in a crude oil impacted soil remediated with cow dung, poultry droppings and NPK. *International Journal of Engineering and Industry, 5*(2), 7–14.

Ahmad, A. A., Muhammad, I., Shah, T., Kalwar, Q., Jianbo, Z., & Zeyi, L. (2020). Remediation methods of crude oil contaminated soil. *World Journal of Agriculture and Soil Science, 4*(3), 1–8.

Akyürek, Z., Çelebi, H., Çakal, G. Ö., & Turgut, S. (2022). Enhanced arsenic (V) removal from aqueous solution by a novel magnetic biochar derived from dairy cattle manure. *Korean Chemical Engineering Research, 60*(3), 423–432.

Ali, M., Song, X., Wang, Q., Zhang, Z., Zhang, M., Chen, X., Tang, Z., & Liu, X. (2023). Thermally enhanced biodegradation of benzo [a] pyrene and benzene co-contaminated soil: Bioavailability and generation of ROS. *Journal of Hazardous Materials, 455*, 131494.

Ambaye, T. G., Vaccari, M., Franzetti, A., Prasad, S., Formicola, F., Rosatelli, A., Hassani, A., Aminabhavi, T. M., & Rtimi, S. (2023). Microbial electrochemical bioremediation of petroleum hydrocarbons (PHCs) pollution: Recent advances and outlook. *Chemical Engineering Journal, 452*, 139372.

Appiagyei Osei-Owusu, B., Baidoo, M. F., Arthur, R., & Oduro-Kwarteng, S. (2023). Physico-chemical characterization of selected feedstocks as co-substrates for household biogas generation in Ghana. *International Journal of Sustainable Engineering, 16*(1), 117–128.

Atai, E., Jumbo, R. B., Cowley, T., Azuazu, I., Coulon, F., & Pawlett, M. (2023). Efficacy of bioadmendments in reducing the influence of salinity on the bioremediation of oil-contaminated soil. *Science of the Total Environment*, 164720.

Aziz, S., Ali, M. I., Farooq, U., Jamal, A., Liu, F. J., He, H., Guo, H., Urynowicz, M., & Huang, Z. (2020). Enhanced bioremediation of diesel range hydrocarbons in soil using biochar made from organic wastes. *Environmental Monitoring and Assessment, 192*(9), 1–14.

Banerjee, S., Kanaujia, A., & Sharma, S. (2023). Importance of Indigenous Cow dung and its microbiota. *International Journal of Current Microbiology and Applied Sciences, 12*(04), 26–36.

BBC. (March 7, 2020). *Chad repaying $100 m debt to Angola with cattle.* Retrieved on 5 October 2020 from https://www.bbc.com/news/world-africa-51925035

Beef Cattle Research Council of Canada. (2023). *Manure and nutrient management.* Retrieved on 27 November 2023 from https://www.beefresearch.ca/topics/manure-nutrient-management/

Behera, S. S., & Ray, R. C. (2021). Bioprospecting of cowdung microflora for sustainable agricultural, biotechnological and environmental applications. *Current Research in Microbial Sciences, 2*, 100018.

Blenis, N., Hue, N., Maaz, T. M., & Kantar, M. (2023). Biochar production, modification, and its uses in soil remediation: A review. *Sustainability, 15*(4), 3442.

Boopathy, S., Appavoo, M. S., & Radhakrishnan, I. (2021). Sunflower seed husk combined with poultry droppings to degrade petroleum hydrocarbons in crude oil-contaminated soil. *Environmental Engineering Research, 26*(5).

Borah, P., Sharma, V., Kashyap, D., Kumar, M., & Sahoo, B. M. (2023). Indigenous techniques to remove metals from contaminated water. In *Metals in water* (pp. 285–303). Elsevier.

Cao, S., Zhan, G., Wei, K., Zhou, B., Zhang, H., Gao, T., & Zhang, L. (2023). Raman spectroscopic and microscopic monitoring of on-site and in-situ remediation dynamics in petroleum contaminated soil and groundwater. *Water Research, 233*, 119777.

Chilvers, B. L., & McClelland, P. J. (2023). Lessons learned for pre-emptive capture management as a tool for wildlife conservation during oil spills and eradication events. *Animals, 13*(5), 833.

Cook, R. (October 7, 2020). *Cattle inventory of United States of America.* Retrieved on 3 January 2023 from https://beef2live.com/story-united-states-cattle-inventory-124-108177%23:%7E:text=The%20cattle%20inventory%20in%20the%20United%20States%20in%202017%20is,2%20million%20head%20of%20cattle

Cox, D. (2019). The planet's prodigious poo problem. In *The Guardian.* Retrieved on 28 November 2023 from https://www.theguardian.com/news/2019/mar/25/animal-waste-excrement-four-billion-tonnes-dung-poo-faecebook

de Souza, D. M., da Silva, J. D. L., Ludwig, L. D. C., Petersen, B. C., Brehm, F. A., Modolo, R. C. E., De Marchi, T. C., Figueiredo, R., & Moraes, C. A. (2023). Study of the phytoremediation potential of native plant species identified in an area contaminated by volatile organic compounds: A systematic review. *International Journal of Phytoremediation*, 1–18.

Dhiman, S., Kumar, S., Baliyan, N., Dheeman, S., & Maheshwari, D. K. (2021). Cattle dung manure microbiota as a substitute for mineral nutrients and growth management practices in plants. *Endophytes: Mineral Nutrient Management, 3*, 77–103.

Ding, Y., Feng, L., Zhang, Z., Zhang, S., Zhang, X., Li, Y., Fang, X., Hu, X., Wang, X., Tam, K. C., Zhou, G., & Zhang, Z. (2023). Lignocellulose nanoparticles extracted from cattle dung as Pickering emulsifiers for microencapsulating phase change materials. *ACS Sustainable Chemistry and Engineering, 11*(38), 14255–14266.

Dumanović, J., Nepovimova, E., Natić, M., Kuča, K., & Jaćević, V. (2021). The significance of reactive oxygen species and antioxidant defense system in plants: A concise overview. *Frontiers in Plant Science, 11*, 552969.

Ehiomogue, P. O., Ahuchaogu, I. I., & Ahaneku, E. I. (2021). A review of bioremediation of hydrocarbon contaminated soils in Niger Delta area of Nigeria. *Poljoprivredna Tehnika, 46*(3), 23–39.

El-Gendy, N. S., & Nassar, H. N. (2021). Biosynthesized magnetite nanoparticles as an environmental opulence and sustainable wastewater treatment. *Science of the Total Environment, 774*, 145610.

Elmi, A., Anwar, S. A., & Al-Dashti, H. (2023). Measurements and regional climate modeling of soil temperature with and without bias correction method under arid environment: Can soil temperature outperform air temperature as a climate change indicator?. *Environmental Modeling and Assessment*, 1–11.

Eltoum, H., & AWad Salih, A. (2022). *Report on the FMD situation in Republic of Sudan*. Consultative seminar on progress made in FMD/PPR roadmaps for East Mediterranean countries Beirut-Lebanon 11–12 September 2022.

Emoyan, O. O. (2020). Bioremediation of in-situ crude oil contaminated soil using selected organic dung. *Egyptian Journal of Chemistry, 63*(8), 8–9.

Environmental Dynamics International (EDI). (2021). *Five barrels of water produced per barrel of oil*. Retrieved on 27 November 2023 from https://www.ediweekly.com/five-barrels-water-produced-per-barrel-oil/

Erickson, P. S., & Kalscheur, K. F. (2020). Nutrition and feeding of dairy cattle. In *Animal agriculture* (pp. 157–180). Academic Press.

Fadugba, O., Ola, S., Babatola, J. O., Ojuri, O., Uduebor, M., & Akande, S. (2023). Assessment of heavy metals in petroleum hydrocarbon contaminated groundwater (A case study of Baruwa community, Lagos State, Nigeria). *Covenant Journal of Engineering Technology, 7*(2), 65–69.

Fan, J., Ai, S., Zheng, G., Yin, T., Zhang, H., Tao, D., & Wang, S. (2023). Insight into effects of pyrolysis products and white-rot fungi on co-composting of pig manure and corn stalk. *Biomass Conversion and Biorefinery*, 1–11.

Fasake, V., & Dashora, K. (2021). A sustainable potential source of ruminant animal waste material (dung fiber) for various industrial applications: A review. *Bioresource Technology Reports, 15*, 100693.

Feepee, E. F., Eludoyin, O. S., & Deekor, T. N. (2023). Assessment of plant species diversity, density and abundance of crude oil post impacted communities in Gokana LGA of rivers state, Nigeria. *Assessment*, 2(2).

Gao, G., Zeng, H., & Zhou, Q. (2022). Biomonitoring atmospheric pollution of polycyclic aromatic hydrocarbons using mosses. *Atmosphere, 14*(1), 26.

Gharekhani, J., Yakhchali, M., & Berahmat, R. (2020). Neospora caninum infection in Iran (2004–2020): A review. *Journal of Parasitic Diseases, 44*, 671–686.

Hanafiah, M. M., Ibraheem, A. J., & Razman, K. K. (October, 2021). Emissions of carbon dioxide and methane from dairy cattle manure. *IOP Conference Series: Earth and Environmental Science, 880*(1), 012037. IOP Publishing.

Haokip, I. C., Devi, M. H., Mishra, R., Kumar, D., & Dey, P. (2023). Remediation of polluted soils for managing toxicity stress in crops of dryland ecosystems. In *Enhancing resilience of dryland agriculture under changing climate: Interdisciplinary and convergence approaches* (pp. 259–303). Springer Nature Singapore.

Hauptvogl, M., Kotrla, M., Prčík, M., Pauková, Ž, Kováčik, M., & Lošák, T. (2020). Phytoremediation potential of fast-growing energy plants: Challenges and perspectives—A review. *Polish Journal of Environmental Studies, 29*(1), 505–516.

Hidalgo, D., Corona, F., & Martín-Marroquín, J. (2022). Manure biostabilization by effective microorganisms as a way to improve its agronomic value. *Biomass Conversion and Biorefinery, 12*, 4649–4664.

Homayounpour, F. (2023). *Thermal treatments of contaminated soils* (Doctoral dissertation, Politecnico di Torino).

International Soil Reference and Information Centre (ISRIC). (2023). *Why are soils important?* Retrieved on 27 November 2023 from https://www.isric.org/discover/about-soils/why-are-soils-important

Iwuozor, K. O., Emenike, E. C., Aniagor, C. O., Iwuchukwu, F. U., Ibitogbe, E. M., Okikiola, T. B., Omuku, P. E., & Adeniyi, A. G. (2022). Removal of pollutants from aqueous media using cow dung-based adsorbents. *Current Research in Green and Sustainable Chemistry, 5*, 100300.

Izah, S. C., & Aigberua, A. O. (2023). Potential threats and possible conservation strategies of biodiversity in Niger delta region of Nigeria. In *Sustainable utilization and conservation of Africa's biological resources and environment* (pp. 151–176). Springer Nature Singapore.

Javed, A., Ali, E., Afzal, K. B., Osman, A., & Riaz, S. (2022). Soil fertility: Factors affecting soil fertility, and biodiversity responsible for soil fertility. *International Journal of Plant, Animal and Environmental Sciences, 12*(1), 21–33.

John, O. U., Umana, S. I., Asuquo, C. E., & Eduok, S. I. (2021). A comparative study on the use of soil-organic and inorganic biostimulants in the remediation of oily waste. *Current Journal of Applied Science and Technology, 40*(17), 21–33.

Khan, N., Ahmad, A., Sharma, V., Saha, A. K., Pandey, A., & Bhargava, P. C. (2022). An integrative study for efficient removal of hazardous azo dye using microbe-immobilized cow dung biochar in a continuous packed bed reactor. *Renewable Energy, 200*, 1589–1601.

King, R. B., Sheldon, J. K., & Long, G. M. (2023). *Practical environmental bioremediation: The field guide*. CRC Press.

Koolivand, A., Saeedi, R., Coulon, F., Kumar, V., Villaseñor, J., Asghari, F., & Hesampoor, F. (2020). Bioremediation of petroleum hydrocarbons by vermicomposting process bioaugmentated with indigenous bacterial consortium isolated from petroleum oily sludge. *Ecotoxicology and Environmental Safety, 198*, 110645.

Kusmiyati, K., Wijaya, D. K., Hartono, B. R., Shidik, G. F., & Fudholi, A. (2023). Harnessing the power of cow dung: Exploring the environmental, energy, and economic potential of biogas production in Indonesia. *Results in Engineering, 20*, 101431.

Kuyukina, M. S., Krivoruchko, A. V., & Ivshina, I. B. (2020). Advanced bioreactor treatments of hydrocarbon-containing wastewater. *Applied Sciences, 10*(3), 831.

Lacalle, R. G., Becerril, J. M., & Garbisu, C. (2020). Biological methods of polluted soil remediation for an effective economically-optimal recovery of soil health and ecosystem services. *Journal of Environmental Science and Public Health, 4*(2), 112–133.

Landaeta-Hernández, A. J., Ungerfeld, R., & Chenoweth, P. J. (2023). Biostimulation and pheromones in livestock: A review. *Animal Reproduction Science, 248*, 107154.

Li, Y., Li, X., Ma, X., Qiu, T., Fu, X., Ma, Z., Ping, H., & Li, C. (2023). Livestock wastes from family-operated farms are potential important sources of potentially toxic elements, antibiotics, and estrogens in rural areas in North China. *Environmental Science and Pollution Research*, 1–12.

Liu, X., & Zhang, L. (2023). Effects of additives on the co-composting of forest residues with cattle manure. *Bioresource Technology, 368*, 128384.

Lombardi, B., Loaiza, S., Trujillo, C., Arevalo, A., Vázquez, E., Arango, J., & Chirinda, N. (2022). Greenhouse gas emissions from cattle dung depositions in two Urochloa forage fields with contrasting biological nitrification inhibition (BNI) capacity. *Geoderma, 406*, 115516.

Manian, R. (2023). Bioremediation of polycyclic aromatic hydrocarbons contaminated soils: Recent progress, perspectives and challenges. *Environmental Monitoring and Assessment, 195*(12), 1–17.

Mansoor, S., Ali Wani, O., Lone, J. K., Manhas, S., Kour, N., Alam, P., Ahmad, A., & Ahmad, P. (2022). Reactive oxygen species in plants: From source to sink. *Antioxidants, 11*(2), 225.

Masotti, F., Garavaglia, B. S., Gottig, N., & Ottado, J. (2023). Bioremediation of the herbicide glyphosate in polluted soils by plant-associated microbes. *Current Opinion in Microbiology, 73*, 102290.

Ministry of Agriculture, Food and Rural Affairs Ontario. (August 28, 2020). *Dietary phosphorus in dairy cattle—How to manage nutrients and protect the environment*. Factsheet: Queens printer for Ontario. Retrieved on 24 January 2023 from http://www.omafra.gov.on.ca/english/livestock/dairy/facts/04-001.htm

Muhammad, R. G., Mohammed, J. N., Muhammad, I. L., & Hamzat, A. (2022). Stimulated bioremediation of soil contaminated with spent engine oil using organic wastes. *Science World Journal, 17*(2), 308–314.

Muneeswari, R., Iyappan, S., Swathi, K. V., Vinu, R., Ramani, K., & Sekaran, G. (2022). Biocatalytic lipoprotein bioamphiphile induced treatment of recalcitrant hydrocarbons in petroleum refinery oil sludge through transposon technology. *Journal of Hazardous Materials, 431*, 128520.

Nishat, F., & Keshav, S. (2023). Accumulation of heavy metals from the combination of different biological wastes by earthworm *Lampito mauritii* Kinberg. *International Journal of Zoological Investigations, 9*(1), 257–271.

Nivetha, N., Srivarshine, B., Sowmya, B., Rajendiran, M., Saravanan, P., Rajeshkannan, R., Rajasimman, M., Pham, T. H. T., Shanmugam, V., & Dragoi, E. N. (2023). A comprehensive review on bio-stimulation and bio-enhancement towards remediation of heavy metals degeneration. *Chemosphere, 312*, 137099.

Noviardi, R., Karuniawan, A., Sofyan, E. T., & Suryatmana, P. (2023). Heavy metals content in sweet potato (*Ipomoea batatas* L.) grown on soil contaminated by gold mine tailings with composted cow manure amendment. *Journal of Degraded and Mining Lands Management, 10*(4).

Obasi, S. E., Obasi, N. A., Nwankwo, E. O., Enemchukwu, B. N., Igbolekwu, R. I., & Nkama, J. O. (2021). Effects of organic manures bioremediation on growth performance of Maize (*Zea mays* L.) in crude oil polluted soil. *International Journal of Recycling Organic Waste in Agriculture, 10*(4), 415–426.

Obi, C. C., Umanu, G., Anozie, C. P., & Umar, H. (2022). Biodegradation of spent automobile engine oil in soil microcosms amended with cow dung. *Journal of Applied Sciences and Environmental Management, 26*(2), 255–263.

Ogoanah, S. O., Ngwoke, U. N., Ohanmu, E. O., Okoye, P. C., & Ikhajiagbe, B. (2020). Enhancement of the soil quality of an oil-polluted ultisol using livestock wastes. *Notulae Scientia Biologicae, 12*(2), 387–398.

Okpashi, V. E., Ushie, O. A., Abeng, F. E., & Inyang, I. H. (2020). Monitoring the in-situ bioremediation of spend engine-oil contaminated soil after irrigation with fermented chicken-droppings. *Journal of Applied Sciences and Environmental Management, 24*(3), 411–416.

Olawale, O., Obayomi, K. S., Dahunsi, S. O., & Folarin, O. (2020). Bioremediation of artificially contaminated soil with petroleum using animal waste: Cow and poultry dung. *Cogent Engineering, 7*(1), 1721409.

Osinowo, I., Abioye, O. P., Oyewole, O. A., & Oyeleke, S. B. (2020). The bioremediation of diesel contaminated soil using bacterial cocktail and organic nutrients. *Journal of Microbiology, Biotechnology and Food Sciences, 10*(2), 150–158.

Otten, J. G., Williams, L., & Refsnider, J. M. (2023). Freshwater turtle populations as bioindicators following an oil spill: Delayed demographic changes reveal long-term impacts. *Ecological Indicators, 154*, 110519.

Pande, V., Pandey, S. C., Sati, D., Pande, V., & Samant, M. (2020). Bioremediation: an emerging effective approach towards environment restoration. *Environmental Sustainability*, 1–13.

Pinto, A. P., Faria, J. M., Dordio, A. V., & Carvalho, A. P. (2023). Organic farming—A sustainable option to reduce soil degradation. *Agroecological Approaches for Sustainable Soil Management*, 83–143.

Pothula, G. K., Vij, R. K., & Bera, A. (2023). An overview of chemical enhanced oil recovery and its status in India. *Petroleum Science*.

Puri-Marza, A. (January 3, 2023). *Number of farm cows in Saudi Arabia from 2014 to 2019 (in 1000 s)*. Retrieved on 27 November 2023 from https://www.statista.com/statistics/976249/saudi-arabia-number-of-cows-in-farms/

Rafeeq, H., Afsheen, N., Rafique, S., Arshad, A., Intisar, M., Hussain, A., Bilal, M., & Iqbal, H. M. (2023). Genetically engineered microorganisms for environmental remediation. *Chemosphere, 310*, 136751.

Raimi, M. O., Sawyerr, O. H., Ezekwe, C. I., & Salako, G. (2022). Many oil wells, one evil: Comprehensive assessment of toxic metals concentration, seasonal variation and human health risk in drinking water quality in areas surrounding crude oil exploration facilities in rivers state, Nigeria. *International Journal of Hydrology, 6*(1), 23–42.

Rajasekaran, M., & Kandasamy, R. (2023). High-throughput bioamphiphile production by ethyl methane sulphonate induced mutant of hydrocarbonoclastic Enterobacter xiangfangensis STP-3: In depth structural elucidation and application to petroleum refinery oil sludge bioremediation. *Journal of Hazardous Materials, 458*, 131961.

Rajput, V. D., Harish, Singh, R. K., Verma, K. K., Sharma, L., Quiroz-Figueroa, F. R., Meena, M., Gour, V. S., Minkina, T., Sushkova, S., & Mandzhieva, S. (2021). Recent developments in enzymatic antioxidant defence mechanism in plants with special reference to abiotic stress. *Biology, 10*(4), 267.

Raspor, M., Berić, T., Ćosić, T., & Stanković, S. (2023). The rhizosphere—A hub of emerging importance for plant biotic interactions. In *New frontiers in plant-environment interactions: Innovative technologies and developments* (pp. 289–327). Springer Nature Switzerland.

Rezanezhad, F., Bieroza, M., Contosta, A., & Van Cappellen, P. (2023). The cold regions in transition: Impacts on soil and groundwater biogeochemistry. *Frontiers in Environmental Science, 11*, 93.

Ridder, M. (March 15, 2023). *Number of dairy cows in Norway from 2010 to 2022.* Retrieved on 27 November 2023 from https://www.statista.com/statistics/690233/number-of-dairy-cows-in-norway/

Rout, P. R., Pandey, D. S., Haynes-Parry, M., Briggs, C., Manuel, H. L. C., Umapathi, R., Mukherjee, S., Panigrahi, S., & Goel, M. (2023). Sustainable valorisation of animal manures via thermochemical conversion technologies: an inclusive review on recent trends. *Waste and Biomass Valorization, 14*(2), 553–582.

Ruley, J. A., Galla, J. O., Massawe, P., Ladu, J. L. C., & Tumuhairwe, J. B. (2023). Pseudomonas as backbone for environmental health. In *Microbiomes for the management of agricultural sustainability* (pp. 239–253). Springer Nature Switzerland.

Ruley, J. A. (2022). *Bioremediation of petroleum contaminated soils using native plants and their rhizobacteria in the Sudd region of South Sudan* (Doctoral dissertation, Makerere University, Kampala-Uganda).

Ruley, J. A., Amoding, A., Tumuhairwe, J. B., & Basamba, T. A. (2022). Rhizoremediation of petroleum hydrocarbon–contaminated soils: A systematic review of mutualism between phytoremediation species and soil living microorganisms. *Phytoremediation*, 263–296.

Ruley, J. A., Tumuhairwe, J. B., Amoding, A., Westengen, O. T., & Vinje, H. (2020a). Rhizobacteria communities of phytoremediation plant species in petroleum hydrocarbon contaminated soil of the Sudd ecosystem, South Sudan. *International Journal of Microbiology, 2020.*

Ruley, J. A., Amoding, A., Tumuhairwe, J. B., Basamba, T. A., Opolot, E., & Oryem-Origa, H. (2020b). Enhancing the phytoremediation of hydrocarbon-contaminated soils in the Sudd wetlands, South Sudan, using organic manure. *Applied and Environmental Soil Science, 2020.*

Ruley, J. A., Amoding, A., Tumuhairwe, J. B., Basamba, T. A., Opolot, E., & Oryem-Origa, H. (2020c). Enhancing the phytoremediation of hydrocarbon-contaminated soils in the Sudd wetlands, South Sudan, using organic manure. *Applied and Environmental Soil Science, 2020*, 1–8.

Saraswat, S. K., Demir, M., & Gosu, V. (2020). Adsorptive removal of heavy metals from industrial effluents using cow dung as the biosorbent: Kinetic and isotherm modeling. *Environmental Quality Management, 30*(1), 51–60.

Sasu, D. D. (June 30, 2023). *Stock of live cattle in Nigeria from 2010 to 2021.* https://www.statista.com/statistics/1297914/stock-of-live-cattle-in-nigeria/

Scully, S. M., & Orlygsson, J. (2023). Thermostable enzymes and their applications. *Thermophilic anaerobes: Phylogeny, physiology and biotechnological applications* (pp. 155–186). Springer International Publishing.

Selva Filho, A. A. P., Converti, A., Soares da Silva, R. D. C. F., & Sarubbo, L. A. (2023). Biosurfactants as multifunctional remediation agents of environmental pollutants generated by the petroleum industry. *Energies, 16*(3), 1209.

Shahbandeh, M. (May 16, 2023). *Cattle inventory in Canada as of January 2023, by type.* Retrieved on 27 November 2023 from https://www.statista.com/statistics/442302/cattle-inventory-by-type-canada/

Sharma, I. (2020). Bioremediation techniques for polluted environment: Concept, advantages, limitations, and prospects. In *Trace metals in the environment-new approaches and recent advances.* IntechOpen.

Statista Research Department. (May 13, 2023). *Cattle livestock population in Russia 2001–2022.* Retrieved on 27 November 2023 from https://www.statista.com/statistics/1037803/russia-cattle-population/

Susianti, B., Warmadewanthi, I. D. A. A., & Tangahu, B. V. (2022). Characterization and experimental evaluation of cow dung biochar+ dolomite for heavy metal immobilization in solid waste from silica sand purification. *Bioresource Technology Reports, 18*, 101102.

Taha, A., Saad, S., Jubara, A., Wani, C., Phiri, A. M., Simuunza, M., Munyeme, M., Hang'ombe, B., & Mumba, C. (2023). Financial losses arising from cattle organ and carcass condemnation at lokoloko abattoir in Wau, South Sudan. *Advances in Preventive Medicine, 2023.*

Textor, C. (2023). *Cattle and buffaloes livestock in China from 2012 to 2022.* Retrieved on 27 November 2023 from https://www.statista.com/statistics/275667/cattle-livestock-in-china/

Thacharodi, A., Hassan, S., Singh, T., Mandal, R., Khan, H. A., Hussain, M. A., & Pugazhendhi, A. (2023). Bioremediation of polycyclic aromatic hydrocarbons: An updated microbiological review. *Chemosphere*, 138498.

Thai, T. D., Lim, W., & Na, D. (2023). Synthetic bacteria for the detection and bioremediation of heavy metals. *Frontiers in Bioengineering and Biotechnology, 11*, 1178680.

Thalassinos, G., Petropoulos, S. A., & Antoniadis, V. (2023). The response of purslane (*Portulaca oleracea*) to soil-added Pb: Is it suitable as a potential phytoremediation species? *Toxics, 11*(2), 153.

Tian, Y., Zhou, J., He, C., He, L., Li, X., & Sui, H. (2022). The formation, stabilization and separation of oil–water emulsions: A review. *Processes, 10*(4), 738.

United States Department of Agriculture (USDA). (2023). *United States cattle inventory down 3%.* Retrieved on 27 November 2023 from https://www.nass.usda.gov/Newsroom/2023/01-31-2023.php

Vaksmaa, A., Guerrero-Cruz, S., Ghosh, P., Zeghal, E., Hernando-Morales, V., & Niemann, H. (2023). Role of fungi in bioremediation of emerging pollutants. *Frontiers in Marine Science, 10*, 1070905.

Vendatu. (November 27, 2023). *Question and answers: Agricultural waste.* Retrieved on 28 November 2023 from https://www.vedantu.com/question-answer/animal-dung-is-waste-a-biodegradable-b-class-8-biology-cbse-5f3f675962176b65adca9583

Verma, S., Thakur, D., Pandey, C. M., & Kumar, D. (2023). Recent prospects of carbonaceous nanomaterials-based laccase biosensor for electrochemical detection of phenolic compounds. *Biosensors, 13*(3), 305.

Wokem, V. C., & Madufuro, C. (2020). Application of cow dung and sawdust as biostimulants for enhanced bioremediation of diesel contaminated soil. *Journal of Applied Sciences and Environmental Management, 24*(1), 49–57.

World Organization for Animal Health (WOAH). (2023). *Iraq.* Retrieved on 26 November 2023 from http://rr-middleeast.woah.org/en/about-us/regional-members-of-woah/iraq/

Wu, Q., Kwak, J. H., Chang, S. X., Han, G., & Gong, X. (2020). Cattle urine and dung additions differently affect nitrification pathways and greenhouse gas emission in a grassland soil. *Biology and Fertility of Soils, 56*, 235–247.

Xinhua. (July 19, 2021). *Customers rarely seen in Khartoum's livestock markets due to high prices.* Retrieved on 27 November 2023 from http://www.xinhuanet.com/english/2021-07/19/c_1310070781.htm

Xue, Y., Chen, L., Xiang, L., Zhou, Y., & Wang, T. (2023). Experimental investigation on electromagnetic induction thermal desorption for remediation of petroleum hydrocarbons contaminated soil. *Journal of Environmental Management, 328*, 117200.

Yang, X., Li, L., Zhao, W., Tian, Y., Zheng, R., Deng, S., & Mu, Y. (2023). The influence of potassium hydroxide concentration and temperature on pulp characteristics and cow dung-based paper performance. *Journal of Natural Fibers, 20*(1), 2164546.

Yuan, C. L., Li, Q., Sun, Z. Y., Zhang, W. J., Chen, J. R., Chen, Z., Na, S., & Sun, H. W. (2023). Chemical oxidation of arsenic in the environment and its application in remediation: A mini review. *Pedosphere, 33*(1), 185–193.

Zhu, Y., Merbold, L., Leitner, S., Pelster, D. E., Okoma, S. A., Ngetich, F., Onyango, A. A., Pellikka, P., & Butterbach-Bahl, K. (2020). The effects of climate on decomposition of cattle, sheep and goat manure in Kenyan tropical pastures. *Plant and Soil, 451*, 325–343.

The manufacturer's authorised representative in the EU is Springer Nature Customer Service Centre GmbH, Europaplatz 3, 69115 Heidelberg, Germany. If you have any concerns regarding our products, please contact ProductSafety@springernature.com

Printed and bound by CPI Group (UK) Ltd, Croydon, CR0 4YY
01/10/2026
02225904-0012